Universitext

Universitext

Universitext is a series of textbooks that presents material from a wide variety of mathematical disciplines at master's level and beyond. The books, often well class-tested by their author, may have an informal, personal, even experimental approach to their subject matter. Some of the most successful and established books in the series have evolved through several editions, always following the evolution of teaching curricula, into very polished texts.

Thus as research topics trickle down into graduate-level teaching, first textbooks written for new, cutting-edge courses may make their way into *Universitext*.

More information about this series at http://www.springer.com/series/223

Pierre Brémaud

Fourier Analysis
and Stochastic Processes

 Springer

Pierre Brémaud
Inria
Paris
France

and

École Polytechnique Fédérale de Lausanne
Lausanne
Switzerland

ISSN 0172-5939   ISSN 2191-6675   (electronic)
ISBN 978-3-319-09589-9   ISBN 978-3-319-09590-5   (eBook)
DOI 10.1007/978-3-319-09590-5

Library of Congress Control Number: 2014946193

Mathematics Subject Classification: 42A38, 42B10, 60G10, 60G12, 60G35, 60G55

Springer Cham Heidelberg New York Dordrecht London

Printed on acid-free paper

Springer is part of Springer Science+Business Media (www.springer.com)

Pour Marion

Contents

Introduction

A unified treatment of all the aspects of Fourier theory relevant to the Fourier analysis of stochastic processes is not only unavoidable, but also intellectually satisfying, and in fact time-saving. This is why this book contains the classical Fourier theory of functions (Fourier series and transforms in L^1 and L^2, z-transforms), the Fourier theory of probability distributions (characteristic functions, convergence in distribution) and the elements of Hilbert space theory (orthogonal projection principle, orthonormal bases, isometric extension theorem) which are indispensable prerequisites for the Fourier analysis of stochastic processes (random fields, time series and point processes) of the last three chapters.

Stochastic processes and random fields are studied from the point of view of their second order properties embodied in two functions: the mean function and the covariance function (or the covariance measure in the case of point process random fields). This limited information suffices to satisfy most demands of signal processing and has made this class of processes a rich source of models: white and coloured noises, ARMA series and, more recently, complex signals based on point processes. Wide-sense stationary processes are, in addition to their usefulness, supported by an elegant theory at the interface of Fourier analysis and Hilbert spaces on one hand, and probability and stochastic processes on the other hand. This book presents the classical results in the field together with more recent ones on the power spectral measure (or Bartlett spectrum) of wide-sense stationary point processes and related stochastic processes, such as (non-Poissonian) shot noises, modulated point processes, random samples, etc.

The mathematical prerequisites on integration and probability theory are reviewed in the Appendix, generally without proof, except for the completeness of L^p-spaces (stating in particular that the square-integrable functions form a Hilbert space), because this result is the foundation of the second-order theory of stochastic processes, together with some results of Hilbert space theory. The latter are given, this time with proofs, in the course of the Chap. 1. No prerequisite on stochastic processes is necessary, since the relevant models (especially Gaussian processes and Brownian motion, but also Markov chains, Poisson processes and renewal processes) are presented in detail in the main text as the need arises.

The Chap. 1 gives the basic Fourier theory of functions. The first section deals with the Fourier theory in L^1, featuring among other topics the Poisson summation formula and the Shannon–Nyquist sampling theorem. The second section gives the theory of the z-transform, which is essential to the Fourier theory of time series in Chap. 4. The third section provides, in a self-contained way, the background in Hilbert spaces which is a prerequisite for the Fourier theory in L^2 of the last section, as well as for the rest of the book.

The Chap. 2 provides the interface between the Fourier analysis of functions and the Fourier analysis of stochastic processes, namely the theory of characteristic functions and the pivotal Bochner theorem, which guarantees under very general conditions the existence and uniqueness of the power spectral measure of wide-sense stationary stochastic processes. This theorem relies on Paul Lévy's theorem of characterization of convergence of probability distributions. A slight extension of the latter to the problem of convergence of finite measures is given since it will be needed in the proof of existence of the power spectral measure, or Bartlett spectrum, of point processes.

Chapters 3–5 concern stochastic processes. There are two types of continuous-time stochastic processes and continuous-space random fields that may be distinguished, as we do in the present text. The classical one and indeed most frequently—if not exclusively—dealt with in textbooks, whether applied or theoretical, concerns mathematical objects, such as Gaussian processes, for which the second-order analysis, in particular the spectral representation theory, is reasonably well-developed and can serve as a starting point for their spectral analysis. The second one concerns point processes. These are underlying many models of the first type (regenerative processes among others) and are in general not studied per se. However, the analysis of signals whose point process structure is essential, such as those arising in biology (the spike trains across the nervous fibers) and in communications (the so-called ultra-wide-band signals in which the information is coded into a sequence of random times) require a special treatment. Chapter 3 concerns the first category of signals, whereas Chap. 5 deals with stochastic processes structured by point processes. Chapter 4 is devoted to discrete-time wide-sense stationary processes, with emphasis on the ARMA models, of interest in signal processing as well as in econometrics, among other fields of application.

The formulation of the mathematical theory, the choice of topics and of examples should allow students and researchers in the applied sciences to recognize the objects and paradigms of their respective trade, the aim being to give a firm mathematical basis on which to develop applications-oriented research. In particular, this book features examples relevant to signal processing and communications: Shannon–Nyquist sampling theorem, transfer functions, white noise, pulse-amplitude modulation, filtering, narrowband signals, sampling in the presence of jitter, spike trains, ultra-wide-band signals, etc. However, this choice of examples in one of the richest domains of application of Fourier theory does not diminish the generality of the mathematical theory developed in the text. It will perhaps remind the reader that Fourier theory was originally a physical theory.

I would like to close this introduction with the expression of my sincere gratitude to Justin Salez who reviewed and corrected important portions of the manuscript.

Paris, March 2014   Pierre Brémaud

Chapter 1
Fourier Analysis of Functions

The classical Fourier theory of functions is an indispensable prerequisite for the Fourier theory of stationary stochastic processes. By "classical" Fourier analysis, we mean Fourier series and Fourier transforms in L^1 and L^2, but also z-transforms which are the backbone of discrete-time signal processing together with the notion of (time-invariant) linear filtering. We spend some time with the famous Poisson summation formula—the bridge between Fourier transforms and Fourier series—which is intimately connected to the celebrated Shannon–Nyquist sampling theorem of signal processing and is of special interest to physicists and engineers in that it justifies the calculations involving the Dirac train of impulses without recourse to distribution theory. For the L^2 Fourier theory of functions and sequences, some background in Hilbert spaces is required. The results obtained, such as the orthogonal projection theorem, the isometric extension theorem and the orthonormal basis theorem, will be recurrently used in the rest of the book.

1.1 Fourier Theory in L^1

1.1.1 Fourier Transform and Fourier Series

The sets of integers, positive integers and relative integers are denoted respectively by $\mathbb{N}$, $\mathbb{N}_+$ and $\mathbb{Z}$. The sets of real numbers, non-negative real numbers and complex numbers are denoted respectively by $\mathbb{R}$, $\mathbb{R}_+$ and $\mathbb{C}$.

The Fourier Transform in L^1

For $p \in \mathbb{N}_+$, $L^p_{\mathbb{C}}(\mathbb{R})$ denotes the collection of measurable functions $f : \mathbb{R} \to \mathbb{C}$ such that $\int_{\mathbb{R}} |f(t)|^p \, dt < \infty$. Similarly, for all intervals $(a, b) \subset \mathbb{R}$, $L^p_{\mathbb{C}}([a, b])$ denotes the collection of measurable functions $f : \mathbb{R} \to \mathbb{C}$ such that $\int_a^b |f(t)|^p \, dt < \infty$.

© Springer International Publishing Switzerland 2014
P. Brémaud, *Fourier Analysis and Stochastic Processes*, Universitext,
DOI 10.1007/978-3-319-09590-5_1

A function $f \in L^1_\mathbb{C}(\mathbb{R})$ is called integrable (with respect to Lebesgue measure). A function $f \in L^2_\mathbb{C}(\mathbb{R})$ is called square-integrable (with respect to Lebesgue measure). $L^p_{\mathbb{C},loc}(\mathbb{R})$ is, by definition, the set of functions f that are in $L^p_\mathbb{C}([a,b])$ for all intervals $(a,b) \subset \mathbb{R}$. A function $f \in L^1_{\mathbb{C},loc}(\mathbb{R})$ is said to be locally integrable. Similar definitions and notation will be used for functions $f : \mathbb{R}^d \to \mathbb{C}$ when $d > 1$.

Definition 1.1.1 The *Fourier transform* (FT) of a function $f \in L^1_\mathbb{C}(\mathbb{R})$ is the function $\widehat{f} : \mathbb{R} \to \mathbb{C}$ defined by:

$$\widehat{f}(v) = \int_\mathbb{R} f(t)\, e^{-2i\pi vt}\, dt.$$

The mapping from the function to its Fourier transform will be denoted by $\mathcal{F} : f \to \widehat{f}$ or more simply, $f \to \widehat{f}$, whenever the context is unambiguous.

Theorem 1.1.1 *The Fourier transform of a function $f \in L^1_\mathbb{C}(\mathbb{R})$ is bounded and uniformly continuous.*

Proof From the definition, we have that

$$|\widehat{f}(v)| \le \int_\mathbb{R} |f(t)\, e^{-2i\pi vt}|\, dt = \int_\mathbb{R} |f(t)|\, dt,$$

where the last term does not depend on v and is finite. Also, for all $h \in \mathbb{R}$,

$$|\widehat{f}(v+h) - \widehat{f}(v)| \le \int_\mathbb{R} |f(t)|\, |e^{-2i\pi(v+h)t} - e^{-2i\pi vt}|\, dt$$

$$= \int_\mathbb{R} |f(t)|\, |e^{-2i\pi ht} - 1|\, dt\,.$$

The last term is independent of v and tends to 0 as $h \to 0$ by dominated convergence. $\square$

From time to time, we will make the following abuse of notation, writing "the function $f(t)$" instead of the more correct "the function f". The context will prevent confusion between the function and its value at a specific value of its domain.

Let f, f_1 and f_2 be in $L^1_\mathbb{C}(\mathbb{R})$, $\lambda_1, \lambda_2 \in \mathbb{C}$, and $a \in \mathbb{R}, a > 0$. The following basic rules are easily derived from the definition of the Fourier transform:

Linearity: $\lambda_1 f_1(t) + \lambda_2 f_2(t) \to \lambda_1 \widehat{f_1}(v) + \lambda_2 \widehat{f_2}(v)$
Conjugation: $f^*(t) \to \widehat{f}(-v)^*$.
Modulation: $e^{2i\pi v_0 t} f(t) \to \widehat{f}(v - v_0)$.
Delay: $f(t - t_0) \to e^{-2i\pi v t_0} \widehat{f}(v)$.
Doppler: $f(at) \to \frac{1}{|a|} \widehat{f}\left(\frac{v}{a}\right)$.

It follows from the conjugation rule that if the function is *real*, its FT is *Hermitian even*, that is:

$$\widehat{f}(\nu) = \widehat{f}(-\nu)^*.$$

(Exercise 1.5.1 features other relations of the same kind.)

Example 1.1.1 (The rectangular pulse) Define the *rectangular pulse* of width $T > 0$ by

$$\text{rec}_T(t) = 1_{[-T/2,+T/2]}(t).$$

An elementary computation gives

$$\mathcal{F} : \text{rec}_T(t) \rightarrow T\text{sinc}(\nu T),$$

where

$$\text{sinc}(x) = \frac{\sin(\pi x)}{\pi x}$$

is the *cardinal sine function*.

Example 1.1.2 (One-sided exponential) For $a > 0$,

$$\mathcal{F} : e^{-at} 1_{\mathbb{R}_+}(t) \rightarrow \frac{1}{a + 2i\pi\nu}.$$

Proof Let $A := \int_0^\infty e^{-at} \cos(2\pi\nu t) \, dt$, $B := \int_0^\infty e^{-at} \sin(2\pi\nu t) \, dt$. Integration by parts gives

$$-1 = \left(e^{-at} \cos(2\pi\nu t)\right)_0^\infty = -aA - 2\pi\nu B$$

and

$$0 = \left(e^{-at} \sin(2\pi\nu t)\right)_0^\infty = -aB + 2\pi\nu A,$$

from which we obtain

$$\int_0^\infty e^{-at} e^{+2i\pi\nu t} \, dt = A + iB = \frac{1}{a - 2i\pi\nu},$$

which, after replacement of ν by $-\nu$, gives the announced result. □

Example 1.1.3 (Two-sided exponential) We deduce from the result of Example 1.1.2 the FT of the two-sided exponential:

$$\mathcal{F}: e^{-a|t|} \rightarrow \frac{2a}{a^2 + 4\pi^2 v^2}.$$

Proof $\int_{-\infty}^0 e^{at} e^{-2i\pi vt}\, dt = \int_0^{+\infty} e^{-at} e^{+2i\pi vt}\, dt = \frac{1}{a-2i\pi v}$, and therefore

$$\int_{-\infty}^{+\infty} e^{-a|t|} e^{-2i\pi vt}\, dt = \frac{1}{a + 2i\pi v} + \frac{1}{a - 2i\pi v} = \frac{2a}{a^2 + 4\pi^2 v^2}. \qquad \square$$

The Riemann–Lebesgue Lemma

Theorem 1.1.2 *The FT $\widehat{f}$ of a function $f \in L_{\mathbb{C}}^1(\mathbb{R})$ is such that*

$$\lim_{|v|\to\infty} |\widehat{f}(v)| = 0.$$

Proof The FT of a rectangular pulse is bounded in absolute value by the function $v \to K/|v|$ for some finite constant K (Example 1.1.1). Hence, every function that is a finite linear combination of indicator functions of intervals satisfies the same property. Such finite combinations are dense in $L_{\mathbb{C}}^1(\mathbb{R})$ (Corollary A.3.1). Therefore there exists a sequence $\{f_n\}_{n\geq 1}$ of integrable functions such that

$$\lim_{n\to\infty} \int_{\mathbb{R}} |f_n(t) - f(t)|\, dt = 0,$$

and

$$|\widehat{f_n}(v)| \leq \frac{K_n}{|v|},$$

for finite numbers K_n. From the inequality

$$|\widehat{f}(v) - \widehat{f_n}(v)| \leq \int_{\mathbb{R}} |f(t) - f_n(t)|\, dt,$$

we deduce that

$$|\widehat{f}(v)| \leq |\widehat{f_n}(v)| + \int_{\mathbb{R}} |f(t) - f_n(t)|\, dt$$

$$\leq \frac{K_n}{|v|} + \int_{\mathbb{R}} |f(t) - f_n(t)|\, dt,$$

and therefore

$$\limsup_{|v| \to \infty} |\widehat{f}(v)| \le \int_{\mathbb{R}} |f(t) - f_n(t)| \, dt$$

from which the conclusion follows by letting $n \uparrow \infty$. □

The Convolution-Multiplication Rule

Let h and f be two integrable functions. Then the right-hand side of

$$g(t) = \int_{\mathbb{R}} h(t - s) f(s) \, ds$$

is defined almost-everywhere, and defines almost-everywhere an integrable function.

Proof By Tonelli's theorem and the integrability assumptions

$$\int_{\mathbb{R}} \int_{\mathbb{R}} |h(t - s)| \, |f(s)| \, dt \, ds = \left(\int_{\mathbb{R}} |h(t)| \, dt \right) \left(\int_{\mathbb{R}} |f(t)| \, dt \right) < \infty.$$

This implies that, for almost all t,

$$\int_{\mathbb{R}} |h(t - s) f(s)| \, ds < \infty.$$

The integral $\int_{\mathbb{R}} h(t - s) f(s) \, ds$ is therefore well-defined for almost all t. Also

$$\int_{\mathbb{R}} |g(t)| \, dt = \int_{\mathbb{R}} \left| \int_{\mathbb{R}} h(t - s) f(s) \, ds \right| dt$$

$$\le \int_{\mathbb{R}} \int_{\mathbb{R}} |h(t - s) f(s)| \, dt \, ds < \infty.$$

Therefore g is integrable. □

The function g is the *convolution* of h with f, and is denoted by $g = h * f$.

Theorem 1.1.3 *The convolution-multiplication rule*

$$\mathcal{F} : h * f \to \widehat{h} \widehat{f}$$

applies whenever h and f are functions in $L_{\mathbb{C}}^1(\mathbb{R})$.

Proof We have

$$\int_{\mathbb{R}} \left(\int_{\mathbb{R}} h(t-s) f(s)\, ds \right) e^{-2i\pi vt}\, dt$$

$$= \int_{\mathbb{R}} \int_{\mathbb{R}} h(t-s) e^{-2i\pi v(t-s)} f(s) e^{-2i\pi vs}\, ds\, dt$$

$$= \int_{\mathbb{R}} f(s) e^{-2i\pi vs} \left(\int_{\mathbb{R}} h(t-s) e^{-2i\pi v(t-s)}\, dt \right) ds = \widehat{h}(v)\widehat{f}(v),$$

by Fubini's theorem, which is applicable here because the function

$$(t,s) \to \left| h(t-s) f(s) e^{-2i\pi vt} \right| = |h(t-s) f(s)|$$

is integrable with respect to the product measure $dt \times ds$, as a previous computation has shown. □

Example 1.1.4 (The triangular pulse) The convolution of the rectangular pulse with itself is the *triangular pulse* of width $2T$ and height T,

$$\Delta_T(t) = (T - |t|) 1_{[-T,+T]}(t),$$

and therefore, from the result of Example 1.1.1 and the convolution-multiplication rule,

$$\mathcal{F}: \Delta_T(t) \to (T\,\mathrm{sinc}(vT))^2 .$$

The Differentiation Rule

Denote by C^n the set of functions form $\mathbb{R}$ to $\mathbb{C}$ that have continuous derivatives up to the nth order.

Theorem 1.1.4 *(a) If the integrable function f is such that $t \to t^k f(t)$ is integrable for all $1 \le k \le n$, then its FT is in C^n, and*

$$\mathcal{F}: (-2i\pi t)^k f(t) \to \widehat{f}^{(k)}(v) \ \text{for all } 1 \le k \le n.$$

(b) If the function $f \in C^n$ is, together with its n first derivatives, integrable, then

$$\mathcal{F}: f^{(k)}(t) \to (2i\pi v)^k \widehat{f}(v) \ \text{for all } 1 \le k \le n.$$

Proof (a) In the right-hand side of the expression

$$\widehat{f}(v) = \int_{\mathbb{R}} e^{-2i\pi vt} f(t)\,dt,$$

we can differentiate k times ($k \leq n$) under the integral sign and obtain

$$\widehat{f}^{(k)}(v) = \int_{\mathbb{R}} (-2i\pi t)^k e^{-2i\pi t} f(t)\,dt.$$

(b) It suffices to prove this for $n = 1$ and iterate the result. We first observe that $\lim_{|a|\uparrow\infty} f(a) = 0$. Indeed, with $a > 0$ for instance,

$$f(a) = f(0) + \int_0^a f'(t)dt,$$

and therefore, since $f' \in L^1_{\mathbb{C}}(\mathbb{R})$, the limit exists and is finite. But this limit must be 0, otherwise f would not be integrable. Now, the FT of the derivative f' is

$$\int_{\mathbb{R}} e^{-2i\pi vt} f'(t)\,dt = \lim_{a\uparrow\infty} \int_{-a}^{+a} e^{-2i\pi vt} f'(t)\,dt.$$

Integrating by parts:

$$\int_{-a}^{+a} e^{-2i\pi vt} f'(t)\,dt = \left(e^{-2i\pi vt} f(t)\right)_{-a}^{+a} + (2i\pi v)\int_{-a}^{+a} e^{-2i\pi vt} f(t)\,dt.$$

It then suffices to let a tend to ∞ to obtain the announced result. □

Example 1.1.5 (The Gaussian pulse) The *Gaussian pulse* is its own FT:

$$\mathcal{F}: e^{-\pi t^2} \to e^{-\pi v^2}.$$

Proof Differentiating the function $f(t) = e^{-\pi t^2}$, we obtain $f'(t) = -2\pi t\, f(t)$ and therefore, from Theorem 1.1.4, $2i\pi v \widehat{f}(v) = -i\widehat{f'}(v)$, that is

$$(2\pi v)\widehat{f}(v) + \widehat{f}'(v) = 0.$$

This differential equation has for solution $\widehat{f}(v) = Ke^{-\pi v^2}$. Since $\widehat{f}(0) = \int_{-\infty}^{+\infty} e^{-\pi t^2} dt = 1$, necessarily $K = 1$. □

From the above expression of the FT of the Gaussian pulse and the Doppler rule, we have that for $\alpha > 0$,

$$\mathcal{F}: e^{-\alpha t^2} \rightarrow \sqrt{\frac{\pi}{\alpha}}\, e^{-\frac{\pi^2}{\alpha} v^2}.$$

From Theorem 1.1.4 and the Riemann–Lebesgue lemma (Theorem 1.1.2), we have:

Theorem 1.1.5 *If the function $f : \mathbb{R} \to \mathbb{C}$ in C^n is, together with its n first derivatives, integrable, then*

$$\widehat{f}(v) = o\left(v^{-n}\right) \quad (v \to \infty),$$

that is: $\lim_{v \to \infty} \widehat{f}(v)\, v^n = 0.$

In other words: "smooth functions have fast-decay Fourier transforms".

Fourier Series of Periodic Functions in L^1_{loc}

A function $f : \mathbb{R} \to \mathbb{C}$ is called *periodic* with period $T > 0$ (or T-periodic) if for all $t \in \mathbb{R}$,

$$f(t + T) = f(t).$$

A periodic function is not integrable unless it is almost everywhere null (Exercise 1.5.8). Clearly, a T-periodic function f is locally integrable if and only if $f \in L^1_{\mathbb{C}}([0, T])$.
The relevant notion now is that of Fourier series.[1]

Definition 1.1.2 The Fourier coefficients sequence $\{\widehat{f}_n\}_{n\in\mathbb{Z}}$ of the locally integrable T-periodic function $f : \mathbb{R} \to \mathbb{C}$ is defined by the formula

$$\widehat{f}_n = \frac{1}{T} \int_0^T f(t) e^{-2i\pi \frac{n}{T} t}\, dt,$$

and $\widehat{f}_n$ is called the nth *Fourier coefficient* of the function f.

The nth Fourier coefficient is the value at frequency $v = \frac{n}{T}$ of the Fourier transform of the (integrable) function $t \to \frac{1}{T} f(t) 1_{[0,T]}(t)$. A number of properties of the Fourier coefficients follow from this remark and the corresponding properties of the Fourier transform.

---

[1] Note however that, historically, Fourier series were introduced before Fourier transforms, in contrast with the order of appearance chosen in this text.

Example 1.1.6 Let f be a $2T$-periodic function equal to the triangular pulse of height T and width $2T$ on $[-T, +T]$. Since the Fourier transform of the triangular pulse of height T and width $2T$ is $\widehat{f}(v) = \left(T \frac{\sin(\pi v T)}{\pi v T}\right)^2$, the Fourier coefficients of f are $\widehat{f_0} = \frac{T}{2}$ and for $n \neq 0$,

$$\widehat{f_n} = \frac{2T}{\pi^2 n^2} 1_{\{n \text{ is odd}\}}$$

The formal Fourier series associated with the locally integrable T-periodic function f is the formal function

$$S_f(t) := \sum_{n \in \mathbb{Z}} \widehat{f_n} e^{+2i\pi \frac{n}{T} t}.$$

The qualification "formal" is used to insist on the necessity of giving a meaning to the (formal) sum on the right (does it converge, and in what sense?). This question will receive some answers in the next subsection.

When f is real, it is common to express the formal Fourier series in terms of sines and cosines. In fact, in the real case, $\widehat{f}_{-n} = \widehat{f}_n^*$ and therefore, denoting $\widehat{f}_n = a'_n + i c'_n$, we have

$$S_f(t) = \widehat{f_0} + \sum_{n=1}^{\infty} \left(\widehat{f_n} e^{2i\pi \frac{n}{T} t} + \widehat{f}_{-n} e^{-2i\pi \frac{n}{T} t}\right)$$

$$= \widehat{f_0} + \sum_{n=1}^{\infty} 2\text{Re}\left(\widehat{f_n} e^{2i\pi \frac{n}{T} t}\right)$$

$$= \widehat{f_0} + 2\sum_{n=1}^{\infty} a'_n \cos\left(2\pi \frac{n}{T} t\right) - 2\sum_{n=1}^{\infty} c'_n \sin\left(2\pi \frac{n}{T} t\right),$$

where $a'_n = \frac{1}{T} \int_0^T f(t) \cos(2\pi \frac{n}{T} t) \, dt$ and $c'_n = \frac{1}{T} \int_0^T f(t) \sin(-2\pi \frac{n}{T} t) \, dt = -\frac{1}{T} \int_0^T f(t) \sin(2\pi \frac{n}{T} t) \, dt$. Therefore

$$S_f(t) = \frac{1}{2} a_0 + \sum_{n=1}^{\infty} a_n \cos\left(2\pi \frac{n}{T} t\right) + \sum_{n=1}^{\infty} b_n \sin\left(2\pi \frac{n}{T} t\right),$$

where

$$a_n := \frac{2}{T} \int_0^T f(t) \cos\left(2\pi \frac{n}{T} t\right) dt \quad \text{and} \quad b_n := \frac{2}{T} \int_0^T f(t) \sin\left(2\pi \frac{n}{T} t\right) dt.$$

In textbooks of pure mathematics (as well as in the more theoretical Sect. 1.1.2), the results are often expressed with the choice $T = 2\pi$ for the period. The Fourier sum is then

$$S_f(t) = \sum_{n \in \mathbb{Z}} c_n(f) e^{inx} ,$$

where

$$c_n(f) = \frac{1}{2\pi} \int\limits_{-\pi}^{+\pi} f(u) e^{inu} \, du ,$$

and if the function f is real,

$$S_f(x) = \frac{1}{2} a_0 + \sum_{n=1}^{\infty} a_n \cos(nx) + \sum_{n=1}^{\infty} b_n \sin(nx) ,$$

where

$$a_n := \frac{1}{\pi} \int\limits_{-\pi}^{+\pi} f(u) \cos(nu) \, du \text{ and } b_n := \frac{1}{\pi} \int\limits_{-\pi}^{+\pi} f(u) \sin(nu) \, du .$$

From the remark following Definition 1.1.2 and the Riemann–Lebesgue lemma (Theorem 1.1.2), we have the following consequence of the latter:

Corollary 1.1.1 *For the Fourier coefficients sequence of a locally integrable periodic function $f : \mathbb{R} \to \mathbb{C}$, it holds that $\lim_{|n| \uparrow \infty} \widehat{f_n} = 0$.*

We shall need later on the following result that may be considered as an extension of the Riemann–Lebesgue lemma.

Theorem 1.1.6 *Let $f : \mathbb{R} \to \mathbb{C}$ be a 2π-periodic locally integrable function, and let $g : [a, b] \to \mathbb{C}$ be in C^1, where $[a, b] \subseteq [-\pi, +\pi]$. Then*

$$\lim_{\lambda \uparrow \infty} \int\limits_a^b f(x - u) g(u) \sin(\lambda u) \, du = 0$$

uniformly in x.

Proof For arbitrary $\varepsilon > 0$ select a 2π-periodic function h (anticipate on Corollary 1.1.8 or see Theorem A.3.1 and Example 1.3.2) in C^1 such that

$$\int\limits_{-\pi}^{+\pi} |f(x) - h(x)| \, dx < \varepsilon .$$

Integrating by parts yields

$$I(\lambda) := \int_a^b h(x-u)g(u) \sin(\lambda u) \, du$$

$$= -h(x-u)g(u) \frac{\cos(\lambda u)}{\lambda} \Big|_a^b + \int_a^b [h(x-u)g(u)]' \frac{\cos(\lambda u)}{\lambda} \, du.$$

Since $h \in C^1$ and is periodic, h and h' are uniformly bounded. The same is true of g, g' (g is in C^1). Therefore

$$\lim_{\lambda \uparrow \infty} I(\lambda) = 0 \quad \text{uniformly in } x.$$

Now,

$$\left| \int_a^b f(x-u)g(u) \sin(\lambda u) \, du \right|$$

$$\leq |I(\lambda)| + \int_a^b |h(x-u) - f(x-u)| \, |g(u)| \sin(\lambda u) \, du$$

$$\leq |I(\lambda)| + \max_{a \leq u \leq b} |g(u)| \int_a^b |h(x-u) - f(x-u)| \sin(\lambda u) \, du$$

$$\leq |I(\lambda)| + \max_{a \leq u \leq b} |g(u)| \, \varepsilon.$$

The conclusion then follows because ε is arbitrary. $\square$

Inversion Formulas

An inversion formula is, in the present context, a formula that allows to pass from the Fourier transform, or the Fourier series, to the function itself. Ideally, for the Fourier transform, we would like a formula like

$$f(t) = \int_{\mathbb{R}} \widehat{f}(v) e^{+2i\pi vt} \, dv$$

and, for the Fourier series,

$$f(t) = \sum_{n\in\mathbb{Z}} \widehat{f}_n e^{+2i\pi\frac{n}{T}t}.$$

These are the *Fourier decompositions* of the function f. In this subsection we shall give rather stringent conditions under which such decompositions are available. We shall see that both hold *almost-everywhere*, the first one under the condition that the Fourier transform is *integrable*, and the second one under the condition that the Fourier coefficients sequence be absolutely summable. These results will then be improved in the next subsection.

In the proofs of the inversion formulas for Fourier transforms and Fourier series, the following notion, of independent interest, will be helpful.

Definition 1.1.3 A *regularization kernel* is a function $h_\sigma : \mathbb{R} \to \mathbb{R}$ depending on a parameter $\sigma > 0$, and such that

(i) $h_\sigma(u) \geq 0$ for all $u \in \mathbb{R}$
(ii) $\int_\mathbb{R} h_\sigma(u)\, du = 1$, for all $\sigma > 0$,
(iii) $\lim_{\sigma\downarrow 0} \int_{-a}^{+a} h_\sigma(u)\, du = 1$, for all $a > 0$.

A regularization kernel h_σ is an approximation of the so-called Dirac's generalized function δ, in that for all $\varphi \in C_c$ (continuous functions with compact support) ,

$$\lim_{\sigma\downarrow 0} \int_\mathbb{R} h_\sigma(t)\,\varphi(t)\, dt = \varphi(0).$$

Example 1.1.7 (The Gaussian regularization kernel) The verification that

$$h_\sigma(t) = \frac{1}{\sigma\sqrt{2\pi}}\, e^{-\frac{t^2}{2\sigma^2}}$$

is a regularization kernel is left as an exercise.

The following result is referred to as the *regularization lemma*.

Theorem 1.1.7 *Let* $h_\sigma : \mathbb{R} \to \mathbb{R}$ *be a regularization kernel. Let* f *be a function in* $L^1_\mathbb{C}(\mathbb{R})$. *Then*

$$\lim_{\sigma\downarrow 0} \int_\mathbb{R} |(f * h_\sigma)(t) - f(t)|\, dt = 0.$$

Proof Observe that

$$\int_\mathbb{R} |(f * h_\sigma)(t) - f(t)|\, dt = \int_\mathbb{R} \left| \int_\mathbb{R} (f(t-u) - f(t)) h_\sigma(u)\, du \right| dt$$

(using condition (ii)), and therefore, defining $\tilde{f}(u) = \int_{\mathbb{R}} |f(t-u) - f(t)| \, dt$,

$$\int_{\mathbb{R}} |(f * h_\sigma)(t) - f(t)| \, dt \leq \int_{\mathbb{R}} \tilde{f}(u) h_\sigma(u) \, du \, .$$

We must therefore prove that

$$\lim_{\sigma \downarrow 0} \int_{\mathbb{R}} \tilde{f}(u) h_\sigma(u) = 0 \, .$$

Suppose that we can prove that $\lim_{|u| \downarrow 0} \tilde{f}(u) = 0$ (to be done in a few lines). Therefore, for any given $\varepsilon > 0$, there exists $a = a(\varepsilon)$ such that (using (i))

$$\int_{-a}^{+a} \tilde{f}(u) h_\sigma(u) du \leq \frac{\varepsilon}{2} \int_{-a}^{+a} h_\sigma(u) du \leq \frac{\varepsilon}{2} \, .$$

Since $\tilde{f}$ is bounded (say, by M)

$$\int_{\mathbb{R} \setminus [-a, +a]} \tilde{f}(u) h_\sigma(u) du \leq M \int_{\mathbb{R} \setminus [-a, +a]} h_\sigma(u) du \, .$$

By (ii) and (iii), the last integral is, for sufficiently small σ, lesser than $\frac{\varepsilon}{2M}$. Therefore, for sufficiently small σ,

$$\int_{\mathbb{R}} \tilde{f}(u) h_\sigma(u) du \leq \frac{\varepsilon}{2} + \frac{\varepsilon}{2} = \varepsilon.$$

It now remains to prove that $\lim_{|u| \downarrow 0} \tilde{f}(u) = 0$.

We begin with the case where f is continuous with compact support. In particular, it is uniformly bounded. Since we are interested in a limit as $|u|$ tends to 0, we may suppose that u is in a bounded interval around 0, and in particular, the function $t \to |f(t-u) - f(t)|$ is bounded uniformly in u by an integrable function. The result then follows by dominated convergence.

Let now f be only integrable. Let $\{f_n\}_{n \geq 1}$ be a sequence of continuous functions with compact support that converges in $L_{\mathbb{C}}^1(\mathbb{R})$ to f (Theorem A.3.3). Writing

$$\tilde{f}(u) \leq d(f(\cdot - u), f_n(\cdot - u)) + \int_{\mathbb{R}} |f_n(t-u) - f_n(t)| \, dt + d(f, f_n)$$

$$= \int_{\mathbb{R}} |f_n(t-u) - f_n(t)|\, dt + 2\, d(f, f_n)$$

where $d(f, f_n) := \int_{\mathbb{R}} |f(t) - f_n(t)|\, dt$, the result easily follows. $\square$

The Gaussian Kernel

In spite of the fact that the FT of an integrable function is uniformly bounded and uniformly continuous, it is not necessarily integrable. For instance, the FT of the rectangular pulse is the cardinal sine, a non-integrable function. However, if its FT is integrable, the function admits a Fourier decomposition:

Theorem 1.1.8 *Let f be a function in $L^1_{\mathbb{C}}(\mathbb{R})$ with Fourier transform $\widehat{f}$. Under the additional condition $\widehat{f} \in L^1_{\mathbb{C}}(\mathbb{R})$, the inversion formula*

$$f(t) = \int_{\mathbb{R}} \widehat{f}(v) e^{+2i\pi vt}\, dv$$

holds for almost all t, and for all t if f is, in addition to the above assumptions, continuous.

(Note that in the inversion formula the exponent of the exponential of the integrand is $+2i\pi vt$.)

Proof We first check (Exercise 1.5.4) that the above result is true for the function

$$e_{\alpha,a}(t) = e^{-\alpha t^2 + at} \qquad (\alpha \in \mathbb{R},\ \alpha > 0,\ a \in \mathbb{C}).$$

Let now f be an integrable function and consider the Gaussian kernel of Example 1.1.7, whose FT is

$$\widehat{h}_\sigma(v) = e^{-2\pi^2\sigma^2 v^2}.$$

We first show that the inversion formula is true for the convolution $f * h_\sigma$. Indeed,

$$(f * h_\sigma)(t) = \int_{\mathbb{R}} f(u) h_\sigma(u) e_{\frac{1}{2\sigma^2}, \frac{u}{\sigma^2}}(t)\, du, \qquad (*)$$

and the FT of this function is, by the convolution–multiplication formula $\widehat{f}(v)\widehat{h}_\sigma(v)$. Computing this FT directly from the right-hand side of $(*)$, we obtain

$$\widehat{f}(v)\widehat{h}_\sigma(v) = \int_{\mathbb{R}} f(u)h_\sigma(u)\left(\int_{\mathbb{R}} e_{\frac{1}{2\sigma^2},\frac{u}{\sigma^2}}(t)e^{-2i\pi vt}\,dt\right)du$$

$$= \int_{\mathbb{R}} f(u)h_\sigma(u)\widehat{e}_{\frac{1}{2\sigma^2},\frac{u}{\sigma^2}}(v)\,du\,.$$

Using the result of Example 1.1.5,

$$\int_{\mathbb{R}} \widehat{f}(v)\widehat{h}_\sigma(v)e^{2i\pi vt}\,dv = \int_{\mathbb{R}}\left(\int_{\mathbb{R}} f(u)h_\sigma(u)\widehat{e}_{\frac{1}{2\sigma^2},\frac{u}{\sigma^2}}(v)\,du\right)e^{2i\pi vt}\,dv$$

$$= \int_{\mathbb{R}} f(u)h_\sigma(u)e_{\frac{1}{2\sigma^2},\frac{u}{\sigma^2}}(t)\,du$$

$$= (f * h_\sigma)(t).$$

Therefore, we have

$$(f * h_\sigma)(t) = \int_{\mathbb{R}} \widehat{f}(v)\widehat{h}_\sigma(v)e^{2i\pi vt}\,dv, \qquad (**)$$

and this is the inversion formula for $f * h_\sigma$.

Since for all $v \in \mathbb{R}$ we have that $\lim_{\sigma\downarrow 0} \widehat{h}_\sigma(v) = 1$, it follows from Lebesgue's dominated convergence theorem that when $\sigma \downarrow 0$ the right-hand side of $(**)$ tends to

$$\int_{\mathbb{R}} \widehat{f}(v)e^{2i\pi vt}\,dv$$

for all $t \in \mathbb{R}$. But, when $\sigma \downarrow 0$, the function on the left-hand side of $(**)$ converges in $L^1_{\mathbb{C}}(\mathbb{R})$ to f, by the regularization lemma (Theorem 1.1.7). The announced equality then follows from Theorem A.1.30 which says that if a given function has an almost-everywhere limit and a limit in $L^1_{\mathbb{C}}(\mathbb{R})$, both limits are almost-everywhere equal.

Suppose that, in addition, f is continuous. The right-hand side of the inversion formula defines a continuous function because $\widehat{f}$ is integrable. The everywhere equality in the inversion formula follows from the fact that two continuous functions that are almost-everywhere equal are in fact *everywhere* equal (see Theorem A.1.11). □

The Fourier transform characterizes an integrable function:

Corollary 1.1.2 *If two integrable functions $f_1 : \mathbb{R} \to \mathbb{C}$ and $f_2 : \mathbb{R} \to \mathbb{C}$ have the same Fourier transform, then they are almost everywhere equal.*

Proof The function $f = f_1 - f_2$ has the FT $\widehat{f} \equiv 0$, which is integrable, and therefore by the inversion formula, $f(t) = 0$ for almost all t. □

Example 1.1.8 (Band-pass functions) The terminology in this example is that of communications theory. A *band-pass* (v_0, B) function, where $B < v_0$, is an integrable function whose FT is null if $|v| \notin [-B + v_0, v_0 + B]$. A *base-band* (B) function f is an integrable function whose FT is null outside the interval $[-B, +B]$. It will be assumed, moreover, that f is real, and hence that its FT is Hermitian even:

$$\hat{f}(-v) = \hat{f}(v)^*.$$

Such function admits the representation

$$f(t) = m(t)\cos(2\pi v_0 t) - n(t)\sin(2\pi v_0 t), \qquad (1.1)$$

where m and n are two functions that are *real* and *base-band* (B), called the *quadrature components* of the band-pass function f. One way of proving (1.1) is to form the *analytic transform* of f

$$f_a(t) = 2\int_0^\infty \hat{f}(v)e^{2i\pi vt}\,dv,$$

and then its *complex envelope*

$$u(t) = \int_{\mathbb{R}} \hat{f}_a(v + v_0)e^{2i\pi vt}\,dv.$$

One remarks that the FT of the function $t \to \mathrm{Re}\{u(t)e^{2i\pi v_0 t}\}$ is $\hat{f}$, and therefore

$$f(t) = \mathrm{Re}\{u(t)e^{2i\pi v_0 t}\}. \qquad (*)$$

Let $m(t)$ and $n(t)$ be the real and imaginary parts of $u(t)$:

$$u(t) = m(t) + in(t). \qquad (**)$$

The quadrature decomposition (1.1) follows from $(*)$ and $(**)$.

Example 1.1.9 (The heat equation) Consider the partial differential equation

$$\frac{\partial \theta}{\partial t} = \kappa\frac{\partial^2 \theta}{\partial^2 x},$$

the so-called *heat equation*, where $\theta(x, t)$ is the temperature at time $t \geq 0$ and at location $x \in \mathbb{R}$ of an infinite rod, and κ is the heat conductance. The initial temperature distribution at time 0 is given:

$$\theta(x, 0) = f(x),$$

where f is assumed integrable. We seek a solution such that for each $t \geq 0$, the function $x \to \theta(x, t)$ is in C^2 and has integrable derivatives up to order 2.

Let $\xi \to \Theta(\xi, t)$ be the FT of $x \to \theta(x, t)$ (the variable ξ is a "spatial frequency" corresponding to the spatial argument x, hence the different notation). In the frequency domain, the heat equation becomes (see part (b) of Theorem 1.1.4)

$$\frac{d\,\Theta(\xi, t)}{dt} = -\kappa(4\pi^2\xi^2)\Theta(\xi, t),$$

with the initial condition $\Theta(\xi, 0) = F(\xi)$, where $F(\xi)$ is the FT of $f(x)$. The solution is

$$\Theta(\xi, t) = F(\xi)e^{-4\pi^2\kappa\xi^2 t}.$$

Since $x \to (4\pi\kappa t)^{-1/2}e^{(4\kappa t)^{-1/2}x^2}$ has the FT $\xi \to e^{-4\pi^2\kappa\xi^2 t}$, we have (by the convolution–multiplication formula and the inversion formula)

$$\theta(x, t) = (4\pi\kappa t)^{-\frac{1}{2}} \int_{\mathbb{R}} f(x - y)e^{-\frac{y^2}{4\kappa t}}\,dy,$$

or equivalently

$$\theta(x, t) = \frac{1}{\sqrt{\pi}} \int_{\mathbb{R}} f(x - 2\sqrt{\kappa t}\,y)e^{-y^2}\,dy.$$

The Poisson Kernel

The *Poisson kernel* will play in the proof of the Fourier series inversion formula a role similar to that of the Gaussian kernel in the proof of the Fourier transform inversion formula.

The Poisson kernel is the family of functions $P_r : \mathbb{R} \mapsto \mathbb{C}$ defined by

$$P_r(t) = \sum_{n \in \mathbb{Z}} r^{|n|}e^{2i\pi\frac{n}{T}t}$$

for $0 < r < 1$. It is T-periodic, and elementary computations reveal that

$$P_r(t) = \sum_{n \geq 0} r^n e^{2i\pi\frac{n}{T}t} + \sum_{n \geq 0} r^n e^{-2i\pi\frac{n}{T}t} - 1 = \frac{1 - r^2}{\left|1 - re^{2i\pi\frac{t}{T}}\right|^2}.$$

and therefore

$$P_r(t) \geq 0.$$

Also

$$\frac{1}{T}\int_{-T/2}^{+T/2} P_r(t)\,dt = 1.$$

In view of the above expression of the Poisson kernel, we have the bound

$$\frac{1}{T}\int_{[-\frac{T}{2},+\frac{T}{2}]\setminus[-\varepsilon,+\varepsilon]} P_r(t)\,dt \le \frac{1-r^2}{\left|1-e^{2i\pi\frac{\varepsilon}{T}}\right|^2},$$

and therefore, for all $\varepsilon > 0$,

$$\lim_{r\uparrow 1}\frac{1}{T}\int_{[-\frac{T}{2},+\frac{T}{2}]\setminus[-\varepsilon,+\varepsilon]} P_r(t)\,dt = 0.$$

The above properties make of the Poisson kernel a regularization kernel, and in particular

$$\lim_{r\uparrow 1}\frac{1}{T}\int_{-\frac{T}{2}}^{+\frac{T}{2}} \varphi(t)P_r(t)\,dt = \varphi(0),$$

for all bounded continuous $\varphi : \mathbb{R} \to \mathbb{C}$ (same proof as in Theorem 1.1.7; see also the remark following Definition 1.1.3).

The following result is similar to the Fourier inversion formula for integrable functions (Theorem 1.1.8).

Theorem 1.1.9 *Let $f : \mathbb{R} \to \mathbb{C}$ be a T-periodic locally integrable complex function with Fourier coefficients $\widehat{f}_n$, $n \in \mathbb{Z}$. If*

$$\sum_{n\in\mathbb{Z}} |\widehat{f}_n| < \infty,$$

then we have the inversion formula

$$f(t) = \sum_{n\in\mathbb{Z}} \widehat{f}_n e^{+2i\pi\frac{n}{T}t}$$

which holds for almost all $t \in \mathbb{R}$, and for ALL t if f is, in addition, a continuous function.

Proof The proof is similar to that of Theorem 1.1.8. We have

$$\sum_{n\in\mathbb{Z}} \widehat{f}_n r^{|n|} e^{2i\pi \frac{n}{T}t} = \frac{1}{T} \int_{-\frac{T}{2}}^{+\frac{T}{2}} f(u) P_r(t-u)\, du, \qquad (*)$$

and (regularization lemma)

$$\lim_{r\uparrow 1} \int_{-\frac{T}{2}}^{+\frac{T}{2}} \left| \int_{-\frac{T}{2}}^{+\frac{T}{2}} f(u) P_r(t-u) \frac{du}{T} - f(t) \right| dt = 0,$$

that is to say: the right-hand side of (*) tends to f in $L^1_{\mathbb{C}}([-T/2, +T/2])$ as $r \uparrow 1$. Since $\sum_{n\in\mathbb{Z}} |\widehat{f}_n| < \infty$, the function of t in the left-hand side of (*) tends to $\sum_{n\in\mathbb{Z}} \widehat{f}_n e^{+2i\pi(n/T)t}$ pointwise. The result then follows again from Theorem A.1.30 which says that if a given function has an almost-everywhere limit and a limit in $L^1_{\mathbb{C}}(\mathbb{R})$, both limits are almost-everywhere equal.

The statement in the case where f is continuous is proved exactly as the corresponding statement in Theorem 1.1.8. □

In particular,

Corollary 1.1.3 *Let $f : \mathbb{R} \to \mathbb{C}$ be a T-periodic locally integrable complex function with Fourier coefficients $\widehat{f}_n$, $n \in \mathbb{Z}$. If f is continuous and $\sum_{n\in\mathbb{Z}} |\widehat{f}_n| < \infty$, then*

$$f(0) = \sum_{n\in\mathbb{Z}} \widehat{f}_n \,.$$

This simple formula may be used to compute the sum of series.

Example 1.1.10 Applying the above result to the triangle function of Example 1.1.6, we obtain $T = \frac{T}{2} + 2\sum_{n=1, n\,odd}^{\infty} \frac{2T}{\pi^2 n^2}$, that is

$$\frac{\pi^2}{8} = \sum_{m\geq 1} \frac{1}{(2m+1)^2}\,.$$

As in the case of integrable functions, we deduce from the inversion formula the *uniqueness theorem*:

Corollary 1.1.4 *If two locally integrable T-periodic functions from $\mathbb{R}$ to $\mathbb{C}$ have the same Fourier coefficients, they are equal almost everywhere.*

The Spatial Case

The definitions and results above extend to the spatial case. For instance let f : $\mathbb{R}^n \to \mathbb{C}$, that is $t = (t_1, \ldots, t_n) \in \mathbb{R}^n \to f(t) \in \mathbb{C}$. The Fourier transform of this function is, when it exists, a function $\widehat{f} : \mathbb{R}^n \to \mathbb{C}$ that is, $v = (v_1, \ldots, v_n) \in \mathbb{R}^n \to \widehat{f}(v) \in \mathbb{C}$. We shall occasionally only quote an important result, but we shall omit the proofs since these are the same, *mutatis mutandis*, as in the univariate case. For instance:

If f is in $L_{\mathbb{C}}^1(\mathbb{R}^n)$ then the Fourier transform $\widehat{f}$ is well defined by

$$\widehat{f}(v) = \int_{\mathbb{R}^n} f(t)\, e^{2i\pi \langle t, v\rangle} dt\,,$$

where $\langle t, v\rangle := \sum_{k=1}^n t_k v_k$. The Fourier transform is then uniformly continuous and bounded, and if moreover $\widehat{f}$ is integrable, then the inversion formula

$$f(t) = \int_{\mathbb{R}^n} \widehat{f}(v)\, e^{2i\pi \langle t, v\rangle} dv$$

holds almost-everywhere (with respect to the Lebesgue measure) and *it holds everywhere if f is continuous.*

The proof of this result is similar to that of Theorem 1.1.8, with the obvious adaptations. For instance the multivariate extension of the function h_σ thereof is

$$h_\sigma(t) = \frac{1}{(2\pi\sigma^2)^{\frac{n}{2}}}\, e^{-\frac{\|t\|^2}{2\sigma^2}}\,,$$

where $\|t\|^2 = \sum_{k=1}^n t_k^2$. We also need to observe that if $f_1, \ldots, f_n$ are functions in $L_{\mathbb{C}}^1(\mathbb{R})$, then $f : \mathbb{R}^n \to \mathbb{C}$ defined by

$$f(t_1, \ldots, t_n) = f_1(t_1) \times \cdots \times f_n(t_n)$$

is in $L_{\mathbb{C}}^1(\mathbb{R}^n)$ and its Fourier transform is

$$\widehat{f}(v_1, \ldots, v_n) = \widehat{f_1}(v_1) \times \cdots \times \widehat{f_n}(v_n).$$

We shall give more details for the extension of the Riemann–Lebesgue lemma and the derivative formulas. For this, recall the following standard notation. For a given function $g : \mathbb{R}^d \to \mathbb{C}$, and $\alpha = (\alpha_1, \ldots, \alpha_d)$, $x = (x_1, \ldots, x_d)$,

$$D^\alpha g(x) := D_1^{\alpha_1} \cdots D_d^{\alpha_d} g(x) := \frac{\partial^{\alpha_1} g}{\partial x_1^{\alpha_1}} \cdots \frac{\partial^{\alpha_d} g}{\partial x_d^{\alpha_d}} (x_1, \ldots, x_d)$$

whenever the above derivative exists. If this is the case for all multi-indices α such that $\alpha_1 + \cdots + \alpha_d \leq n$, one says that g is C^n.

For $x = (x_1, \ldots, x_d) \in \mathbb{R}^d$, denote $|x| = \max\{|x_1|, \ldots, |x_d|\}$.

Theorem 1.1.10 *Let* $f : \mathbb{R}^d \to \mathbb{C}$ *be in* $L^1_{\mathbb{C}}(\mathbb{R}^d)$, *and let* $\widehat{f} : \mathbb{R}^d \to \mathbb{C}$ *be its Fourier transform. Then* $\lim_{|v|\uparrow\infty} \widehat{f}(v) = 0$.

Proof The proof of Theorem 1.1.2 can be adapted. However we choose a slightly different one. We first prove the theorem for $g \in \mathcal{D}(\mathbb{R}^d)$, that is, a function that is infinitely differentiable and with compact support. Let T be such that $[-T, +T]^d$ contains the support of g. In particular

$$\widehat{g}(v) = \int_{[-T,+T]^d} g(x)e^{-2i\pi \langle v,x \rangle} \, dx \, .$$

An integration by parts using the fact that $g(\pm T, x_2, \ldots, x_d) = 0$ gives

$$\int_{[-T,+T]^d} D_1 g(x) e^{-2i\pi \langle v,x \rangle} \, dx = (-2i\pi v_1)\widehat{g}(v) \, .$$

Repeating the above operation of derivation on u_2, ..., u_d gives, using Fubini's theorem,

$$\int_{[-T,+T]^d} D_1 \cdots D_d g(x) e^{-2i\pi \langle v,x \rangle} \, dx = (-2i\pi v_1) \cdots (-2i\pi v_d)\widehat{g}(v) \, ,$$

and therefore

$$|\widehat{g}(v)| \leq \frac{1}{(2\pi)^d |v_1| \cdots |v_d|} ||D_1 \cdots D_d g||_1 \overset{|v| \to \infty}{\to} 0 \, .$$

Let now f be as in the statement of the theorem. For any $\varepsilon > 0$, one can find $g \in \mathcal{D}(\mathbb{R}^d)$ such that $||f - g||_1 \leq \varepsilon$ since $\mathcal{D}(\mathbb{R}^d)$ is dense in $L^1_{\mathbb{C}}(\mathbb{R}^d)$. Therefore,

$$|\widehat{f}(v) - \widehat{g}(v)| = |(\widehat{f - g})(v)| \leq ||f - g||L^1_{\mathbb{C}}(\mathbb{R}^d) \leq \frac{\varepsilon}{2} \, .$$

But for $|v|$ sufficiently large, $|\widehat{g}(v)| \leq \frac{\varepsilon}{2}$. Therefore, for $|v|$ sufficiently large,

$$|\widehat{f}(v)| \leq \frac{\varepsilon}{2} + \frac{\varepsilon}{2} = \varepsilon \, . \qquad \square$$

Theorem 1.1.11 *Let $f : \mathbb{R}^d \to \mathbb{C}$ be of class C^n, such that $\lim_{|x|\uparrow\infty} f(x) = 0$. Suppose moreover that $D^\alpha f \in L^1_{\mathbb{C}}(\mathbb{R}^d)$ for all multi-indices $\alpha = (\alpha_1, \ldots, \alpha_d)$ such that $\alpha_1 + \cdots + \alpha_d \leq n$. Then*

$$\widehat{D^\alpha f}(v) = (2i\pi v)^\alpha \widehat{f}(v),$$

where $(2i\pi v)^\alpha := (2i\pi)^{|\alpha|} v_1^{\alpha_1} \cdots v_d^{\alpha_d}$.

Proof Let $f : \mathbb{R}^d \to \mathbb{C}$ be of class C^1, such that $\lim_{|x|\uparrow\infty} f(x) = 0$. Suppose moreover that $D_j f \in L^1_{\mathbb{C}}(\mathbb{R}^d)$. If we can show that for all $1 \leq j \leq d$,

$$\widehat{D_j f}(v) = (2i\pi v_j)\widehat{f}(v), \tag{*}$$

we are done, since the announced result follows from the special case by recurrence. We prove the special case for $j = 1$ (of course without loss of generality). If $x = (x_1, \ldots, x_d)$ write $x = (x_1, x')$ where $x' = (x_2, \ldots, x_d)$. Since f and $D_1 f$ are integrable, by Fubini, the functions $x_1 \to f(x_1, x')$ and $x_1 \to D_1 f(x_1, x')$ are integrable for almost all $x' \in \mathbb{R}^{d-1}$. For fixed x' for which the just stated integrability conditions are satisfied, an integration by parts gives:

$$\int_{\mathbb{R}} D_1 f(x_1, x') e^{2i\pi v_1 x_1}\, dx_1 = \lim_{T\uparrow\infty} \int_{-T}^{+T} D_1 f(x_1, x') e^{2i\pi v_1 x_1}\, dx_1$$

$$= \lim_{T\uparrow\infty} \int_{-T}^{+T} f(x_1, x')(2i\pi v_1) e^{2i\pi v_1 x_1}\, dx_1$$

$$= (2i\pi v_1) \int_{-T}^{+T} f(x_1, x') e^{2i\pi v_1 x_1}\, dx_1,$$

where we have taken into account the fact that, by hypothesis, $\lim_{T\uparrow\infty} f(\pm T, x') = 0$. From this, we have (*) by successive integrations. □

Next result is displayed for future reference.

Corollary 1.1.5 *Any function $f \in \mathcal{D}(\mathbb{R}^d)$ is the Fourier transform of an integrable function.*

Proof Just observe that the FT of such function f is integrable since for all $\alpha > 0$, $\widehat{D^\alpha f}(v) = (2i\pi v)^\alpha \widehat{f}(v)$ behaves at infinity as $o(v)^{-\alpha}$. Therefore the inversion formula holds true. □

1.1.2 Convergence Theory

The inversion formula given in the preceding subsection requires the rather strong condition of summability of the Fourier coefficients series, which implies in particular that the function itself is an almost-everywhere continuous function. How far can the class of functions for which the inversion formula holds be extended?

In fact, Kolmogorov has proven that there exist locally integrable periodic functions whose Fourier series diverges everywhere. This disquieting result urges one to detect reasonable conditions which a locally integrable periodic function must satisfy in order for its Fourier series to converge to the function itself. Recall that the Fourier series associated with a 2π-periodic locally integrable function $f : \mathbb{R} \to \mathbb{C}$ is

$$\sum_{n\in\mathbb{Z}} c_n(f)\, e^{inx} ,$$

where

$$c_n(f) = \frac{1}{2\pi} \int_{-\pi}^{+\pi} f(u) e^{-inu} \, du$$

is the nth Fourier coefficient. As we mentioned before, the above Fourier series is *formal* as long as one does not say anything about its convergence in some sense (pointwise, almost everywhere, in L^1, etc.).

Dirichlet's Integral

However, the truncated Fourier series function

$$S_n^f(x) := \sum_{k=-n}^{+n} c_k(f) e^{ikx}$$

is well-defined. We have to specify when and in what sense convergence of the series in the right-hand side takes place, and what the limit is. Ideally, the convergence should be pointwise and to f itself. Exercise 1.5.12 gives a simple instance where this is true.

This truncated series function will be put in a form suitable for analysis. Define the *Dirichlet kernel*

$$D_n(t) := \sum_{k=-n}^{+n} e^{ikt} = \frac{\sin((n+\frac{1}{2})t)}{\sin(\frac{1}{2}t)} .$$

Then

$$
S_n^f(x) = \sum_{k=-n}^{+n} \left\{ \frac{1}{2\pi} \int_{-\pi}^{+\pi} f(s) e^{-iks} \, ds \right\} e^{ikx}
$$

$$
= \frac{1}{2\pi} \int_{-\pi}^{+\pi} \left\{ \sum_{k=-n}^{+n} e^{ik(x-s)} \right\} f(s) \, ds = \frac{1}{2\pi} \int_{-\pi}^{+\pi} D_n(x-s) f(s) \, ds .
$$

Performing the change of variable $x - s = u$ and taking into account the fact that f and the Dirichlet kernel are 2π-periodic, we obtain

$$
S_n^f(x) = \frac{1}{2\pi} \int_{-\pi}^{+\pi} D_n(u) f(x+u) \, du. \tag{1.2}
$$

The right-hand side of the above equality is called the *Dirichlet integral*. With $f \equiv 1$, we obtain

$$
\frac{1}{2\pi} \int_{-\pi}^{+\pi} D_n(u) \, du = 1 .
$$

Therefore, for any complex number A,

$$
S_n^f(x) - A = \frac{1}{2\pi} \int_{-\pi}^{+\pi} D_n(u) \left(f(x+u) - A \right) du , \tag{1.3}
$$

or, equivalently,

$$
S_n^f(x) - A = \frac{1}{2\pi} \int_0^{\pi} D_n(u) \{ f(x+u) + f(x-u) - 2A \} \, du .
$$

Therefore, in order to show that, for a given $x \in \mathbb{R}$, $S_n^f(x)$ tends to A as $n \uparrow \infty$, we must show that the Dirichlet integral in the right-hand side of (1.3) converges to zero.

Example 1.1.11 Let $a \in \mathbb{R}_+ \backslash \{0\}$. Consider the 2π-periodic function equal to e^{iax} on $(-\pi, +\pi]$. We show that its Fourier series converges uniformly on any interval $[-c, +c] \subset (-\pi, +\pi)$ to the function itself. By (1.3), it suffices to show that

$$
\lim_{n\uparrow\infty} \int_{-\pi}^{+\pi} \frac{\sin((n+\frac{1}{2})u)}{\sin(u/2)} \left| e^{ia(x-u)} - e^{iax} \right| du = 0,
$$

uniformly on $[-c, +c]$. Equivalently,

$$\lim_{n \uparrow \infty} \int_{-\pi}^{+\pi} \left| \frac{\sin(au/2)}{\sin(u/2)} \right| |\sin((n+1/2)u)| \, du = 0.$$

This is guaranteed by the extended Riemann–Lebesgue lemma (Theorem 1.1.6).

The Localization Principle

The convergence of the Fourier series is a *local* property. More precisely:

Theorem 1.1.12 *If f and g are two locally integrable 2π-periodic complex-valued functions such that, for a given $x \in \mathbb{R}$, f and g are equal in a neighborhood of x, then*

$$\lim_{n \uparrow \infty} \{S_n^f(x) - S_n^g(x)\} = 0.$$

Proof By assumption, for any given $x \in \mathbb{R}$, there exists some $\delta > 0$ (possibly depending on x) such that $f(t) = g(t)$ whenever $t \in [x - \delta, x + \delta]$. By (1.2) we have

$$S_n^f(x) - S_n^g(x) = \frac{1}{2\pi} \int_{-\pi}^{+\pi} \sin\left(\left(n + \frac{1}{2}\right)u\right) 1_{\{|u| \geq \delta\}} \frac{f(x+u) - g(x+u)}{\sin(\frac{1}{2}u)} \, du \,.$$

Since the function

$$w(u) = 1_{\{|u| \geq \delta\}} \frac{f(x+u) - g(x+u)}{\sin(\frac{1}{2}u)}$$

is integrable over $[-\pi, +\pi]$, the last integral tends to zero by the Riemann–Lebesgue lemma. □

Inspection of the proof reveals that the condition that f and g are equal in a neighborhhood of x can be replaced by condition that for some $\delta > 0$ (possibly depending on x)

$$\lim_{n \uparrow \infty} \int_{-\delta}^{+\delta} \sin\left(\left(n + \frac{1}{2}\right)u\right) \frac{f(x+u) - g(x+u)}{\sin(\frac{1}{2}u)} \, du = 0 \,.$$

We now state the general *pointwise convergence theorem.*

Theorem 1.1.13 *Let f be a locally integrable 2π-periodic complex-valued function, and let $x \in \mathbb{R}$ and $A \in \mathbb{R}$ be given. If for some $\delta \in (0, \pi]$ (possibly depending on x),*

$$\lim_{n\uparrow\infty} \int_0^\delta \sin\left(\left(n+\frac{1}{2}\right)u\right) \frac{\phi(u)}{\frac{1}{2}u}\, du = 0\,, \tag{1.4}$$

where $\phi(u) := f(x+u) + f(x-u) - 2A$, then $\lim_{n\uparrow\infty} S_n^f(x) = A$.

Proof Taking for g a constant function equal to A, we have $S_n(g) = A$, and therefore we are looking for a sufficient condition guaranteeing that $S_n(f) - S_n(g)$ tends to 0. By the localization principle, Theorem 1.1.12, or rather, the remark following its proof, it is enough to show that

$$\lim_{n\uparrow\infty} \int_0^\delta \sin\left(\left(n+\frac{1}{2}\right)u\right) \frac{\phi(u)}{\sin(\frac{1}{2}u)}\, du = 0\,. \tag{*}$$

The integrals in (1.4) and (*) differ by

$$\int_0^\delta \sin\left(\left(n+\frac{1}{2}\right)u\right) v(u)\, du\,, \tag{†}$$

where

$$v(u) = \phi(u)\left\{\frac{1}{u/2} - \frac{1}{\sin(\frac{1}{2}u)}\right\}$$

is integrable on $[0, \delta]$. Therefore, by the Riemann–Lebesgue lemma, the quantity (†) tends to zero as $n \uparrow \infty$.   □

Dini's Theorem

We are now in a position to discover larger classes of periodic functions whose Fourier series indeed converges to the function itself. We begin with *Dini's theorem*.

Theorem 1.1.14 *Let f be a 2π-periodic locally integrable complex-valued function and let $x \in \mathbb{R}$. If for some $\delta \in (0, \pi]$ (possibly depending on x) and some $A \in \mathbb{R}$, the function*

$$t \ \rightarrow \ \frac{f(x+t) + f(x-t) - 2A}{t}$$

is integrable on $[0, \delta]$, then

$$\lim_{n\uparrow\infty} S_n^f(x) = A\,.$$

Proof This is an immediate corollary of Theorem 1.1.13. □

Corollary 1.1.6 *If a 2π-periodic locally integrable complex-valued function f is Lipschitz continuous of order $\alpha > 0$ at $x \in \mathbb{R}$, that is, if*

$$|f(x+h) - f(x)| = O(|h|^\alpha) \quad as \ h \to 0,$$

then, $\lim_{n\uparrow\infty} S_n^f(x) = f(x)$.

Proof Indeed, with $A = f(x)$

$$\left| \frac{f(x+t) + f(x-t) - 2A}{t} \right| \le K \frac{1}{|t|^{1-\alpha}}$$

for some constant K. We conclude the proof with Theorem 1.1.14 because for all t in some neighborhood of zero, $t \to 1/|t|^{1-\alpha}$ is integrable in this neighborhood $(1 - \alpha < 1)$. □

Corollary 1.1.7 *Let f be a 2π-periodic locally integrable complex-valued function, and let $x \in \mathbb{R}$ be such that*

$$f(x+0) = \lim_{h\downarrow 0} f(x+h) \quad and \quad f(x-0) = \lim_{h\downarrow 0} f(x-h)$$

exist and are finite, and further assume that the derivatives to the left and to the right at x exist. Then

$$\lim_{n\uparrow\infty} S_n^f(x) = \frac{f(x+0) + f(x-0)}{2}.$$

Proof By definition, one says that the derivative to the right exists if

$$\lim_{t\downarrow 0} \frac{f(x+t) - f(x+0)}{t}$$

exists and is finite, with a similar definition for the derivative to the left. The differentiability assumptions imply that

$$\lim_{t\downarrow 0} \frac{f(x+t) - f(x+0) + f(x-t) - f(x-0)}{t}$$

exists and is finite, and therefore that

$$\frac{\phi(t)}{t} = \frac{f(x+t) + f(x-t) - 2A}{t}$$

is integrable in a neighborhood of zero, where $2A := f(x+0) + f(x-0)$. We conclude the proof with Theorem 1.1.14. □

Example 1.1.12 (The saw) Applying the previous corollary to the 2π-periodic function

$$f(t) = t \quad \text{when } 0 < t \leq 2\pi$$

gives

$$t = \pi - \sum_{n \geq 1} \frac{2\sin(nt)}{n} \quad \text{when } 0 < t < 2\pi.$$

At the discontinuity point $t = 0$, we can check directly that the sum of the Fourier series is

$$\frac{1}{2}(f(0+) + f(0-)) = \frac{1}{2}(0 + 2\pi) = \pi,$$

as announced in the last corolary. For $t = \pi/2$, we obtain the remarkable identity

$$\frac{\pi}{4} = \frac{1}{1} - \frac{1}{3} + \frac{1}{5} - \frac{1}{7} + \cdots.$$

Jordan's Theorem

Recall the definition and properties of *bounded variation* functions.

Definition 1.1.4 A real-valued function $\varphi : \mathbb{R} \to \mathbb{R}$ is said to have *bounded variation* on the interval $[a, b] \subset \mathbb{R}$ if

$$\sup_{\mathcal{D}} \sum_{i=0}^{n-1} |\varphi(x_{i+1}) - \varphi(x_i)| < \infty,$$

where the supremum is over all subdivisions $\mathcal{D} = \{a = x_0 < x_1 < \cdots < x_n = b\}$.

The fundamental result on the structure of bounded variation functions says that a real-valued function $\varphi : \mathbb{R} \to \mathbb{R}$ has bounded variation over $[a, b]$ if and only if there exist two *non-decreasing* real-valued functions φ_1, φ_2 such that for all $t \in [a, b]$

$$\varphi(t) = \varphi_1(t) - \varphi_2(t).$$

In particular, for all $x \in [a, b)$, φ has a limit to the right $\varphi(x+0)$ and for all $x \in (a, b]$ it has a limit to the left $\varphi(x - 0)$, and the discontinuity points of φ in $[a, b]$ form a denumerable set, and therefore a set of Lebesgue measure zero.

Theorem 1.1.15 *Let f be a 2π-periodic locally integrable real-valued function, of bounded variation in a neighborhood of a given $x \in \mathbb{R}$. Then*

$$\lim_{n\uparrow\infty} S_n^f(x) = \frac{f(x+0) + f(x-0)}{2}.$$

This is *Jordan's theorem*, whose proof is omitted.

Example 1.1.13 Integration of Fourier series. Let f be a real valued 2π-periodic locally integrable function and let c_n be its nth Fourier coefficient. Since f is real-valued, $c_{-n} = c_n^*$, and therefore the Fourier series of f can be written as

$$\frac{1}{2}a_0 + \sum_{n=1}^{\infty}\{a_n\cos(nx) + b_n\sin(nx)\}, \qquad (*)$$

where, for $n \geq 1$,

$$a_n = \frac{1}{\pi}\int_0^{2\pi} f(t)\cos(nt)\,dt, \qquad b_n = \frac{1}{\pi}\int_0^{2\pi} f(t)\sin(nt)\,dt.$$

Of course, the series in $(*)$ is purely formal when no additional constraints are put on f in order to guarantee its convergence. But the function F defined for $t \in [0, 2\pi)$ by

$$F(t) = \int_0^t \left(f(x) - \frac{1}{2}a_0\right)dx$$

is 2π-periodic, continuous (observe that $F(0) = F(2\pi) = 0$), and it is of bounded variation on finite intervals. Therefore, by Jordan's theorem, its Fourier series converges everywhere, and for all $x \in \mathbb{R}$

$$F(x) = \frac{1}{2}A_0 + \sum_{n=1}^{\infty}\{A_n\cos(nx) + B_n\sin(nx)\},$$

where, for $n \geq 1$,

$$A_n = \frac{1}{\pi}\int_0^{2\pi} F(t)\cos(nt)\,dt$$

$$= \frac{1}{\pi}\left[F(x)\frac{\sin(nx)}{n}\right]_0^{2\pi} - \frac{1}{n\pi}\int_0^{2\pi}(f(t) - \frac{1}{2}a_0)\sin(nt)\,dt$$

$$= -\frac{1}{n\pi}\int_0^{2\pi} f(t)\sin(nt)\,dt = -\frac{b_n}{n},$$

and, with a similar computation,

$$B_n = \frac{1}{\pi} \int_0^{2\pi} F(t)\sin(nt)\,dt = \frac{a_n}{n}.$$

Therefore for all $x \in \mathbb{R}$

$$F(x) = \frac{1}{2}A_0 + \sum_{n=1}^{\infty}\left\{\frac{a_n}{n}\sin(nx) - \frac{b_n}{n}\cos(nx)\right\}.$$

The constant A_0 is identified by setting $x = 0$ in the above identity:

$$\frac{1}{2}A_0 = \sum_{n=1}^{\infty}\frac{b_n}{n}.$$

Since A_0 is finite, we have shown in particular that $\sum_{n=1}^{\infty} b_n/n$ converges for any sequence $\{b_n\}_{n\geq 1}$ of the form

$$b_n = \frac{1}{\pi} \int_0^{2\pi} f(t)\sin(nt)\,dt,$$

when f is a real function integrable over $[0, 2\pi]$.

Example 1.1.14 (Gibbs overshoot phenomenon) This phenomenon is typical of the behaviour of a Fourier series at a discontinuity of the function. It has nothing to do with the failure of the Fourier series to converge at a point of discontinuity of the corresponding function. It concerns the overshoot of the partial sums at such a point of discontinuity. An example will demonstrate this effect.

Consider the 2π-periodic function defined in the interval $[0, +2\pi)$ by

$$f(x) = \frac{1}{2}(\pi - x)$$

The partial sums of its Fourier series is

$$S_n^f(x) = \sum_{k=1}^{n} \frac{\sin(nx)}{n}.$$

By Dini's Theorem, $S_n^f(0)$ converges to $(1/2)(f(0+) + f(0-)) = \pi/2$. However, we shall see that for some $c > 0$ and sufficiently large n

$$S_n^f\left(\frac{\pi}{n}\right) \geq S_n^f(0) + c. \qquad (\dagger)$$

There is indeed convergence at 0 to some value, but there are points approaching 0 which are at a fixed difference apart from the limit. This constitutes Gibbs's overshoot phenomenon, which can be observed whenever the function has a point of discontinuity. The proof of ($\dagger$) for this special case retains most features of the general proof which is left for the reader. In this special case

$$S_n^f(x) = \int_0^x \frac{\sin((n+\frac{1}{2})t)}{2\sin(\frac{1}{2}t)}\, dt - \frac{x}{2}.$$

Now,

$$\int_0^x \frac{\sin((n+\frac{1}{2})t)}{2\sin(\frac{1}{2}t)}\, dt$$

$$= \int_0^x \left(\frac{\sin(nt)\cos(\frac{1}{2}t)}{2\sin(\frac{1}{2}t)} + \frac{1}{2}\cos(nt)\right) dt$$

$$= \int_0^x \frac{\sin(nt)}{t}\, dt + \int_0^x \sin(nt)\left(\frac{\cos(\frac{1}{2}t)}{2\sin(\frac{1}{2}t)} - \frac{1}{t}\right) dt$$

$$+ \frac{1}{2}\int_0^x \cos(nt)\, dt.$$

The last two integrals converge uniformly to zero (by Corollary 1.1.1). Also,

$$\int_0^{\frac{\pi}{n}} \frac{\sin(nt)}{t}\, dt = \int_0^{\pi} \frac{\sin(t)}{t}\, dt \simeq 1.28\frac{\pi}{2} > \frac{\pi}{2}.$$

Féjer's Theorem

For the Fourier series to converge pointwise to the function, certain conditions are required, for instance those in Dini's and Jordan's theorem. However, Cesaro convergence of the series requires much milder conditions. This is because, for a 2π-periodic locally stable function f, Féjer's sum

$$\sigma_n^f(x) = \frac{1}{n} \sum_{k=0}^{n-1} S_k^f(x)$$

behaves more nicely than the Fourier series itself. In particular, for continuous functions, it converges pointwise to the function itself. Therefore, Féjer's theorem is a kind of weak inversion formula, in that it shows that a continuous periodic function is characterized by its Fourier coefficients (see the precise statement in Theorem 1.1.19 below).

Take the imaginary part of the identity

$$\sum_{k=0}^{n-1} e^{i(k+1/2)u} = e^{iu/2} \frac{1 - e^{inu}}{1 - e^{iu}},$$

to obtain

$$\sum_{k=0}^{n-1} \sin((k + \frac{1}{2})u) = \frac{\sin^2(\frac{1}{2}nu)}{\sin(\frac{1}{2}u)}.$$

Starting from Dirichlet's integral expression for S_n^f (Eq. (1.2)), we therefore obtain Féjer's integral representation,

$$\sigma_n^f(x) = \int_{-\pi}^{+\pi} K_n(u) f(x - u) \, du = \int_{-\pi}^{+\pi} K_n(x - u) f(x) \, du.$$

where

$$K_n(t) = \frac{1}{2n\pi} \frac{\sin^2(\frac{1}{2}nt)}{\sin^2(\frac{1}{2}t)}$$

is, by definition, *Féjer's kernel*. It has the following properties:

(i) $K_n(t) \geq 0$,
(ii) $\int_{-\pi}^{+\pi} K_n(u) du = 1$ (letting $f \equiv 1$ in Féjer's integral representation), and
(iii) for all $\delta \leq \pi$, $\lim_{n \uparrow \infty} \int_{-\delta}^{+\delta} K_n(u) \, du = 1$ (the proof is left as exercise).

These properties make of Féjer's kernel a *regularization kernel* on $[-\pi, +\pi]$.
    We then have *Féjer's theorem*:

Theorem 1.1.16 *Let f be a 2π-periodic continuous function. Then*

$$\lim_{n \uparrow \infty} \sup_{x \in [-\pi, +\pi]} |\sigma_n^f(x) - f(x)| = 0. \tag{1.5}$$

Proof From Féjer's integral representation, we have

$$|\sigma_n^f(x) - f(x)| \leq \int_{-\pi}^{+\pi} K_n(u)\,|f(x-u) - f(x)|\,du$$

$$= \int_{-\delta}^{+\delta} + \int_{[-\pi,+\pi]\setminus[-\delta,+\delta]} = A + B.$$

For a given $\varepsilon > 0$, choose δ such that $|f(x-u) - f(x)| \leq \varepsilon/2$ when $|u| \leq \delta$. Note that f is uniformly continuous and uniformly bounded (being a periodic and continuous function), and therefore δ can be chosen independent of x. We have by (i) and (ii)

$$A \leq \frac{\varepsilon}{2} \int_{-\delta}^{+\delta} K_n(u)\,du \leq \frac{\varepsilon}{2},$$

and, calling M the uniform bound of f,

$$B \leq 2M \int_{[-\pi,+\pi]\setminus[-\delta,+\delta]} K_n(u)\,du.$$

By (iii), $B \leq \frac{\varepsilon}{2}$ for n sufficiently large. Therefore, for n sufficiently large, $A + B \leq \varepsilon$. $\qquad\square$

Féjer's theorem for continuous periodic functions is the key to important approximation theorems. The first one is for free. We call *trigonometric polynomial* any finite trigonometric sum of the form

$$p(x) = \sum_{-n}^{+n} c_k e^{ikx}.$$

Corollary 1.1.8 *Let f be a 2π-periodic continuous function. For any $\varepsilon > 0$, there exists a trigonometric polynomial p such that*

$$\sup_{x\in[-\pi,+\pi]} |f(x) - p(x)| \leq \varepsilon.$$

Proof This is a restatement of Theorem 1.1.16 after observing that σ_n^f is a trigonometric polynomial. See also the usual proof of this result *via* the Stone–Weierstrass theorem (Example 1.3.2). $\qquad\square$

From this, follows the *Weierstrass approximation theorem* (see Example 1.3.2).

Theorem 1.1.17 *Let* $f : [a, b] \to \mathbb{C}$ *be a continuous function. Select an arbitrary* $\varepsilon > 0$. *There exists a polynomial P such that*

$$\sup_{x \in [a,b]} |f(x) - P(x)| \leq \varepsilon.$$

If, moreover, f is real-valued, then P can be chosen with real coefficients.

Proof First, suppose that $a = 0, b = 1$. One can then extend $f : [0, 1] \to \mathbb{C}$ to a function still denoted by f, $f : [-\pi, +\pi]] \to \mathbb{C}$, that is continuous and such that $f(+\pi) = f(-\pi) = 0$. By Theorem 1.1.8, there exists a trigonometric polynomial p such that

$$\sup_{x \in [0,1]} |f(x) - p(x)| \leq \sup_{x \in [-\pi,+\pi]} |f(x) - p(x)| \leq \frac{\varepsilon}{2}.$$

Now replace each function $x \to e^{ikx}$ in p by a sufficiently large portion of its Taylor series expansion, to obtain a polynomial P such that

$$\sup_{x \in [0,1]} |P(x) - p(x)| \leq \frac{\varepsilon}{2}.$$

Then

$$\sup_{x \in [0,1]} |f(x) - P(x)| \leq \frac{\varepsilon}{2}.$$

To treat the general case $f : [a, b] \to \mathbb{C}$, apply the result just proven to $\varphi : [0, 1] \to \mathbb{C}$ defined by $\varphi(t) = f(a + (b - a)t)$ to obtain the approximating polynomial $\pi(x)$, and take $P(x) = \pi(x - a)/(b - a))$. Finally, in order to prove the last statement of the theorem, observe that

$$|f(x) - \operatorname{Re} P(x)| \leq |f(x) - P(x)|. \qquad \square$$

There is for the Féjer sum a result analogous to Theorem 1.1.13. First,

$$\sigma_n^f(x) = \frac{1}{2n\pi} \int_0^{\pi} \frac{\sin^2(\frac{1}{2}nu)}{\sin^2(\frac{1}{2}u)} \{f(x + u) - f(x - u)\} \, du,$$

and therefore for any number A

$$\sigma_n^f(x) - A = \frac{1}{2n\pi} \int_0^{\pi} \frac{\sin^2(\frac{1}{2}nu)}{\sin^2(\frac{1}{2}u)} \{f(x + u) + f(x - u) - 2A\} \, du.$$

Theorem 1.1.18 *For any $x \in \mathbb{R}$ and any constant A*

$$\lim_{n \uparrow \infty} \sigma_n^f(x) = A$$

if for some $\delta > 0$

$$\lim_{n \uparrow \infty} \frac{1}{n} \int_0^\delta \sin^2\left(\frac{1}{2}nu\right) \frac{\varphi(u)}{u^2} \, du = 0, \tag{1.6}$$

where $\varphi(u) = f(x+u) + f(x-u) - 2A$.

Proof The quantity

$$\left| \frac{1}{n} \int_\delta^\pi \frac{\sin^2(\frac{1}{2}nu)}{\sin^2(\frac{1}{2}u)} \varphi(u) \, du \right| \le \frac{1}{n} \int_\delta^\pi \frac{|\varphi(u)|}{\sin^2(\frac{1}{2}u)} \, du$$

tends to 0 as $n \uparrow \infty$. We must therefore show that

$$\frac{1}{n} \int_0^\delta \frac{\sin^2(\frac{1}{2}nu)}{\sin^2(\frac{1}{2}u)} \varphi(u) \, du$$

tends to 0 as $n \uparrow \infty$. But (1.6) guarantees this since

$$\left| \frac{1}{n} \int_0^\delta \sin^2\left(\frac{1}{2}nu\right) \left\{ \frac{1}{\sin^2(\frac{1}{2}u)} - \frac{1}{\frac{1}{4}u^2} \right\} \varphi(u) \, du \right|$$

$$\le \frac{1}{n} \int_0^\delta \left\{ \frac{1}{\sin^2(\frac{1}{2}u)} - \frac{1}{\frac{1}{4}u^2} \right\} |\varphi(u)| \, du$$

tends to 0 as $n \uparrow \infty$ (the expression in curly brackets is bounded in $[0, \delta]$ and therefore the integral is finite). $\qquad\square$

Theorem 1.1.19 *Let f be a 2π-periodic locally integrable function and assume that at some point $x \in \mathbb{R}$, the limits to the right and to the left, denoted by $f(x+0)$ and $f(x-0)$ respectively, exist. Then*

$$\lim_{n \uparrow} \sigma_n^f(x) = \frac{f(x+0) + f(x-0)}{2}.$$

Proof Fix $\delta > 0$. In view of the last result, it suffices to prove (1.6) with

$$\varphi(u) = \{f(x+u) - f(x+0)\} + \{f(x-u) - f(x-0)\}.$$

Since $\varphi(u)$ tends to zero as $n \uparrow \infty$, for any given $\varepsilon > 0$ there exists $\eta = \eta(\varepsilon)$, $0 < \eta \le \delta$, such that $|\varphi(u)| \le \varepsilon$ when $0 < u \le \eta$. Now,

$$\left| \frac{1}{n} \int_0^\delta \frac{\sin^2(\frac{1}{2}nu)}{u^2} \varphi(u)\, du \right|$$

$$\le \frac{1}{n} \int_0^\eta \frac{\sin^2(\frac{1}{2}nu)}{u^2} \varepsilon\, du + \frac{1}{n} \int_\eta^\delta \frac{\sin^2(\frac{1}{2}nu)}{u^2} |\varphi(u)|\, du$$

$$\le \frac{\varepsilon}{n} \int_0^\eta \frac{\sin^2(\frac{1}{2}nu)}{u^2}\, du + \frac{1}{n} \int_\eta^\delta \frac{|\varphi(u)|}{u^2}\, du.$$

The last integral is bounded and therefore the last term goes to 0 as $n \uparrow \infty$. As for the penultimate term, it is bounded by $A\varepsilon$, where

$$A = \int_0^\infty \frac{\sin^2(\frac{1}{2}v)}{v^2}\, dv < \infty. \qquad \square$$

Corollary 1.1.9 *Let f be a 2π-periodic locally integrable function and suppose that for some $x \in \mathbb{R}$:*

(a) the function f is continuous at x; and
(b) $S_n^f(x)$ converges to some number A.

Then $A = f(x)$.

Proof From (b) and Cesaro's mean convergence theorem,

$$\lim_{n \uparrow \infty} \sigma_n^f(x) = A.$$

From Theorem 1.1.19 and (a),

$$\lim_{n \uparrow \infty} \sigma_n^f(x) = f(x). \qquad \square$$

1.1.3 The Poisson Summation Formula

This formula connects Fourier transforms and Fourier series. It plays an important role in the theory of series and in signal processing.

Weak Version

Conditions under which the strong version of the Poisson summation formula

$$T \sum_{n \in \mathbb{Z}} f(nT) = \sum_{n \in \mathbb{Z}} \widehat{f}\left(\frac{n}{T}\right) \qquad (1.7)$$

holds true will be given in the next subsection. Right now, we give the *weak* Poisson summation formula, which is sufficient for most purposes in applications.

Theorem 1.1.20 *If $f \in L^1_{\mathbb{C}}(\mathbb{R})$ and $T > 0$, the series $\sum_{n \in \mathbb{Z}} f(t + nT)$ converges absolutely almost-everywhere to a T-periodic locally integrable function Φ, the nth Fourier coefficient of which is $\frac{1}{T} \widehat{f}\left(\frac{n}{T}\right)$.*

Proof We first show that Φ is almost-everywhere well-defined. In fact,

$$\int_0^T \sum_{n \in \mathbb{Z}} |f(t + nT)|\, dt = \sum_{n \in \mathbb{Z}} \int_0^T |f(t + nT)|\, dt$$

$$= \sum_{n \in \mathbb{Z}} \int_{nT}^{(n+1)T} |f(t)|\, dt = \int_{\mathbb{R}} |f(t)|\, dt < \infty,$$

which implies that

$$\sum_{n \in \mathbb{Z}} |f(t + nT)| < \infty \quad \text{almost-everywhere}.$$

In particular, Φ is almost-everywhere well-defined and finite. It is clearly T-periodic. It is integrable since

$$\int_0^T |\Phi(t)|\, dt = \int_0^T \left| \sum_{n \in \mathbb{Z}} f(t + nT) \right| dt$$

$$\leq \int_0^T \sum_{n \in \mathbb{Z}} |f(t + nT)|\, dt = \int_{\mathbb{R}} |f(t)|\, dt < \infty.$$

Its nth Fourier coefficient is

$$c_n(\Phi) = \frac{1}{T} \int_0^T \Phi(t) e^{-2i\pi \frac{n}{T} t} \, dt$$

$$= \frac{1}{T} \int_0^T \left\{ \sum_{k \in \mathbb{Z}} f(t + kT) \right\} e^{-2i\pi \frac{n}{T} t} \, dt$$

$$= \frac{1}{T} \int_0^T \left\{ \sum_{k \in \mathbb{Z}} f(t + kT) e^{-2i\pi \frac{n}{T} (t + kT)} \right\} dt$$

$$= \frac{1}{T} \int_{\mathbb{R}} f(t) e^{-2i\pi \frac{n}{T} t} \, dt = \frac{1}{T} \widehat{f}\left(\frac{n}{T}\right). \qquad \square$$

We may paraphrase Theorem 1.1.20 as follows: Under the above conditions, the function

$$\Phi(t) := \sum_{n \in \mathbb{Z}} f(t + nT)$$

is T-periodic and locally integrable, and its *formal* Fourier series is

$$S_\Phi(t) := \frac{1}{T} \sum_{n \in \mathbb{Z}} \widehat{f}\left(\frac{n}{T}\right) e^{2i\pi \frac{n}{T} t}.$$

Recall once again that we speak of a "formal" Fourier series because nothing is said about the convergence of the sum of the right-hand side of the above expression. However:

Corollary 1.1.10 *Under the conditions of Theorem 1.1.20, and if moreover* $\sum_{n \in \mathbb{Z}} |\widehat{f}(\frac{n}{T})| < \infty$, *then*

$$\sum_{n \in \mathbb{Z}} f(t + nT) = \frac{1}{T} \sum_{n \in \mathbb{Z}} \widehat{f}\left(\frac{n}{T}\right) e^{2i\pi \frac{n}{T} t} \qquad \textit{a.e.with respect to } \ell.$$

Corollary 1.1.11 *Let* $g : \mathbb{R} \to \mathbb{C}$ *be a continuous and integrable function whose* FT $\widehat{g}$ *is integrable. The two following conditions are equivalent:*

(a) $g(jT) = 0$ *for all* $j \in \mathbb{Z}, j \neq 0$;
(b) *The function* $v \to \sum_{n \in \mathbb{Z}} \widehat{g}\left(v + \frac{n}{T}\right)$ *is constant almost everywhere.*

Proof The FT $\widehat{\widehat{g}}$ of the (integrable) function $\widehat{g}$ is

$$\widehat{\widehat{g}}(t) = \int_{\mathbb{R}} \widehat{g}(v)\, e^{-2i\pi vt}\, dt\,.$$

By the inversion formula, the last integral is equal to $g(-t)$. Since g is assumed continuous, this equality holds *everywhere*, and in particular for $t = nT$.

By the weak version of the Poisson sum formula, $T\widehat{\widehat{g}}(nT) = Tg(-nT)$ is the nth Fourier coefficient of $\sum_{n\in\mathbb{Z}} \widehat{g}(v + n/T)$. Therefore if (b) is true, then (a) is necessarily true. Conversely, if (a) is true, then the sequence $\{Tg(-nT)\}_{n\in\mathbb{Z}}$ is the sequence of Fourier coefficients of two functions, the constant function equal to $Tg(0)$, and $\sum_{n\in\mathbb{Z}} \widehat{g}(v+n/T)$, and therefore the two functions must be equal almost everywhere.   □

Example 1.1.15 (Intersymbol interference) In a certain type of digital communications system one transmits 'discrete' information, consisting of a sequence of complex numbers $\mathbf{a} := \{a_n\}_{n\in\mathbb{Z}}$, in the form of an analog signal

$$f(t) = \sum_{n\in\mathbb{Z}} a_n\, g(t - nT), \tag{1.8}$$

where g is a complex function (the "pulse") such that $g(0) \neq 0$, assumed continuous, integrable and with an integrable FT $\widehat{g}$. Such a 'coding' of the information sequence is referred to as *pulse amplitude modulation*. Here, $T > 0$ determines the rate of transmission of information, and also the rate at which the information is extracted at the receiver. At time kT the receiver extracts the sample

$$f(kT) = a_k g(0) + \sum_{\substack{j\in\mathbb{Z}\\ j\neq 0}} a_{k-j}\, g(jT)\,.$$

If one only seeks to recover $\mathbf{a}$ from the samples $f(kT)$, $k \in \mathbb{Z}$, the term $\sum_{\substack{j\in\mathbb{Z}\\ j\neq 0}} a_{k-j} g(jT)$ is a parasitic term which disappears for *every* sequence $\mathbf{a}$ if and only if

$$g(jT) = 0 \quad \text{for all } j \neq 0.$$

By Corollary 1.1.11, this is equivalent to condition

$$\sum_{n\in\mathbb{Z}} \widehat{g}\left(v - \frac{n}{T}\right) = Tg(0).$$

This condition is the *Nyquist condition* for the absence of intersymbol interference.

Strong Version

Let us return to the situation of Theorem 1.1.20. In the case where we are able to show that the Fourier series S_Φ represents the function Φ at $t = 0$, that is, if $\Phi(0) = S_\Phi(0)$, then we obtain the strong Poisson summation formula (1.7).

Theorem 1.1.21 *Let $f : \mathbb{R} \to \mathbb{C}$ be an integrable function, and suppose that for some $T \in \mathbb{R}_+\backslash\{0\}$,*

(1) $\sum_{n\in\mathbb{Z}} f(\cdot + nT)$ converges everywhere to a continuous function;
(2) $\sum_{n\in\mathbb{Z}} \widehat{f}\left(\frac{n}{T}\right) e^{2i\pi\frac{n}{T}t}$ converges for all t.

Then, the strong Poisson summation formula holds, that is

$$\sum_{n\in\mathbb{Z}} f(t + nT) = \frac{1}{T}\sum_{n\in\mathbb{Z}} \widehat{f}\left(\frac{n}{T}\right) e^{2i\pi\frac{n}{T}t} \quad \text{for all } t \in \mathbb{R}.$$

Proof The result is an immediate consequence of Theorem 1.1.20 and of Corollary 1.1.9 (substituting Φ for f therein). □

Here are two important cases for which the conditions of Theorem 1.1.21 are satisfied.

Corollary 1.1.12 *Let $f : \mathbb{R} \to \mathbb{C}$ be an integrable function, and let $T \in \mathbb{R}_+\backslash\{0\}$. If, in addition, $\sum f(t + nT)$ converges everywhere to a continuous function with bounded variation then the strong Poisson summation formula holds true.*

Proof We must verify conditions (1) and (2) of Theorem 1.1.21. Condition (1) is part of the hypothesis. Condition (2) is a consequence of Jordan's theorem (Theorem 1.1.15). □

Corollary 1.1.13 *Let $f : \mathbb{R} \to \mathbb{C}$ be an integrable function such that for some $\alpha > 1$,*

$$f(t) = O\left(\frac{1}{1 + |t|^\alpha}\right) \quad (|t| \to \infty), \text{ and } \widehat{f}(v) = O\left(\frac{1}{1 + |v|^\alpha}\right) \quad (|v| \to \infty),$$

then the strong Poisson summation formula holds for all $T \in \mathbb{R}_+\backslash\{0\}$.

Proof This is an immediate consequence of Theorem 1.1.21. □

Example 1.1.16 (Acceleration of convergence) The Poisson summation formula can be used to replace a series with slow convergence by another one (with the same limit however!) with rapid convergence, or to obtain some remarkable formulas. Here is a typical example. Let $a > 0$. The FT of the function $f(t) = e^{-2\pi a|t|}$ is $\widehat{f}(v) = \frac{a}{\pi(a^2+v^2)}$. Since

$$\sum_{n\in\mathbb{Z}} f(t + n) = \sum_{n\in\mathbb{Z}} e^{-2\pi a|t+n|}$$

is a continuous function with bounded variation, we have by the strong Poisson summation formula

$$\sum_{n\in\mathbb{Z}} e^{-2\pi a|n|} = \sum_{n\in\mathbb{Z}} \frac{a}{\pi(a^2+n^2)}.$$

The left-hand side is equal to $\frac{2}{1-e^{-2\pi a}} - 1$, whereas the right-hand side can be written as $\frac{1}{\pi a} + 2\sum_{n\geq1} \frac{a}{\pi(a^2+n^2)}$. Therefore

$$\sum_{n\geq1} \frac{1}{a^2+n^2} = \frac{\pi}{2a} \frac{1+e^{-2\pi a}}{1-e^{-2\pi a}} - \frac{1}{2a^2}.$$

This example seems useless as far as acceleration of convergence is concerned since we know explicitly the limit. However: letting $a\downarrow0$, we obtain the equality

$$\sum_{n\geq1} \frac{1}{n^2} = \frac{\pi^2}{6}.$$

This remarkable formula can be used to compute π. But now, the slow convergence of the series therein is a problem. This series is therefore best replaced by a series with exponential rate of convergence.

The general feature of the above example is the following. We have a series which is obtained by sampling a very regular function (in the example: C^∞) but unfortunately slowly decreasing. However, because of its strong regularity, its FT has a fast decay, by the Riemann–Lebesgue lemma. The series obtained by sampling the FT is therefore rapidly converging.

Sampling Theorem

How much information about a given function f is contained in its samples $f(nT)$, $n\in\mathbb{Z}$? The forthcoming discussion is based on the following general result.

Theorem 1.1.22 *Let $f:\mathbb{R}\to\mathbb{C}$ be an integrable and continuous function with Fourier transform $\widehat{f}\in L^1_{\mathbb{C}}(\mathbb{R})$, and assume in addition that for some $B\in\mathbb{R}_+\backslash\{0\}$*

$$\sum_{n\in\mathbb{Z}} \left|f\left(\frac{n}{2B}\right)\right| < \infty. \tag{1.9}$$

Let $h:\mathbb{R}\to\mathbb{C}$ be a function of the form

$$h(t) = \int_R \widehat{h}(v)e^{2i\pi vt}\,dv,$$

where $\widehat{h} \in L^1_{\mathbb{C}}(\mathbb{R})$. Then

$$\frac{1}{2B} \sum_{n \in \mathbb{Z}} f\left(\frac{n}{2B}\right) h\left(t - \frac{n}{2B}\right) = \int_{\mathbb{R}} \left\{ \sum_{j \in \mathbb{Z}} \widehat{f}(v + j2B) \right\} \widehat{h}(v) e^{2i\pi vt} dv \quad a.e. \ (1.10)$$

Note that it is not required that $\widehat{h}$ be the FT of h in the L^1 sense. It would be if h were assumed integrable, by Theorem 1.1.8. However we do not make this assumption since we shall take in the Shannon–Nyquist sampling theorem (Theorem 1.1.23) for function $\widehat{h}$ a rectangular pulse, whose FT is not integrable.

Proof We start by proving that

$$\Phi(v) := \sum_{j \in \mathbb{Z}} \widehat{f}(v + j2B) = \frac{1}{2B} \sum_{n \in \mathbb{Z}} f\left(\frac{n}{2B}\right) e^{-2i\pi v \frac{n}{2B}}, \ \text{a.e.} \qquad (\dagger)$$

By Theorem 1.1.20, the $2B$-periodic function Φ is locally integrable, and its nth Fourier coefficient is

$$\frac{1}{2B} \int_{\mathbb{R}} \widehat{f}(v) e^{-2i\pi \frac{n}{2B} v} dv .$$

The Fourier inversion formula for f holds ($\widehat{f}$ is integrable) and it holds *everywhere* (f is continuous). In particular, the nth Fourier coefficient of Φ is equal to

$$f\left(-\frac{n}{2B}\right).$$

The formal Fourier series of Φ is therefore

$$\frac{1}{2B} \sum_{n \in \mathbb{Z}} f\left(\frac{n}{2B}\right) e^{-2i\pi \frac{n}{2B} v}.$$

In view of condition (1.9), the Fourier inversion formula holds a.e. (Theorem 1.1.9), that is Φ is almost everywhere equal to its Fourier series. This proves ($\dagger$).

Define now

$$\tilde{f}(t) := \frac{1}{2B} \sum_{n \in \mathbb{Z}} f\left(\frac{n}{2B}\right) h\left(t - \frac{n}{2B}\right) \qquad (1.11)$$

Since $\widehat{h} \in L^1_{\mathbb{C}}(\mathbb{R})$, the function h is bounded and uniformly continuous, and therefore $\tilde{f}(t)$ is bounded and continuous (the right-hand side of (1.11) is, in view of (1.9), a normally convergent series of bounded and continuous functions). Upon substituting the expression of h in (1.11), we obtain

$$\tilde{f}(t) = \frac{1}{2B} \sum_{n\in\mathbb{Z}} f\left(\frac{n}{2B}\right) \int_{\mathbb{R}} \widehat{h}(\nu) e^{2i\pi\nu(t-\frac{n}{2B})}\, d\nu$$

$$= \int_{\mathbb{R}} \left\{ \sum_{n\in\mathbb{Z}} \frac{1}{2B} f\left(\frac{n}{2B}\right) e^{-2i\pi\nu\frac{n}{2B}} \right\} \widehat{h}(\nu) e^{2i\pi\nu t}\, d\nu.$$

The interchange of integration and summation is justified by Fubini's theorem because

$$\int_{\mathbb{R}} \sum_{n\in\mathbb{Z}} \left| f\left(\frac{n}{2B}\right)\right| |\widehat{h}(\nu)|\, d\nu = \left(\sum_{n\in\mathbb{Z}} \left| f\left(\frac{n}{2B}\right)\right|\right) \left(\int_{\mathbb{R}} |\widehat{h}(\nu)|\, d\nu\right) < \infty.)$$

Therefore,

$$\tilde{f}(t) = \int_{\mathbb{R}} g(\nu) e^{2i\pi\nu t}\, d\nu,$$

where

$$g(\nu) = \frac{1}{2B} \left\{ \sum_{n\in\mathbb{Z}} f\left(\frac{n}{2B}\right) e^{-2i\pi\nu\frac{n}{2B}} \right\} \widehat{h}(\nu).$$

The result (1.10) then follows from (†). $\qquad\square$

A direct consequence of the previous theorem is the so-called *Shannon–Nyquist sampling theorem*.

Theorem 1.1.23 *Let B be a positive real number and let* $f : \mathbb{R} \to \mathbb{C}$ *be a continuous and integrable base-band (B) function, and assume condition (1.9) satisfied. We can then recover f from its samples* $f(n/2B)$, $n \in \mathbb{Z}$, *by the formula*

$$f(t) = \sum_{n\in\mathbb{Z}} f\left(\frac{n}{2B}\right) \text{sinc}(2Bt - n), \text{ a.e.} \qquad (1.12)$$

Proof Set $\widehat{h}(\nu) = 1_{[-B,+B]}(\nu)$ in Theorem 1.1.22. Then

$$\left\{ \sum_{j\in\mathbb{Z}} \widehat{f}(\nu + j2B) \right\} \widehat{h}(\nu) = \widehat{f}(\nu) 1_{[-B,+B]}(\nu) = \widehat{f}(\nu),$$

and therefore

$$\int_{\mathbb{R}} \left\{ \sum_{j\in\mathbb{Z}} \widehat{f}(\nu + j2B) \right\} \widehat{h}(\nu) e^{2i\pi\nu t}\, d\nu = \int_{\mathbb{R}} \widehat{f}(\nu) e^{2i\pi\nu t}\, d\nu = f(t).$$

The second equality holds everywhere because f is a continuous function. We conclude the proof with (1.10).   □

The function

$$h(t) = \int_{\mathbb{R}} 1_{[-B,+B]}(v)e^{2i\pi vt}\, dt$$

is called the impulse response of the *base-band (B)* (filter) of transmittance T. Define (informally) the so-called *Dirac comb* by

$$\Delta_B(t) = \frac{1}{2B} \sum_{n \in \mathbb{Z}} \delta\left(t - \frac{n}{2B}\right),$$

where δ is the so-called Dirac pseudo-function. If we interpret the function $t \to f(n/2B)h(t - n/2B)$ as the response of the base-band (B) when a Dirac impulse of height $f(n/2B)$ is applied at time $n/2B$, the right-hand side of equation (1.12) is the response of the base-band (B) to the *Dirac sampling comb*:

$$\Delta_{B,f}(t) = \frac{1}{2B} \sum_{n \in \mathbb{Z}} f\left(\frac{n}{2B}\right)\delta\left(t - \frac{n}{2B}\right). \tag{1.13}$$

The adaptation of Theorem 1.1.23 to the spatial case is straightforward.

Theorem 1.1.24 *Let* $B_1, \ldots, B_n \in \mathbb{R}_+ \backslash \{0\}$ *and let* $f : \mathbb{R}^n \to \mathbb{C}$ *be an integrable continuous function whose* FT $\widehat{f}$ *vanishes outside* $[-B_1, +B_1] \times \cdots \times [-B_n, +B_n]$, *and assume that*

$$\sum_{k_1 \in \mathbb{Z}} \cdots \sum_{k_n \in \mathbb{Z}} \left| f\left(\frac{k_1}{2B_1}, \ldots, \frac{k_n}{2B_n}\right) \right| < \infty.$$

Then, we can then recover f *from its samples* $f\left(\frac{k_1}{2B_1}, \ldots, \frac{k_n}{2B_n}\right)$, $k_1 \in \mathbb{Z}, \ldots,$ $k_n \in \mathbb{Z}$, *by the formula*

$$f(t_1, \ldots, t_n) = \sum_{k_1 \in \mathbb{Z}} \cdots \sum_{k_n \in \mathbb{Z}} f\left(\frac{k_1}{2B_1}, \ldots, \frac{k_n}{2B_n}\right) \operatorname{sinc}(2B_1 t - k_1) \times \cdots \times \operatorname{sinc}(2B_n t - k_n), \ a.e.$$

Aliasing

What happens in the Shannon–Nyquist sampling theorem if one supposes that the function is base-band (B), although it is not the case in reality?

Suppose that an integrable function f is sampled at the frequency $2B$ and that the resulting Dirac sampling comb $\Delta_{B,f}$ given by (1.13) is applied to the base-band (B) filter whose impulse response is $h(t) = 2B\operatorname{sinc}(2Bt)$, to obtain, after division by $2B$, the function

$$\tilde{f}(t) = \sum_{n \in \mathbb{Z}} f\left(\frac{n}{2B}\right) \operatorname{sinc}(2Bt - n).$$

What is the FT of this function? The answer is given by the theorem below which is a direct consequence of Theorem 1.1.22 with $\widehat{h}(v) = 1_{[-B,+B]}(v)$.

Theorem 1.1.25 *Let $B \in \mathbb{R}_+ \backslash \{0\}$ and let $f : \mathbb{R} \to \mathbb{C}$ be an integrable continuous function such that condition (1.9) is satisfied. The function*

$$\tilde{f}(t) = \sum_{n \in \mathbb{Z}} f\left(\frac{n}{2B}\right) \operatorname{sinc}(2Bt - n)$$

admits the representation

$$\tilde{f}(t) = \int_{\mathbb{R}} \widehat{\tilde{f}}(v) e^{2i\pi vt} \, dv,$$

where

$$\widehat{\tilde{f}}(v) = \left\{ \sum_{k \in \mathbb{Z}} \widehat{f}(v + j2B) \right\} 1_{[-B,+B]}(v).$$

If $\tilde{f}$ is integrable, then $\widehat{\tilde{f}}$ is its FT, by the Fourier inversion theorem. This FT is obtained by superposing, in the frequency band $[-B, +B]$, the translates by multiples of $2B$ of the initial spectrum $\widehat{f}$. This constitutes the phenomenon of *spectrum folding*, and the distortion that it creates is called *aliasing*.

1.2   Z-Transforms

1.2.1   Fourier Sums

Sampling naturally leads to discrete-time functions (sequences) which are studied in this section. From the Fourier point of view, their theory parallels that of continuous-time functions, and is in fact much easier. The important result of this section is Féjer's lemma for rational transfer functions which will be central to the study of the ARMA processes of Chap. 4.

Definition 1.2.1 Let $\mathbf{x} := \{x_n\}_{n \in \mathbb{Z}}$ be a sequence of complex numbers such that

$$\sum_{n \in \mathbb{Z}} |x_n| < \infty,$$

that is, in $\ell_{\mathbb{C}}^1(\mathbb{Z})$. Its *Fourier sum* is the function $\tilde{x} : [-\pi, +\pi] \to \mathbb{C}$ defined by

$$\tilde{x}(\omega) = \sum_{k \in \mathbb{Z}} x_k e^{-ik\omega}. \tag{1.14}$$

It is a 2π-periodic function and it is continuous, bounded by $\sum_{n \in \mathbb{Z}} |x_n|$ (repeat the arguments that show that the FT of an integrable function is uniformly continuous and bounded, Theorem 1.1.1).

An inversion formula is available:

Theorem 1.2.1 x_n *is the nth Fourier coefficient of $\tilde{x}$:*

$$x_n = \frac{1}{2\pi} \int_{-\pi}^{+\pi} \tilde{x}(\omega) e^{in\omega} \, d\omega.$$

Proof Multiply (1.14) by $e^{in\omega}$ and integrate from $-\pi$ to $+\pi$, using dominated convergence in order to exchange the order of summation and integration.   □

Definition 1.2.2 The operation which associates to the discrete-time function $\mathbf{x} := \{x_n\}_{n \in \mathbb{Z}} \in \ell^1_{\mathbb{C}}(\mathbb{Z})$ the discrete-time function $\mathbf{y} := \{y_n\}_{n \in \mathbb{Z}}$ defined by

$$y_n = \sum_{k \in \mathbb{Z}} x_k h_{n-k}, \tag{1.15}$$

where $\mathbf{h} := \{h_n\}_{n \in \mathbb{Z}} \in \ell^1_{\mathbb{C}}(\mathbb{Z})$, is called *convolutional filtering*. The function $\mathbf{y}$ is the *output* of the convolutional filter with *impulse response* $\mathbf{h}$, and $\mathbf{x}$ is the *input*.

When $\mathbf{x} \in \ell^1_{\mathbb{C}}(\mathbb{Z})$ and $\mathbf{h} \in \ell^1_{\mathbb{C}}(\mathbb{Z})$, the right-hand side of (1.15) has a meaning. In fact,

$$\sum_{n \in \mathbb{Z}} \sum_{n \in \mathbb{Z}} |x_k| \, |h_{n-k}| = \sum_{n \in \mathbb{Z}} \left\{ |x_k| \sum_{n \in \mathbb{Z}} |h_{n-k}| \right\}$$

$$= \sum_{k \in \mathbb{Z}} |x_k| \sum_{\ell \in \mathbb{Z}} |h_\ell| < \infty,$$

and in particular

$$\sum_{k \in \mathbb{Z}} |x_k| \, |h_{n-k}| < \infty \quad \text{for all } n \in \mathbb{Z}.$$

This also shows that $\mathbf{y} \in \ell^1_{\mathbb{C}}(\mathbb{Z})$.

When the input function $\mathbf{x}$ is the unit impulse at 0, that is,

$$x_n = \delta_n := 1_{\{n=0\}},$$

the output is $\mathbf{y} = \mathbf{h}$, hence the appelation "impulse response" for $\mathbf{h}$.

Definition 1.2.3 A *causal*, or *physically realizable*, filter is one for which the impulse response satisfies the condition

$$h_n = 0 \quad \text{for all } n < 0.$$

The filter is called causal because if $x_n = 0$ for $n \leq n_0$, then $y_n = 0$ for $n \leq n_0$. The input–output relation (1.15) takes, for a causal filter, the form

$$y_n = \sum_{k=-\infty}^{n} x_k h_{n-k}.$$

Definition 1.2.4 The *frequency response* of the convolutional filter with impulse response $\mathbf{h} \in \ell_{\mathbb{C}}^1(\mathbb{Z})$ is, by definition, the Fourier sum

$$\tilde{h}(\omega) = \sum_{n \in \mathbb{Z}} h_n e^{-in\omega}.$$

Let $\tilde{x}$ and $\tilde{y}$ be the Fourier sums of the input $\mathbf{x} \in \ell_{\mathbb{C}}^1(\mathbb{Z})$ and the output $\mathbf{y} \in \ell_{\mathbb{C}}^1(\mathbb{Z})$ of the convolutional filter with impulse response $\mathbf{h} \in \ell_{\mathbb{C}}^1(\mathbb{Z})$, the input–output relation (1.15) reads

$$\tilde{y}(\omega) = \tilde{h}(\omega)\tilde{x}(\omega).$$

Indeed,

$$\tilde{y}(\omega) = \sum_{n \in \mathbb{Z}} y_n e^{-in\omega} = \sum_{n \in \mathbb{Z}} \sum_{k \in \mathbb{Z}} x_k h_{n-k} e^{-in\omega}$$

$$= \sum_{k \in \mathbb{Z}} \left\{ x_k e^{-ik\omega} \sum_{n \in \mathbb{Z}} h_{n-k} e^{-i(n-k)\omega} \right\}$$

$$= \left(\sum_{k \in \mathbb{Z}} x_k e^{-ik\omega} \right) \left(\sum_{\ell \in \mathbb{Z}} h_\ell e^{-i\ell\omega} \right).$$

The exchange of the orders of summations is justified by Fubini's theorem in a way similar to what was done after Definition 1.2.4.

Example 1.2.1 (Pure delay and smoothing filter) (a) The pure delay. The input–output relation defined by $y_n = x_{n-k}$ is a convolutional filtering with impulse response $h_n = \delta_{n-k} = 1_{\{n=k\}}$ and frequency response $\tilde{h}(\omega) = e^{-ik\omega}$.
(b) The smoothing filter. This is the convolutional filter defined by the input–output relation

$$y_n = \frac{1}{2N+1} \sum_{k=-N}^{+N} x_{n-k}.$$

Its frequency response is

$$\tilde{h}(\omega) = \frac{1}{2N+1} \frac{\sin\{(N+\frac{1}{2})\omega\}}{\sin\{\omega/2\}}.$$

Example 1.2.2 (Equivalence of analog and digital filtering) The operation of continuous-time filtering followed by sampling can be performed, under certain conditions, with the same result if one chooses first to sample and then to operate in the sampled domain. More precisely, let $f : \mathbb{R} \to \mathbb{C}$ be an integrable continuous function, base-band (B), sampled at the Nyquist rate $2B$. We obtain the samples

$$f\left(\frac{n}{2B}\right) = \int_{-B}^{+B} e^{2i\pi v \frac{n}{2B}} \widehat{f}(v)\, dv = \frac{1}{2\pi} \int_{-\pi}^{+\pi} e^{in\omega} 2B \widehat{f}\left(\frac{B}{\pi}\omega\right) d\omega$$

(the inversion formula can be applied because $\widehat{f}$ is integrable, being continuous with a bounded support; on the other hand the equality of $f(t)$ and $\int_{\mathbb{R}} \widehat{f}(v) e^{2i\pi vt}\, dt$ holds for all t, since both functions of t are continuous). If it is further assumed that

$$\sum_{n\in\mathbb{Z}} |f\left(\frac{n}{2B}\right)| < \infty, \tag{†}$$

and therefore, the Fourier sum of $\{f(n/2B)\}_{n\in\mathbb{Z}}$ is the function $\omega \to 2B\widehat{f}((B/\pi)\omega)$. Let now $x : \mathbb{R} \to \mathbb{C}$ and $h : \mathbb{R} \to \mathbb{C}$ be continuous functions in $L^2_{\mathbb{C}}(\mathbb{R})$, base-band (B), and both satisfying the condition of type (†). Then

$$y\left(\frac{n}{2B}\right) = \frac{1}{2B} \sum_{n\in\mathbb{Z}} h\left(\frac{k}{2B}\right) x\left(\frac{k}{2B}\right).$$

Proof The discrete-time function $y_n := \frac{1}{2B} \sum_{k\in\mathbb{Z}} h\left(\frac{k}{2B}\right) x\left(\frac{n-k}{2B}\right)$ is in $\ell^1_{\mathbb{C}}(\mathbb{Z})$ and its Fourier sum is

$$\tilde{y}(\omega) = 2B\widehat{h}\left(\frac{B}{\pi}\omega\right) \widehat{x}\left(\frac{B}{\pi}\omega\right).$$

Hence we have

$$y_n = \frac{1}{2\pi} \int_{-\pi}^{+\pi} 2B\widehat{h}\left(\frac{B}{\pi}\omega\right) \widehat{x}\left(\frac{B}{\pi}\omega\right) e^{in\omega}\, d\omega$$

$$= \int\limits_{-B}^{+B} \widehat{h}(v)\widehat{x}(v)e^{2i\pi v \frac{n}{2B}}\,dv.$$

On the other hand, for the continuous-time function $y(t) = \int_{\mathbb{R}} h(t - s)x(s)\,ds$, we have the inversion formula (which holds for all $t \in \mathbb{R}$ since its FT $\widehat{y} = \widehat{h}\widehat{x}$ is integrable; check this)

$$y(t) = \int\limits_{-B}^{+B} e^{2i\pi vt}\widehat{x}(v)\widehat{h}(v)\,dv$$

and therefore $y_n = y(n/2B)$.   □

1.2.2 Transfer Functions

To every discrete-time function $\mathbf{x} = \{x_n\}_{n\in\mathbb{Z}} \in \ell_{\mathbb{C}}^1(\mathbb{Z})$ is associated its *formal z-transform*, which is the formal series

$$X(z) = \sum_{n\in\mathbb{Z}} x_n z^n\,.$$

The qualification "formal" refers to the fact that the z-transform of a function takes a meaning as a function of $z \in \mathbb{C}$ only if one gives the domain of convergence of the series defining it. The formal z-transform of the impulse response $\mathbf{h} = \{h_n\}_{n\in\mathbb{Z}} \in \ell_{\mathbb{C}}^1(\mathbb{Z})$ of a convolutional filter is the *formal transfer function* of the said filter:

$$H(z) = \sum_{n\in\mathbb{Z}} h_n z^n\,.$$

The input–output relation (1.15) reads, in terms of the z-transforms of $\mathbf{x}$, $\mathbf{y}$ and $\mathbf{h}$,

$$Y(z) = H(z)X(z)\,.$$

We use the *unit delay* operation z defined symbolically by

$$z^k x_n = x_{n-k}\,.$$

With this notation, the input-output relation $\sum_{n\in\mathbb{Z}} h_k x_{n-k}$ takes the form

$$y_n = \sum_{n\in\mathbb{Z}} h_k(z^k x_n)$$

$$= \left(\sum_{n \in \mathbb{Z}} h_k z^k \right) x_n = H(z) x_n .$$

In some cases (see the examples below) a function H holomorphic in a *donut* $\{r_1 < |z| < r_2\}$ *containing the unit circle* $\{|z| = 1\}$ is given. This function defines a convolutional filter whose impulse response $\mathbf{h}$ is given by the Laurent expansion

$$H(z) = \sum_{n \in \mathbb{Z}} h_n z^n \qquad (r_1 < |z| < r_2).$$

In particular, the Laurent expansion at $z = 1$ is absolutely convergent, and thus the impulse response $\mathbf{h}$ is in $\ell^1_{\mathbb{C}}(\mathbb{Z})$. The frequency response of the filter is

$$\tilde{h}(\omega) = H(e^{-i\omega}).$$

Example 1.2.3 (Pure delay) The input–output relation $y_n = x_{n-k}$ is a convolutional filter with impulse response $h_n = 1_{\{k\}}(n)$ and transfer function $H(z) = z^k$. (Here the domain of definition of H is the whole complex plane.) The transmittance of the pure delay is $H(e^{-i\omega}) = e^{-ik\omega}$.

The coefficients of the Laurent expansion are explicitly given by the Cauchy formula

$$h_n = \frac{1}{2i\pi} \oint_C \frac{H(z)}{z^{n+1}} \, dz, \qquad (1.16)$$

where C is a closed path without multiple points that lies within the interior of the donut of convergence, for example the unit circle, taken in the anti-clockwise sense. This equality also takes the form

$$h_n = \frac{1}{2\pi} \int_0^{2\pi} H(e^{-i\omega}) e^{in\omega} \, d\omega .$$

Example 1.2.4 (Rational transfer functions) Let

$$P(z) = 1 + \sum_{j=1}^{p} a_j z^j, \qquad Q(z) = 1 + \sum_{\ell=1}^{q} b_\ell z^\ell ,$$

be two polynomials of the complex variable z with complex coefficients. We shall assume that P has *no roots on the unit circle* $\{|z| = 1\}$.

Let $r_1 = \max\{|z| : P(z) = 0 \text{ and } |z| < 1\}$; $r_1 = 0$ if there is no root of $P(z)$ with modulus strictly smaller than 1. Let $r_2 = \inf\{|z| : P(z) = 0 \text{ and } |z| > 1\}$; $r_2 = +\infty$ if there is no root of $P(z)$ with modulus strictly larger than 1.

The function

$$H(z) = \frac{Q(z)}{P(z)}$$

is holomorphic in the donut $C_{r_1,r_2} := \{r_1 < |z| < r_2\}$ (in the open disk $\{|z| < r_2\}$ if $r_1 = 0$) which contains the unit circle since $r_2 > 1$. We therefore have a Laurent expansion in C_{r_1,r_2}

$$H(z) = \sum_{n \in \mathbb{Z}} h_n z^n$$

which defines a filter with impulse response $\mathbf{h} \in \ell^1_{\mathbb{C}}(\mathbb{Z})$ and frequency response

$$H(e^{i\omega}) = \frac{Q(e^{-i\omega})}{P(e^{-i\omega})}.$$

Stable Filters and Causal Filters

If $\mathbf{y}$ is the output of the filter with transfer function $H(z) = \frac{Q(z)}{P(z)}$ corresponding to the input $\mathbf{x} \in \ell^1_{\mathbb{C}}(\mathbb{Z})$, we have $\tilde{y}(\omega) = H(e^{-i\omega})\tilde{x}(\omega)$, that is

$$P(e^{-i\omega})\tilde{y}(\omega) = Q(e^{-i\omega})\tilde{x}(\omega).$$

Now $P(e^{-i\omega})\tilde{y}(\omega)$ is the Fourier sum of the function $y_n + \sum_{j=1}^{p} a_j y_{n-j}$ and $Q(e^{-i\omega})\tilde{x}(\omega)$ is the Fourier sum of $x_n + \sum_{\ell=1}^{q} b_\ell x_{n-\ell}$. Therefore

$$y_n + \sum_{j=1}^{p} a_j y_{n-j} = x_n + \sum_{\ell=1}^{q} b_\ell x_{n-\ell}, \qquad (*)$$

or, symbolically, $P(z)y_n = Q(z)x_n$. Given the input $\mathbf{x}$ and the initial conditions $y_0, y_{-1}, \ldots, y_{-p+1}$, the solution of $(*)$ is completely determined.

The general solution of this recurrence equation is the sum of an arbitrary solution and of the general solution of the equation without right-hand side

$$y_n + \sum_{j=1}^{p} a_j y_{n-j} = 0. \qquad (**)$$

The general solution of $(**)$ is a weighted sum of terms of the form $r(n)\rho^{-n}$, where ρ is a root of P and $r(n)$ is a polynomial of degree equal to the multiplicity of this root minus one.

For this solution never to blow up (it is said to blow up if $\lim_{|n|\uparrow\infty}|y_n| = \infty$) whatever the input $\mathbf{x} \in \ell^1_{\mathbb{C}}(\mathbb{Z})$ and the initial conditions $y_{-p+1}, \ldots, y_{-1}, y_0$, it is necessary and sufficient that all the roots of P have modulus > 1.

A particular solution of $(*)$ is

$$y_n = \sum_{k \geq 0} h_k x_{n-k} \, .$$

If the input $\mathbf{h}$ is in $\ell^1_{\mathbb{C}}(\mathbb{Z})$, then also the output $\mathbf{y}$ is in $\ell^1_{\mathbb{C}}(\mathbb{Z})$ since the impulse response $\mathbf{h} \in \ell^1_{\mathbb{C}}(\mathbb{Z})$, and therefore the particular solution $\mathbf{y} \in \ell^1_{\mathbb{C}}(\mathbb{Z})$ does not blow up.

Therefore, in order for the general solution of $(*)$ with input $\mathbf{x} \in \ell^1_{\mathbb{C}}(\mathbb{Z})$ not to blow up, it is necessary and sufficient that the polynomial P has all its roots with modulus strictly greater than 1.

Definition 1.2.5 The rational filter $Q(z)/P(z)$ is said to be *stable and causal* if P has all its roots outside the closed unit disk $\{|z| \leq 1\}$.

Causality arises from the property that if P has roots with modulus > 1, $Q/P = H$ is holomorphic inside $\{|z| < r_2\}$ where $r_2 > 1$. The Laurent expansion of $H(z)$ is then an expansion as an entire series $H(z) = \sum_{k \geq 0} h_k z^k$, and this means that the filter is causal ($h_k = 0$ when $k < 0$).

Definition 1.2.6 The stable and causal rational filter Q/P is said to be *causally invertible* if Q has all its roots outside the closed unit disk $\{|z| \leq 1\}$.

In fact, writing the analytic expansion of $P(z)/Q(z)$ in the neighborhood of zero as $\sum_{k \geq 0} w_k z^k$ we have

$$\tilde{x}(\omega) = \frac{P(e^{-i\omega})}{Q(e^{-i\omega})} \tilde{y}(\omega) = \left(\sum_{k \geq 0} w_k e^{-ik\omega} \right) \tilde{y}(\omega),$$

that is, $x_n = \sum_{k \geq 0} w_k y_{n-k}$.

All-Pass Lemma

A particular case of rational filter is the *all-pass filter*.

Theorem 1.2.2 *Let z_i ($1 \leq i \leq L$) be complex numbers with modulus strictly greater than 1. Then, the transfer function*

$$H(z) = \prod_{i=1}^{L} \frac{z z_i^* - 1}{z - z_i}$$

satisfies

$$|H(z)| < 1 \text{ if } |z| < 1,$$
$$= 1 \text{ if } |z| = 1,$$
$$> 1 \text{ if } |z| > 1.$$

Proof Let

$$H_i(z) = \frac{zz_i^* - 1}{z - z_i}$$

be an arbitrary factor of $H(z)$. If $|z| = 1$, we observe that $|H_i(z)| = 1$, using *Féjer's identity*

$$(z - \beta)(z - \frac{1}{\beta^*}) = -\frac{1}{\beta^*}z|z - \beta|^2, \qquad (1.17)$$

which is true for $|z| = 1$, $\beta \in \mathbb{C}$, $\beta \neq 0$. On the other hand, $H_i(z)$ is holomorphic on $|z| < |z_i|$ and $|H_i(0)| = |z_i|^{-1} < 1$. Therefore we must have $|H_i(z)| < 1$ on $\{|z| < 1\}$, otherwise the maximum modulus theorem for holomorphic functions would be contradicted.[2] Observing that

$$\left| H_i\left(\frac{1}{z^*}\right) \right| = \frac{1}{|H_i(z)|}$$

we see that the result just obtained implies that $|H_i(z)| > 1$ if $|z| > 1$. $\qquad \square$

Therefore, the all-pass filter has gain of amplitude unity: it is a pure phase filter, or *all-pass filter*, by definition.

Féjer's Lemma

Consider the discrete-time function $\mathbf{x} \in \ell_{\mathbb{C}}^2(\mathbb{Z})$ with z-transform $X(z)$. Define its nth autocorrelation coefficient

$$c_n = \sum_{k \in \mathbb{Z}} x_{n+k} x_k^* .$$

Then (exercise) $\mathbf{c} := \{c_n\}_{n \in \mathbb{Z}} \in \ell_{\mathbb{C}}^2(\mathbb{Z})$ and its Fourier sum is

$$\tilde{c}(\omega) = |\tilde{x}(\omega)|^2 = R(e^{-i\omega}) := X(e^{-i\omega})X(e^{-i\omega})^* .$$

---

[2] Recall the maximum modulus theorem: If f is analytic in a bounded domain D and $|f|$ is continuous in the closure of D, then $|f|$ takes its maximum only at a point on the boundary of D; see K. Kodaira (1984).

The 2π-periodic function $\omega \to R(e^{-i\omega})$ has the following properties:

(a) $R(e^{-i\omega}) \geq 0$, for all $\omega \in [-\pi, +\pi]$
(b) $\int_{-\pi}^{+\pi} R(e^{-i\omega}) < \infty$, and

Moreover, if X is a rational fraction,

$$R(e^{-i\omega}) \text{ is a rational fraction of } e^{-i\omega}.$$

The next result is *Féjer's lemma*, and it is also called the *spectral factorization* theorem.

Theorem 1.2.3 *Let R be a rational fraction in $z \in \mathbb{C}$ with complex coefficients, such that properties (a) and (b) are verified. Then there exist two polynomials P and Q with complex coefficients, and a constant $c \geq 0$, such that $P(0) = Q(0) = 1$ and*

$$R(e^{-i\omega}) = c \left| \frac{Q(e^{-i\omega})}{P(e^{-i\omega})} \right|^2.$$

Moreover, one can choose P without roots inside the closed *unit disk centered at the origin, and Q without roots inside the* open *unit disk centered at the origin.*

Proof $R(z)$ can be factored as

$$R(z) = a z^{m_0} \prod_{k \in K} (z - z_k)^{m_k},$$

where $a \in \mathbb{C}$, K is a finite subset of $\mathbb{N}$, the z_k are non-null distinct complex numbers, and the $m_k \in \mathbb{Z}$. When $|z| = 1$, $R(z)$ is real (by assumption (a)) and therefore, for $|z| = 1$,

$$R(z) = R(z)^* = a^* (z^*)^{m_0} \prod_{k \in K} (z^* - z_k^*)^{m_k} = a^* (z^{-1})^{m_0} \prod_{k \in K} (z^{-1} - z_k^*)^{m_k}.$$

Therefore, when $|z| = 1$, there exist $b \in \mathbb{C}$ and $r_0 \in \mathbb{Z}$ such that

$$R(z) = b z^{r_0} \prod_{k \in K} \left(z - \frac{1}{z_k^*} \right)^{m_k}.$$

Therefore, if $|z| = 1$,

$$a z^{m_0} \prod_{k \in K} (z - z_k)^{m_k} = b z^{r_0} \prod_{k \in K} \left(z - \frac{1}{z_k^*} \right)^{m_k}.$$

Two rational fractions that coincide when $|z| = 1$ coincide for all $z \in \mathbb{C}$. In particular $a = b$, and whenever we have in $R(z)$ the factor $(z - z_k)$ with $|z_k| \neq 1$, then we also have the factor $(z - \frac{1}{z_k^*})$. Therefore

$$\prod_{k \in K}(z - z_k)^{m_k} = \prod_{j \in J}(z - z_j)^{s_j}\left(z - \frac{1}{z_j^*}\right)^{s_j}\prod_{\ell \in L}(z - z_\ell)^{r_\ell},$$

where $|z_\ell| = 1$ for all $\ell \in L$, and $|z_j| \neq 1$ for all $j \in J$. We show that $r_\ell = 2s_\ell$ where $s_\ell \in \mathbb{N}$ for all $\ell \in L$. For this, we write $z_\ell = e^{-i\omega_\ell}$, and observe that in the neighborhood of ω_ℓ, $R(e^{-i\omega})$ is equivalent to a constant times $(\omega - \omega_\ell)^{r_\ell}$, and therefore can remain non-negative if and only if $r_\ell = 2s_\ell$. Since $R(e^{-i\omega})$ is locally integrable, necessarily $s_\ell \in \mathbb{N}$. Therefore

$$R(z) = bz^{r_0}\prod_{j \in J}(z - z_j)^{s_j}\left(z - \frac{1}{z_j^*}\right)^{s_j}\prod_{\ell \in L}(z - z_\ell)^{2s_\ell}.$$

Using Féjer's identity (1.17), we therefore find that $R(z)$ can be put under the form

$$R(z) = cz^d|G(z)|^2,$$

where

$$G(z) = \prod_{j \in J}(z - z_j)^{s_j}\prod_{\ell \in L}(z - z_\ell)^{s_\ell}.$$

The function $R(e^{-i\omega})$ can remain real and non-negative if and only if $c \geq 0$ and $d = 0$. Finally, we can always suppose that $|z_j| < 1$ for all $j \in J$ (any root z_j of modulus > 1 is paired with another root $\frac{1}{z_j^*}$ of modulus < 1). □

1.3 Hilbert Spaces

1.3.1 Isometric Extension

Hilber spaces play a fundamental role in the Fourier theory of functions, but also of stochastic processes. This section gives all that is needed in the sequel, namely the projection principle, the isometry extension theorem and the theory of Hilbert bases. It will find immediate applications in the next section devoted to the Fourier theory of functions of $L_{\mathbb{C}}^2(\mathbb{R})$.

Let H be a vector space with scalar field $K = \mathbb{C}$ or $\mathbb{R}$, endowed with an application $(x, y) \in H \times H \to \langle x, y \rangle \in K$ such that for all $x, y, z \in H$ and all $\lambda \in K$,

1. $\langle y, x \rangle = \langle x, y \rangle^*$
2. $\langle \lambda y, x \rangle = \lambda \langle y, x \rangle$

3. $\langle x, y + z \rangle = \langle x, y \rangle + \langle x, z \rangle$
4. $\langle x, x \rangle \geq 0$; and $\langle x, x \rangle = 0$ if and only if $x = 0$.

Then H is called a *pre-Hilbert space* over K and $\langle x, y \rangle$ is called the *inner product* of x and y. For any $x \in E$, denote

$$\|x\|^2 = \langle x, x \rangle.$$

The *parallelogram identity*

$$\|x\|^2 + \|y\|^2 = \frac{1}{2}(\|x + y\|^2 + \|x - y\|^2)$$

is obtained by expanding the right-hand side and using the equality

$$\|x + y\|^2 = \|x\|^2 + \|y\|^2 + 2\mathrm{Re}\left\{\langle x, y \rangle\right\}.$$

The *polarization identity*

$$\langle x, y \rangle = \frac{1}{4}\left\{\|x + y\|^2 - \|x - y\|^2 + i\|x + iy\|^2 - i\|x - iy\|^2\right\},$$

is checked by expanding the right-hand side. It shows in particular that two inner products $\langle \cdot, \cdot \rangle_1$ and $\langle \cdot, \cdot \rangle_2$ on E such that $\|\cdot\|_1 = \|\cdot\|_2$ are identical.

Schwarz's Inequality

Theorem 1.3.1 *For all $x, y \in H$,*

$$|\langle x, y \rangle| \leq \|x\| \times \|y\|.$$

Equality occurs if and only if x and y are colinear.

Proof Say $K = \mathbb{C}$. If x and y are colinear, that is $x = \lambda y$ for some $\lambda \in \mathbb{C}$, the inequality is obviously an equality. If x and y are linearly independent, then for all $\lambda \in \mathbb{C}, x + \lambda y \neq 0$. Therefore

$$0 < \|x + \lambda y\|^2 = \|x\|^2 + |\lambda y|^2\|\lambda y\|^2 + \lambda^*\langle x, y \rangle + \lambda\langle x, y \rangle^*$$
$$= \|x\|^2 + |\lambda|^2\|y\|^2 + 2\mathrm{Re}(\lambda^*\langle x, y \rangle).$$

Take $u \in \mathbb{C}, |u| = 1$, such that $u^*\langle x, y \rangle = |\langle x, y \rangle|$. Take any $t \in \mathbb{R}$ and put $\lambda = tu$. Then

$$0 < \|x\|^2 + t^2\|y\|^2 + 2t|\langle x, y \rangle|.$$

This is true for all $t \in \mathbb{R}$. Therefore the discriminant of this second degree polynomial in t of the right-hand side must be strictly negative, that is, $4|\langle x, y \rangle|^2 - 4\|x\|^2 \times \|y\|^2 < 0$. □

Theorem 1.3.2 *The mapping* $x \to \|x\|$ *is a* norm *on* E, *that is to say, for all* $x, y \in E$, *all* $\alpha \in \mathbb{C}$,

(a) $\|x\| \geq 0$; *and* $\|x\| = 0$ *if and only if* $x = 0$,
(b) $\|\alpha x\| = |\alpha| \|x\|$, *and*
(c) $\|x + y\| \leq \|x\| + \|y\|$ (*triangle inequality*).

Proof The proof of (a) and (b) is immediate. For (c) write

$$\|x + y\|^2 = \|x\|^2 + \|y\|^2 + \langle x, y \rangle + \langle y, x \rangle$$

and

$$(\|x\| + \|y\|)^2 = \|x\|^2 + \|y\|^2 + 2\|x\|\|y\|.$$

It therefore suffices to prove

$$\langle x, y \rangle + \langle y, x \rangle = 2\text{Re}(\langle x, y \rangle) \leq 2\|x\|\|y\|,$$

which follows from Schwarz's inequality. □

The norm $\| \cdot \|$ induces a *distance* $d(\cdot, \cdot)$ on H by

$$d(x, y) = \|x - y\|.$$

Recall that a mapping $d : E \times E \to \mathbb{R}_+$ is called a *distance* on E if, for all $x, y, z \in E$,

(a′) $d(x, y) \geq 0$; and $d(x, y) = 0$ if and only if $x = y$,
(b′) $d(x, y) = d(y, x)$, and
(c′) $d(x, y) \geq d(x, z) + d(z, y)$.

The above properties are immediate consequences of (a), (b), and (c) of Theorem 1.3.2. When endowed with a distance, a space H is called a *metric space*.

Definition 1.3.1 The pre-Hilbert space H is called a *Hilbert space* if the distance d makes of it a *complete* metric space.

By this, the following is meant: If $\{x_n\}_{n \geq 1}$ is a Cauchy sequence in H, that is, if $\lim_{m,n \uparrow \infty} d(x_m, x_n) = 0$, then there exists $x \in H$ such that $\lim_{n \uparrow \infty} d(x_n, x) = 0$.

Theorem 1.3.3 *Let* $\{x_n\}_{n \geq 1}$ *and* $\{y_n\}_{n \geq 1}$ *be sequences in a Hilbert space* H *that converge to* x *and* y *respectively. Then,*

$$\lim_{m,n \uparrow \infty} \langle x_n, y_m \rangle = \langle x, y \rangle.$$

In other words, the inner product of a Hilbert space is *bicontinuous*. In particular, the norm $x \mapsto \|x\|$ is a continuous function from H to $\mathbb{R}_+$.

Proof We have for all h_1, h_2 in H,

$$|\langle x + h_1, y + h_2\rangle - \langle x, y\rangle| = |\langle x, h_2\rangle + \langle h_1, y\rangle + \langle h_1, h_2\rangle|.$$

By Schwarz's inequality $|\langle x, h_2\rangle| \leq \|x\|\|h_2\|, |\langle h_1, y\rangle| \leq \|y\|\|h_1\|,$ and $|\langle h_1, h_2\rangle| \leq \|h_1\|\|h_2\|$. Therefore

$$\lim_{\|h_1\|, \|h_2\| \downarrow 0} |\langle x + h_1, y + h_2\rangle - \langle x, y\rangle| = 0. \qquad \square$$

Of special interest is the space $L^2_{\mathbb{C}}(\mu)$ of (equivalence classes of) complex measurable functions $f : X \to \mathbb{R}$ such that

$$\int_X |f(x)|^2 \, \mu(dx) < \infty.$$

We have (Theorem A.1.28):

Theorem 1.3.4 $L^2_{\mathbb{C}}(\mu)$ *is a vector space with scalar field* $\mathbb{C}$, *and when endowed with the inner product*

$$\langle f, g\rangle = \int_X f(x)g(x)^* \, \mu(dx)$$

it is a Hilbert space.

The norm of a function $f \in L^2_{\mathbb{C}}(\mu)$ is

$$\|f\| = \left(\int_X |f(x)|^2 \, \mu(dx)\right)^{\frac{1}{2}}$$

and the distance between two functions f and g in $L^2_{\mathbb{C}}(\mu)$ is

$$d(f, g) = \left(\int_X |f(x) - g(x)|^2 \, \mu(dx)\right)^{\frac{1}{2}}.$$

The completeness property of $L^2_{\mathbb{C}}(\mu)$ reads, in this case, as follows. If $\{f_n\}_{n \geq 1}$ is a sequence of functions in $L^2_{\mathbb{C}}(\mu)$ such that

$$\lim_{m, n \uparrow \infty} \int_X |f_n(x) - f_m(x)|^2 \, \mu(dx) = 0,$$

then, there exists a function $f \in L^2_{\mathbb{C}}(\mu)$ such that

$$\lim_{n \uparrow \infty} \int_X |f_n(x) - f(x)|^2 \, \mu(dx) = 0.$$

In $L^2_{\mathbb{C}}(\mu)$, Schwarz's inequality reads:

$$\left| \int_X f(x)g(x)^* \mu(dx) \right| \leq \left(\int_X |f(x)|^2 \, \mu(dx) \right)^{\frac{1}{2}} \left(\int_X |g(x)|^2 \, \mu(dx) \right)^{\frac{1}{2}}$$

Recall the following simple and often used observation.

Theorem 1.3.5 *Let p and q be positive real numbers such that $p \geq q$. If the measure μ on $(X, \mathcal{X}, \mu)$ is finite, then $L^p_{\mathbb{C}}(\mu) \subseteq L^q_{\mathbb{C}}(\mu)$. In particular, $L^2_{\mathbb{C}}(\mu) \subseteq L^1_{\mathbb{C}}(\mu)$.*

Proof From the inequality $|a|^q \leq 1 + |a|^p$, true for all $a \in \mathbb{C}$, it follows that $\mu(|f|^q) \leq \mu(1) + \mu(|f|^p)$. Since $\mu(1) = \mu(\mathbb{R}) < \infty$, $\mu(|f|^q) < \infty$ whenever $\mu(|f|^p) < \infty$. $\qquad\square$

Example 1.3.1 The space $\ell^2_{\mathbb{C}}(\mathbb{Z})$ of complex sequences $\mathbf{a} = \{a_n\}_{n \in \mathbb{Z}}$ such that $\sum_{n \in \mathbb{Z}} |a_n|^2 < \infty$, with the inner product $\langle \mathbf{a}, \mathbf{b} \rangle = \sum_{n \in \mathbb{Z}} a_n b_n^*$ is a Hilbert space.

This is indeed a particular case of a Hilbert space $L^2_{\mathbb{C}}(\mu)$, with $X = \mathbb{N}$, and μ is the counting measure. In this example, Schwarz's inequality reads

$$\left| \sum_{n \in \mathbb{Z}} a_n b_n^* \right| \leq \left(\sum_{n \in \mathbb{Z}} |a_n|^2 \right)^{\frac{1}{2}} \times \left(\sum_{n \in \mathbb{Z}} |b_n|^2 \right)^{\frac{1}{2}}.$$

The following inclusion will be used recurrently in the Fourier analysis of stochastic processes, and in particular of time series.

Theorem 1.3.6 $\ell^1_{\mathbb{C}}(\mathbb{Z}) \subset \ell^2_{\mathbb{C}}(\mathbb{Z})$.

Proof See Exercise 1.5.20. $\qquad\square$

Of special interest is the case $(X, \mathcal{X}, \mu) = (\Omega, \mathcal{F}, P)$, where P is a probability measure. We briefly summarize the results collected previously in the terminology and notation of probability theory.

The space $L^2_{\mathbb{C}}(P)$ is the space of (equivalence classes of) complex random variables such that

$$E\left[|X|^2 \right] < \infty$$

with the inner product

$$\langle X, Y \rangle = E\left[XY^*\right].$$

The norm of a random variable X is

$$\|X\| = E\left[|X|^2\right]^{\frac{1}{2}}$$

and the distance between two variables X and Y is

$$d(X, Y) = E\left[|X - Y|^2\right]^{\frac{1}{2}}.$$

The completeness property of $L^2_{\mathbb{C}}(P)$ reads, in this case, as follows. If $\{X_n\}_{n \geq 1}$ is a sequence of random variables in $L^2_{\mathbb{C}}(P)$ such that $\lim_{m,n \uparrow \infty} E\left[|X_m - X_n|^2\right] = 0$, then, there exists a random variable $X \in L^2_{\mathbb{C}}(P)$ such that $\lim_{m,n \uparrow \infty} E\left[|X - X_n|^2\right] = 0$. In other words, the sequence $\{X_n\}_{n \geq 1}$ converges in quadratic mean to X.

Isometric Extension

Definition 1.3.2 Let H and K be two Hilbert spaces with inner products denoted by $\langle \cdot, \cdot \rangle_H$ and $\langle \cdot, \cdot \rangle_K$ respectively, and let $\phi : H \mapsto K$ be a linear mapping such that for all $x, y \in H$

$$\langle \phi(x), \phi(y) \rangle_K = \langle x, y \rangle_H.$$

Then, ϕ is called a *linear isometry* from H into K. If, moreover, ϕ is from H *onto* K, then H and K are said to be *isomorphic*.

Note that a linear isometry is necessarily injective, since $\phi(x) = \phi(y)$ implies $\phi(x - y) = 0$, and therefore

$$0 = \|\phi(x - y)\|_K = \|x - y\|_H,$$

which implies $x = y$. In particular, if the linear isometry is *onto*, it is bijective.

Recall that a subset $A \in E$ where (E, d) is a metric space, is said to be *dense* in E, if for all $x \in E$, there exists a sequence $\{x_n\}_{n \geq 1}$ in A converging to x.

Theorem 1.3.7 *Let H and K be two Hilbert spaces with inner products denoted by $\langle \cdot, \cdot \rangle_H$ and $\langle \cdot, \cdot \rangle_K$ respectively. Let V be a vector subspace of H that is dense in H, and $\phi : V \mapsto K$ be a linear isometry from V to K. Then, there exists a unique linear isometry $\tilde{\phi} : H \mapsto K$ whose restriction to V is ϕ.*

Proof We shall first define $\tilde{\phi}(x)$ for $x \in H$. Since V is dense in H, there exists a sequence $\{x_n\}_{n \geq 1}$ in V converging to x. Since ϕ is isometric,

$$\|\phi(x_n) - \phi(x_m)\|_K = \|x_n - x_m\|_H \quad \text{for all } m, n \geq 1.$$

In particular $\{\phi(x_n)\}_{n\geq 1}$ is a Cauchy sequence in K and therefore it converges to some element of K which we denote $\tilde{\phi}(x)$.

The definition of $\tilde{\phi}(x)$ is independent of the sequence $\{x_n\}_{n\geq 1}$ converging to x. Indeed, for another such sequence $\{y_n\}_{n\geq 1}$

$$\lim_{n\uparrow\infty} \|\phi(x_n) - \phi(y_n)\|_K = \lim_{n\uparrow\infty} \|x_n - y_n\|_H = 0.$$

The mapping $\tilde{\phi} : H \mapsto K$ so constructed is clearly an extension of ϕ (for $x \in V$ one can as the approximating sequence of x the sequence $\{x_n\}_{n\geq 1}$ such that $x_n \equiv x$).

The mapping $\tilde{\phi}$ is linear. Indeed, let $x, y \in H$, $\alpha, \beta \in \mathbb{C}$, and let $\{x_n\}_{n\geq 1}$ and $\{y_n\}_{n\geq 1}$ be two sequences in V converging to x and y, respectively. Then $\{\alpha x_n + \beta y_n\}_{n\geq 1}$ converges to $\alpha x + \beta y$. Therefore

$$\lim_{n\uparrow\infty} \phi(\alpha x_n + \beta y_n) = \tilde{\phi}(\alpha x + \beta y).$$

But

$$\phi(\alpha x_n + \beta y_n) = \alpha\phi(x_n) + \beta\phi(y_n) \rightarrow \alpha\tilde{\phi}(x) + \beta\tilde{\phi}(y)$$

tends to $\tilde{\phi}(\alpha x + \beta y) = \alpha\tilde{\phi}(x) + \beta\tilde{\phi}(y)$.

The mapping $\tilde{\phi}$ is isometric since, in view of the bicontinuity of the inner product and of the isometricity of ϕ, if $\{x_n\}_{n\geq 1}$ and $\{y_n\}_{n\geq 1}$ are two sequences in V converging to x and y respectively, then

$$\langle \tilde{\phi}(x), \tilde{\phi}(y) \rangle_K = \lim_{n\uparrow\infty} \langle \phi(x_n), \phi(y_n) \rangle_K$$
$$= \lim_{n\uparrow\infty} \langle x_n, y_n \rangle_H = \langle x, y \rangle_H. \qquad \square$$

1.3.2 Orthogonal Projection

Let H be a Hilbert space. A set $G \subseteq H$ is said to be closed in H if every convergent sequence of G has a limit in G.

Theorem 1.3.8 *Let $G \subseteq H$ be a vector subspace of the Hilbert space H. Endow G with the inner product which is the restriction to G of the inner product on H. Then, G is a Hilbert space if and only if G is closed in H.*

G is then called a *Hilbert subspace* of H.

Proof (i) Assume that G is closed. Let $\{x_n\}_{n\in\mathbb{N}}$ be a Cauchy sequence in G. It is a fortiori a Cauchy sequence in H, and therefore it converges in H to some x, and this x must be in G, because it is a limit of elements of G and G is closed.

(ii) Assume that G is a Hilbert space with the inner product induced by the inner product of H. In particular every convergent sequence $\{x_n\}_{n\in\mathbb{N}}$ of elements of G converges to some element of G. Therefore G is closed.   □

Definition 1.3.3 Two elements x, y of the Hilbert space H are said to be *orthogonal* if $\langle x, y \rangle = 0$. Let G be a Hilbert subspace of the Hilbert space H. The *orthogonal complement* of G in H, denoted $G^\perp$, is defined by

$$G^\perp = \{z \in H : \langle z, x \rangle = 0 \text{ for all } x \in G\}.$$

Clearly, $G^\perp$ is a vector space over $\mathbb{C}$. Moreover, it is closed in H since if $\{z_n\}_{n\geq 1}$ is a sequence of elements of $G^\perp$ converging to $z \in H$ then, by continuity of the inner product,

$$\langle z, x \rangle = \lim_{n\uparrow\infty} \langle z_n, x \rangle = 0 \quad \text{for all } x \in H.$$

Therefore $G^\perp$ is a Hilbert subspace of H.

Observe that a decomposition $x = y + z$ where $y \in G$ and $z \in G^\perp$ is necessarily unique. Indeed, let $x = y' + z'$ be another such decomposition. Then, letting $a = y - y'$, $b = z - z'$, we have that $0 = a + b$ where $a \in G$ and $b \in G^\perp$. Therefore, in particular, $0 = \langle a, a \rangle + \langle a, b \rangle$. But $\langle a, b \rangle = 0$, and therefore $\langle a, a \rangle = 0$, which implies that $a = 0$. Similarly, $b = 0$.

Theorem 1.3.9 *Let G be a Hilbert subspace of H. For all $x \in H$, there exists a unique element $y \in G$ such that $x - y \in G^\perp$. Moreover,*

$$\|y - x\| = \inf_{u\in G} \|u - x\|. \tag{1.18}$$

Proof Let $d(x, G) = \inf_{z\in G} d(x, z)$ and let $\{y_n\}_{n\geq 1}$ be a sequence in G such that

$$d(x, G)^2 \leq d(x, y_n)^2 \leq d(x, G)^2 + \frac{1}{n}. \tag{*}$$

The parallelogram identity gives, for all $m, n \geq 1$,

$$\|y_n - y_m\|^2 = 2(\|x - y_n\|^2 + \|x - y_m\|^2) - 4\|x - \frac{1}{2}(y_m + y_n)\|^2.$$

Since $\frac{1}{2}(y_n + y_m) \in G$,

$$\|x - \frac{1}{2}(y_m + y_n)\|^2 \geq d(x, G)^2,$$

and therefore

$$\|y_n - y_m\|^2 \leq 2\left(\frac{1}{n} + \frac{1}{m}\right).$$

The sequence $\{y_n\}_{n\geq 1}$ is therefore a Cauchy sequence in G and consequently it converges to some $y \in G$ since G is closed. Passing to the limit in (∗) gives (1.18).

Uniqueness of y satisfying (1.18): Let $y' \in G$ be another such element. Then

$$\|x - y'\| = \|x - y\| = d(x, G),$$

and from the parallelogram identity

$$\|y - y'\|^2 = 2\|y - x\|^2 + 2\|y' - x\|^2 - 4\|x - \frac{1}{2}(y + y')\|^2$$
$$= 4d(x, G)^2 - 4\|x - \frac{1}{2}(y + y')\|^2.$$

Since $\frac{1}{2}(y + y') \in G$

$$\|x - \frac{1}{2}(y + y')\|^2 \geq d(x, G)^2,$$

and therefore $\|y - y'\|^2 \leq 0$, which implies $\|y - y'\|^2 = 0$ and therefore $y = y'$.

It now remains to show that $x - y$ is orthogonal to G, that is, $\langle x - y, z \rangle = 0$ for all $z \in G$. Since this is trivially true if $z = 0$, we may assume $z \neq 0$. Because $y + \lambda z \in G$ for all $\lambda \in \mathbb{R}$

$$\|x - (y + \lambda z)\|^2 \geq d(x, G)^2,$$

that is,

$$\|x - y\|^2 + 2\lambda \mathrm{Re}\,\{\langle x - y, z \rangle\} + \lambda^2\|z\|^2 \geq d(x, G)^2.$$

Since $\|x - y\|^2 = d(x, G)^2$, we have

$$-2\lambda \mathrm{Re}\,\{\langle x - y, z \rangle\} + \lambda^2\|z\|^2 \geq 0 \quad \text{for all } \lambda \in \mathbb{R},$$

which implies $\mathrm{Re}\,\{\langle x - y, z \rangle\} = 0$. The same type of calculation with $\lambda \in i\mathbb{R}$ (pure imaginary) leads to $\Im\,\{\langle x - y, z \rangle\} = 0$. Therefore $\langle x - y, z \rangle = 0$.

That y is the unique element of G such that $y - x \in G^\perp$ follows the remark preceding Theorem 1.3.9.   □

Definition 1.3.4 The element y in Theorem 1.3.9 is called the *orthogonal projection* of x on G and is denoted by $P_G(x)$.

The projection theorem states, in particular, that for any $x \in G$ there is a unique decomposition

$$x = y + z, \quad y \in G, \ z \in G^{\perp},$$

and that $y = P_G(x)$, the (unique) element of G closest to x. Therefore

Theorem 1.3.10 *The orthogonal projection* $y = P_G(x)$ *is characterized by the two following properties:*

(1) $y \in G$;
(2) $\langle y - x, z \rangle = 0$ *for all* $z \in G$.

This characterization is called the *projection principle*.

Let C be a collection of vectors in the Hilbert space H. The linear span of C, denoted $span(C)$ is, by definition, the set of all finite linear combinations of vectors of C. This is a vector space. The closure of this vector space, $\overline{span(C)}$, is called the Hilbert subspace generated by C. By definition, x belongs to this subspace if and only if there exists a sequence of vectors $\{x_n\}_{n \geq 1}$ such that

(i) for all $n \geq 1$, x_n is a finite linear combination of vectors of C, and
(ii) $\lim_{n \uparrow \infty} x_n = x$.

Theorem 1.3.11 *An element* $\widehat{x} \in H$ *is the projection of* x *onto* $G = \overline{span(C)}$ *if and only if*
(α) $\widehat{x} \in G$, *and*
(β) $\langle x - \widehat{x}, z \rangle = 0$ *for all* $z \in C$.

Note that we have to satisfy requirement not for all $z \in G$, but only for all $z \in C$.

The proof is easy. We have to show that $\langle x - \widehat{x}, z \rangle = 0$ for all $z \in G$. But $z = \lim_{n \uparrow \infty} z_n$ where $\{z_n\}_{n \geq 1}$ is a sequence of vectors of $span(C)$ such that $\lim_{n \uparrow \infty} z_n = z$. By hypothesis, for all $n \geq 1$, $\langle x - \widehat{x}, z_n \rangle = 0$. Therefore, by continuity of the inner product,

$$\langle x - \widehat{x}, z \rangle = \lim_{n \uparrow \infty} \langle x - \widehat{x}, z_n \rangle = 0.$$

Example 1.3.2 (Linear regression of random variables on random vectors) Let Y and $X_1, \ldots, X_N$ be square-integrable real random variables, that we assume, to begin with, centered. We seek the random variable $\widehat{Y} = a_1 X_1 + \cdots + a_N X_N = a^T X$ which minimizes the *quadratic error*

$$E[(Z - Y)^2]$$

amongst all the random variables Z of the form $b_1 X_1 + \cdots b_N X_N = b^T X$. The random variable $\widehat{Y}$ achieving the minimum is called the *linear regression* of Y on $X_1, \ldots, X_N$, or the best *linear-quadratic approximation* of Y as a function of $X_1, \ldots, X_N$. The vector a is called the *regression vector* (of X on Y).

This minimization problem is a special case of the general projection problem, with $H = L^2_{\mathbb{R}}(P)$ and G the Hilbert space consisting of finite real combinations of $X_1, \ldots, X_N$. By the projection principle, the projection $\widehat{Y}$ of $Y \in H$ is of the form $X^T a$ and verifies

$$E\left[X_i(Y - \widehat{Y})\right] = 0$$

for all $i \in \{1, \ldots, N\}$, that is

$$E\left[X(Y - X^T a)\right] = \Gamma_{XY} - \Gamma_X a = 0,$$

where $\Gamma_{XY} = E[XY]$ and Γ_X is the covariance matrix of X. We assume that $\Gamma_X > 0$ and is therefore invertible. The regression vector is then given by

$$a = \Gamma_X^{-1} \Gamma_{XY}$$

so that

$$\widehat{Y} = \Gamma_{YX} \Gamma_X^{-1} X . \tag{1.19}$$

The quadratic error $d^2 = E\left[|Y - \widehat{Y}|^2\right]$ is given by

$$d^2 = \langle Y - \widehat{Y}, Y \rangle$$

since $Y - \widehat{Y}$ and $\widehat{Y}$ are orthogonal elements of H. Expliciting, we find

$$d^2 = \sigma_Y^2 - \Gamma_{YX} \Gamma_X^{-1} \Gamma_{XY},$$

where σ_Y^2 is the variance of Y.

We now consider the case where the random variables Y and $X_1, \ldots, X_N$ are no longer assumed to be centered. The problem is now to find the *affine* combination of $X_1, \ldots, X_N$ which best approximates Y in the least-squares sense. In other words, we seek to minimize

$$E[(Y - b_0 - b_1 X_1 - \cdots - b_N X_N)^2]$$

with respect to the scalars $b_0, \ldots, b_N$. This problem can be reduced to the preceding one as follows. In fact, for every square integrable random variable U with mean m,

$$E[(U - c)^2] \geq E[(U - m)^2] \quad \text{for all } c.$$

Therefore

$$E[Y - b^T X - b_0]^2 \geq E[(Y - b^T X - E[Y - b^T X])^2],$$

where

$$b^T = (b_1, \ldots, b_N).$$

This shows that b_0 is necessarily of the form $b_0 = m_Y - b^T m_X$. Therefore we have reduced the original problem to that of minimizing with respect to b the quantity $E[((Y - m_Y) - b^T (X - m_X))^2]$, and for this we can use the result obtained in the case of centred random variables. In summary:

If $\Gamma_X > 0$, the best quadratic approximation of Y as an affine function of X is

$$\widehat{Y} = m_Y + \Gamma_{YX} \Gamma_X^{-1} (X - m_X).$$

The quadratic error is given by

$$E[(\widehat{Y} - Y)^2] = \sigma_Y^2 - \Gamma_{YX} \Gamma_X^{-1} \Gamma_{XY}.$$

Theorem 1.3.12 *Let G be a Hilbert subspace of the Hilbert space H.*
(α) The mapping $x \rightarrow P_G(x)$ is linear and continuous, and furthermore

$$\|P_G(x)\| \leq \|x\| \quad \text{for all } x \in H.$$

(β) If F is a Hilbert subspace of H such that $F \subseteq G$ then $P_F \circ P_G = P_F$. In particular $P_G^2 = P_G$ (P_G is then called idempotent).

Proof (α) Let $x_1, x_2 \in H$. They admit the decomposition

$$x_i = P_G(x_i) + w_i \quad (i = 1, 2),$$

where $w_i \in G^\perp$ ($i = 1, 2$). Therefore

$$\begin{aligned} x_1 + x_2 &= P_G(x_1) + P_G(x_2) + w_1 + w_2 \\ &= P_G(x_1) + P_G(x_2) + w, \end{aligned}$$

where $w \in G^\perp$. Now, $x_1 + x_2$ admits an unique decomposition of the type

$$x_1 + x_2 = y + w,$$

where $w \in G^\perp$, $y \in G$, namely: $y = P_G(x_1 + x_2)$. Therefore

$$P_G(x_1 + x_2) = P_G(x_1) + P_G(x_2).$$

One proves similarly that

$$P_G(\alpha x) = \alpha P_G(x) \quad \text{for all } \alpha \in G, \ x \in H.$$

Thus P_G is linear.

From Pythagoras' theorem applied to $x = P_G(x) + w$,

$$\|P_G(x)\| + \|w\|^2 = \|x\|^2,$$

and therefore

$$\|P_G(x)\|^2 \leq \|x\|^2.$$

Hence, P_G is continuous.

(β) The unique decompositions of x on G and $G^\perp$ and of $P_G(x)$ on F and $F^\perp$ are respectively

$$x = P_G(x) + w, \quad P_G(x) = P_F(P_G(x)) + z.$$

From these two equalities, we obtain

$$x = P_F(P_G(x)) + z + w. \tag{*}$$

But $z \in G^\perp$ implies $z \in F^\perp$ since $F \subseteq G$, and therefore $v = z + w \in F^\perp$. On the other hand, $P_F(P_G(x)) \in F$. Therefore (*) is the unique decomposition of x on F and $F^\perp$; in particular, $P_F(x) = P_F(P_G(x))$. $\qquad\square$

The next result says that the projection operator P_G is "continuous" with respect to G.

Theorem 1.3.13 *(i) Let $\{G_n\}_{n\geq 1}$ be a non-decreasing sequence of Hilbert subspaces of H. Then, the closure G of $\bigcup_{n\geq 1} G_n$ is a Hilbert subspace of H and for all $x \in H$*

$$\lim_{n\uparrow\infty} P_{G_n}(x) = P_G(x).$$

(ii) Let $\{G_n\}$ be a non-increasing sequence of Hilbert subspaces of H. Then $G := \bigcap_{n\geq 1} G_n$ is a Hilbert subspace of H and for all $x \in H$

$$\lim_{n\uparrow\infty} P_{G_n}(x) = P_G(x).$$

Proof (i) The set $\bigcup_{n\geq 1} G_n$ is evidently a vector subspace of H (however, it is not closed in general). Its closure G is a Hilbert subspace (Theorem 1.3.8). To any $y \in G$ one can associate a sequence $\{y_n\}_{n\geq 1}$, where $y_n \in G_n$, and

$$\lim_{n\to\infty} \|y - y_n\| = 0.$$

Take $y = P_G(x)$. By the parallelogram identity

$$\|P_{G_n}(x) - P_G(x)\|^2 = \|(x - P_G(x)) - (x - P_{G_n}(x))\|^2$$
$$= 2\|x - P_{G_n}(x)\|^2 + 2\|x - P_G(x)\|^2$$
$$- 4\|x - \frac{1}{2}(P_{G_n}(x) + P_G(x))\|.$$

But since $P_{G_n}(x) + P_G(x)$ is a vector in G,

$$\|x - \frac{1}{2}(P_{G_n}(x) + P_G(x))\|^2 \geq \|x - P_G(x)\|^2,$$

and therefore

$$\|P_{G_n}(x) - P_G(x)\|^2 \leq 2\|x - P_{G_n}(x)\|^2 - 2\|x - P_G(x)\|^2$$
$$\leq 2\|x - y_n\|^2 - 2\|x - P_G(x)\|^2.$$

By the continuity of the norm

$$\lim_{n \uparrow \infty} \|x - y_n\|^2 = \|x - P_G(x)\|^2,$$

and finally

$$\lim_{n \uparrow \infty} \|P_{G_n}(x) - P_G(x)\|^2 = 0.$$

(ii) Devise a direct proof in the spirit of (i) or use the fact that $G^\perp$ is the closure of $\bigcup_{n \geq 1} G_n^\perp$. $\square$

If G_1 and G_2 are orthogonal Hilbert subspaces of the Hilbert space H,

$$G_1 \oplus G_2 := \{z = x_1 + x_2 : x_1 \in G, \ x_2 \in G_2\}$$

is called the *orthogonal sum* of G_1 and G_2.

Riesz's Representation

Definition 1.3.5 Let H be a Hilbert space over $\mathbb{C}$ and let $f : H \to \mathbb{C}$ be a linear mapping; f is then called a (complex) *linear form* on H. It is said to be continuous if there exists $A \geq 0$ such that

$$|f(x_1) - f(x_2)| \leq A\|x_1 - x_2\| \quad \text{for all } x_1, x_2 \in H.$$

The infimum of such constants A is called the *norm* of f.

Example 1.3.3 Let $y \in H$. The mapping $f : H \to \mathbb{C}$ by $f(x) = \langle x, y \rangle$ is a linear form, and by Schwarz's inequality

$$|f(x_1) - f(x_2)| = |f(x_1 - x_2)| = |\langle x_1 - x_2, y \rangle| \leq \|y\| \, \|x_1 - x_2\| \, .$$

Therefore f is continuous. Its norm is $\|y\|$. To prove this it remains to show that if K is such that $|\langle x_1 - x_2, y \rangle| \leq K \|x_1 - x_2\|$ for all $x_1, x_2 \in H$, then $\|y\| \leq K$. It suffices to take $x_1 = x_2 = y$ in the last before last inequality, which gives $\|y\|^2 \leq K\|y\|$.

Theorem 1.3.14 *Let $f : H \mapsto \mathbb{C}$ be a continuous linear form on the Hilbert space H. Then there exists an unique $y \in H$ such that*

$$f(x) = \langle x, y \rangle \, for \, all \, x \in H \, .$$

Proof Uniqueness: Let $y, y' \in H$ be such that

$$f(x) = \langle x, y \rangle = \langle x, y' \rangle \quad \text{for all } x \in H.$$

In particular,

$$\langle x, y - y' \rangle = 0 \quad \text{for all } x \in H.$$

The choice $x = y - y'$ leads to $\|y - y'\|^2 = 0$, that is, $y - y'$.
Existence: The kernel $N := \{u \in H : f(u) = 0\}$ of f is a Hilbert subspace of H. Eliminating the trivial case $f \equiv 0$, it is strictly included in H. In particular, $N^{\perp}$ does not reduces to the singleton $\{0\}$ and therefore, there exists $z \in N^{\perp}, z \neq 0$. Define y by

$$y = f(z)^* \frac{z}{\|z\|^2} \, .$$

For all $x \in N$, $\langle x, y \rangle = 0$, and therefore, in particular, $\langle x, y \rangle = f(x)$. Also,

$$\langle z, y \rangle = \langle\langle z, f(z)^* \frac{z}{\|z\|^2} \rangle\rangle = f(z).$$

Therefore the mappings $x \to f(x)$ and $x \to \langle x, y \rangle$ coincide on the Hilbert subspace generated by N and z. But this subspace is H itself. In fact, for all $x \in H$,

$$x = \left(x - \frac{f(x)}{f(z)} z \right) + \frac{f(x)}{f(z)} z = u + w,$$

where $u \in N$ and w is colinear to z. $\qquad\qquad \square$

1.3.3 Orthonormal Expansions

Definition 1.3.6 A sequence $\{e_n\}_{n \geq 0}$ in a Hilbert space H is called an *orthonormal system* of H if it satisfies the conditions

(α) $\langle e_n, e_k \rangle = 0$ for all $n \neq k$; and

(β) $\|e_n\| = 1$ for all $n \geq 0$.

If only requirement (α) is met, the sequence is called an *orthogonal system*

Example 1.3.4 (Gram–Schmidt's orthonormalization procedure) Let $\{f_n\}_{n \geq 0}$ be a sequence of vectors of a Hilbert space H. Construct $\{e_n\}_{n \geq 0}$ by the *Gram–Schmidt orthonormalization procedure*:

- Set $p(0) = 0$ and $e_0 = f_0 / \|f_0\|$ (assuming $f_0 \neq 0$ without loss of generality);
- $e_0, \ldots, e_n$ and $p(n)$ being defined, let $p(n+1)$ be the first index $p > p(n)$ such that f_p is independent of $e_0, \ldots, e_n$, and define, with $p = p(n+1)$,

$$e_{n+1} = \frac{f_p - \sum_{i=1}^{n} \langle f_p, e_i \rangle e_i}{\left\| f_p - \sum_{i=1}^{n} \langle f_p, e_i \rangle e_i \right\|}.$$

One verifies that $\{e_n\}_{n \geq 0}$ is an orthonormal system.

An orthonormal system $\{e_n\}_{n \geq 0}$ is *free* in the sense that an arbitrary *finite* subset of it is linearly independent. Indeed, taking $(e_1, \ldots, e_k)$ for example, the relation $\sum_{i=1}^{k} \alpha_i e_i = 0$ implies that $\alpha_\ell = \langle e_\ell, \sum_{i=1}^{k} \alpha_i e_i \rangle = 0$ for all $1 \leq \ell \leq k$.

Let $\{e_n\}_{n \geq 0}$ be an orthonormal system of H and let G be the Hilbert subspace of H generated by it.

Theorem 1.3.15 *(a) For an arbitrary sequence $\{\alpha_n\}_{n \geq 0}$ of complex numbers the series $\sum_{n \geq 0} \alpha_n e_n$ is convergent in H if and only if $\{\alpha_n\}_{n \geq 0} \in \ell_{\mathbb{C}}^2$, in which case*

$$\left\| \sum_{n \geq 0} \alpha_n e_n \right\|^2 = \sum_{n \geq 0} |\alpha_n|^2. \tag{1.20}$$

(b) For all $x \in H$, we have Bessel's inequality:

$$\sum_{n \geq 0} |\langle x, e_n \rangle|^2 \leq \|x\|^2. \tag{1.21}$$

(c) For all $x \in H$ the series $\sum_{n \geq 0} \langle x, e_n \rangle e_n$ converges, and

$$\sum_{n \geq 0} \langle x, e_n \rangle e_n = P_G(x), \tag{1.22}$$

where P_G is the projection on G.

(d) For all $x, y \in H$ the series $\sum_{n\geq 0}\langle x, e_n\rangle\langle y, e_n\rangle$ is absolutely convergent, and

$$\sum_{n\geq 0}\langle x, e_n\rangle\langle y, e_n\rangle^* = \langle P_G(x), P_G(y)\rangle. \tag{1.23}$$

Proof (a) From Pythagoras' theorem we have

$$\left\|\sum_{j=m+1}^{n} \alpha_j e_j\right\|^2 = \sum_{j=m+1}^{n} |\alpha_j|^2,$$

and therefore the sequence $\{\sum_{j=0}^{n} \alpha_j e_j\}_{n\geq 0}$ is a Cauchy sequence in H if and only if $\{\sum_{j=0}^{n} |\alpha_j|^2\}_{n\geq 0}$ is a Cauchy sequence in $\mathbb{R}$. In other words, $\sum_{n\geq 0}\alpha_n e_n$ converges if and only if $\sum_{n\geq 0}|\alpha_n|^2 < \infty$. In this case equality (1.20) follows from the continuity of the norm, by letting n tend to ∞ in the last display.

(b) According to (α) of Theorem 1.3.12, $\|x\| \geq \|P_{G_n}(x)\|$, where G_n is the Hilbert subspace spanned by $\{e_1, \ldots, e_n\}$. But

$$P_{G_n}(x) = \sum_{i=0}^{n}\langle x, e_i\rangle e_i,$$

and by Pythagoras' theorem

$$\left\|P_{G_n}(x)\right\|^2 = \sum_{i=0}^{n} |\langle x, e_i\rangle|^2.$$

Therefore

$$\|x\|^2 \geq \sum_{i=0}^{n} |\langle x, e_i\rangle|^2,$$

from which Bessel's inequality follows by letting $n \uparrow \infty$.

(c) From (1.21) and (a), the series $\sum_{n\geq 0}\langle x, e_n\rangle e_n$ converges. For any $m \geq 0$ and for all $N \geq m$

$$\langle x - \sum_{n=0}^{N}\langle x, e_n\rangle e_n, e_m\rangle = 0,$$

and therefore, by continuity of the inner product,

$$\langle x - \sum_{n\geq 0}\langle x, e_n\rangle e_n, e_m\rangle = 0 \quad \text{for all } m \geq 0.$$

This implies that $x - \sum_{n \geq 0} \langle x, e_n \rangle e_n$ is orthogonal to G. Also $\sum_{n \geq 0} \langle x, e_n \rangle e_n \in G$. Therefore, by the projection principle,

$$P_G(x) = \sum_{n \geq 0} \langle x, e_n \rangle e_n.$$

(d) By Schwarz's inequality in $\ell_{\mathbb{C}}^2$, for all $N \geq 0$

$$\left(\sum_{n=0}^{N} |\langle x, e_n \rangle \langle y, e_n \rangle^*| \right)^2 \leq \left(\sum_{n=0}^{N} |\langle x, e_n \rangle|^2 \right) \left(\sum_{n=0}^{N} |\langle y, e_n \rangle|^2 \right)$$

$$\leq \|x\|^2 \|y\|^2.$$

Therefore the series $\sum_{n=0}^{\infty} \langle x, e_n \rangle \langle y, e_n \rangle^*$ is absolutely convergent. Also, by an elementary computation,

$$\left\langle \sum_{n=0}^{N} \langle x, e_n \rangle e_n, \sum_{n=0}^{N} \langle y, e_n \rangle e_n \right\rangle = \sum_{n=0}^{N} \langle x, e_n \rangle \langle y, e_n \rangle^*.$$

Letting $N \uparrow \infty$ we obtain (1.23) (using (1.22) and the continuity of the inner product).   □

Definition 1.3.7 A sequence $\{w_n\}_{n \geq 0}$ in H is said to be *total* in H if it generates H. (That is, the finite linear combinations of the elements of $\{w_n\}_{n \geq 0}$ form a dense subset of H.)

A sequence $\{w_n\}_{n \geq 0}$ in H is total in H if and only if

$$\langle z, w_n \rangle = 0 \text{ for all } n \geq 0 \Rightarrow z = 0. \tag{†}$$

The proof is left as an exercise.

Theorem 1.3.16 *Let $\{e_n\}_{n \geq 0}$ be an orthonormal system of H. The following properties are equivalent:*
(a) $\{e_n\}_{n \geq 0}$ is total in H;
(b) For all $x \in H$, the Plancherel–Parseval identity holds true:

$$\|x\|^2 = \sum_{n \geq 0} |\langle x, e_n \rangle|^2;$$

(c) For all $x \in H$,

$$x = \sum_{n \geq 0} \langle x, e_n \rangle e_n.$$

Proof (a)⇒(c)   According to (c) of Theorem 1.3.15,

$$\sum_{n\geq 0} \langle x, e_n \rangle e_n = P_G(x),$$

where G is the Hilbert subspace generated by $\{e_n\}_{n\geq 0}$. Since $\{e_n\}_{n\geq 0}$ is total, it follows by (†) that $G^\perp = \{0\}$, and therefore $P_G(x) = x$.
(c)⇒(b) This follows from (a) of Theorem 1.3.15.
(b)⇒(a) From (1.20) and (1.22),

$$\sum_{n\geq 0} |\langle x, e_n \rangle|^2 = \|P_G(x)\|^2,$$

and therefore (b) implies $\|x\|^2 = \|P_G(x)\|^2$. From Pythagoras' theorem

$$\|x\|^2 = \|P_G(x) + x - P_G(x)\|^2$$
$$= \|P_G(x)\|^2 + \|x - P_G(x)\|^2$$
$$= \|x\|^2 + \|x - P_G(x)\|^2,$$

and therefore $\|x - P_G(x)\|^2 = 0$, which implies $x = P_G(x)$. Since this is true for all $x \in H$ we must have $G \equiv H$, *i.e.* $\{e_n\}_{n\geq 0}$ is total in H.   □

A sequence $\{e_n\}_{n\geq 0}$ satisfying one (and then all) of the conditions of Theorem 1.3.16 is called a (denumerable) *Hilbert basis* of H.

1.4 Fourier Theory in L^2

1.4.1 Fourier Transform in L^2

Recall the notation $f(.)$ for a function f; in particular, $f(a + .)$ is the function $f_a : \mathbb{R} \mapsto \mathbb{C}$ defined by $f_a(t) = f(a + t)$.

Theorem 1.4.1 *For any $f \in L^2_{\mathbb{C}}(\mathbb{R})$, the mapping from $\mathbb{R}$ into $L^2_{\mathbb{C}}(\mathbb{R})$ defined by*

$$t \rightarrow f(t + \cdot)$$

is uniformly continuous.

Proof We have to prove that the quantity

$$\int_{\mathbb{R}} |f(t + h + u) - f(t + u)|^2 \, du = \int_{\mathbb{R}} |f(h + u) - f(u)|^2 \, du$$

tends to 0 when $h \rightarrow 0$ (the uniformity in t of convergence then follows, since this quantity is independent of t). When f is continuous with compact support the result follows by dominated convergence. The general case is obtained by approximating f in $L^2_{\mathbb{C}}(\mathbb{R})$ by continuous functions with compact support as in the proof of Theorem 1.1.8.   □

It follows from Schwarz's inequality that the function

$$t \to \langle f(t + \cdot), f(\cdot) \rangle_{L^2_{\mathbb{C}}(\mathbb{R})} = \int_{\mathbb{R}} f(t + x) f^*(x)\, dx$$

is uniformly continuous on $\mathbb{R}$, and bounded by $\int_{\mathbb{R}} |f(t)|^2\, dt$. This function is called the *autocorrelation function* of f. Note that it is the convolution $f * \tilde{f}$, where $\tilde{f}(t) = f(-t)^*$.

Theorem 1.4.2 *If $f \in L^1_{\mathbb{C}}(\mathbb{R}) \cap L^2_{\mathbb{C}}(\mathbb{R})$, then $\widehat{f} \in L^2_{\mathbb{C}}(\mathbb{R})$ and*

$$\int_{\mathbb{R}} |f(t)|^2\, dt = \int_{\mathbb{R}} |\widehat{f}(v)|^2\, dv.$$

Proof The function $\tilde{f}$ admits $\widehat{f^*}$ for FT, and therefore, by the convolution–multiplication rule,

$$\mathcal{F} : (f * \tilde{f}) \to |\widehat{f}|^2.$$

Consider the Gaussian regularization kernel

$$h_\sigma(t) = \frac{1}{\sigma\sqrt{2\pi}}\, e^{-\frac{t^2}{2\sigma^2}}.$$

Applying (∗∗) of the proof of Theorem 1.1.8 with $(f * \tilde{f})$ instead of f, and observing that h_σ is an even function, we obtain

$$\int_{\mathbb{R}} |\widehat{f}(v)|^2 \widehat{f_\sigma}(v)\, dv = \int_{\mathbb{R}} (f * \tilde{f})(x) h_\sigma(x)\, dx.$$

Since $\widehat{f_\sigma}(v) = e^{-2\pi^2\sigma^2 x^2} \uparrow 1$ when $\sigma \downarrow 0$, the left-hand side of the previous equality tends to $\int_{\mathbb{R}} |\widehat{f}(v)|^2\, dv$, by dominated convergence. On the other hand, since the autocorrelation function $f * \tilde{f}$ is continuous and bounded, the quantity

$$\int_{\mathbb{R}} (f * \tilde{f})(x) h_\sigma(x)\, dx = \int_{\mathbb{R}} (f * \tilde{f})(\sigma y) h_1(y)\, dy$$

tends when $\sigma \downarrow 0$ towards

$$\int_{\mathbb{R}} (f * \tilde{f})(0) h_1(y) \, dy = (f * \tilde{f})(0) = \int_{\mathbb{R}} |f(t)|^2 \, dt,$$

by dominated convergence. □

The mapping $\varphi : f \to \widehat{f}$ from $L^1_{\mathbb{C}}(\mathbb{R}) \cap L^2_{\mathbb{C}}(\mathbb{R})$ into $L^2_{\mathbb{C}}(\mathbb{R})$ is therefore isometric and linear. Since $L^1 \cap L^2$ is dense in L^2, this linear isometry can be uniquely extended into a linear isometry from $L^2_{\mathbb{C}}(\mathbb{R})$ into itself (Theorem 1.3.7). We continue to denote by $\widehat{f}$ the image of f by this extended isometry, and to call it the Fourier transform of f.

The above isometry is expressed by the *Plancherel–Parseval identity*:

Theorem 1.4.3 *For f_1 and f_2 in $L^2_{\mathbb{C}}(\mathbb{R})$,*

$$\int_{\mathbb{R}} f_1(t) f_2(t)^* \, dt = \int_{\mathbb{R}} \widehat{f_1}(v) \widehat{f_2}(v)^* \, dv.$$

The Convolution-Multiplication Rule

Theorem 1.4.4 *For $h \in L^1_{\mathbb{C}}(\mathbb{R})$ and $f \in L^2_{\mathbb{C}}(\mathbb{R})$, the function*

$$g(t) = \int_{\mathbb{R}} h(t - s) f(s) \, ds$$

is almost-everywhere well-defined and in $L^2_{\mathbb{C}}(\mathbb{R})$. Furthermore, its FT *is*

$$\widehat{g} = \widehat{h} \widehat{f}.$$

Proof First, observe that on the one hand

$$\int_{\mathbb{R}} |h(t - s)| \, |f(s)| \, ds \leq \int_{\mathbb{R}} |h(t - s)| (1 + |f(s)|^2) \, ds$$

$$= \int_{\mathbb{R}} |h(t)| \, dt + \int_{\mathbb{R}} |h(t - s)| \, |f(s)|^2 \, ds,$$

and on the other hand, for almost all t,

$$\int_{\mathbb{R}} |h(t - s)| \, |(f(s)|^2 \, ds < \infty,$$

since $|h|$ and $|f|^2$ are in $L^1_{\mathbb{C}}(\mathbb{R})$. Therefore, for almost all t,

$$\int_{\mathbb{R}} |h(t-s|\,|f(s)\,ds < \infty,$$

and $g(t)$ is almost-everywhere well defined. Let us now show that $g \in L^2_{\mathbb{C}}(\mathbb{R})$. Using Fubini's theorem and Schwarz's inequality we have:

$$\int_{\mathbb{R}} \left| \int_{\mathbb{R}} h(t-s)f(s)\,ds \right|^2 dt$$

$$= \int_{\mathbb{R}} \left| \int_{\mathbb{R}} h(u)f(t-u)\,du \right|^2 dt$$

$$= \int_{\mathbb{R}} \int_{\mathbb{R}} \left\{ \int_{\mathbb{R}} f(t-u)f(t-v)^*\,dt \right\} h(u)h(v)^*\,du\,dv$$

$$\leq \left(\int_{\mathbb{R}} |f(s)|^2\,ds \right) \left(\int_{\mathbb{R}} |h(u)|\,du \right)^2 < \infty.$$

For future reference, we rewrite this

$$\|h * f\|_{L^2_{\mathbb{C}}(\mathbb{R})} \leq \|h\|_{L^1_{\mathbb{C}}(\mathbb{R})} \|f\|_{L^2_{\mathbb{C}}(\mathbb{R})} \qquad (1.24)$$

The function $g = h * f$ is therefore in $L^2_{\mathbb{C}}(\mathbb{R})$ when $h \in L^1_{\mathbb{C}}(\mathbb{R})$ and $f \in L^2_{\mathbb{C}}(\mathbb{R})$. If, furthermore, f is in $L^1_{\mathbb{C}}(\mathbb{R})$, then g is in $L^1_{\mathbb{C}}(\mathbb{R})$. Therefore

$$f \in L^1_{\mathbb{C}}(\mathbb{R}) \cap L^2_{\mathbb{C}}(\mathbb{R}) \Rightarrow g = h * f \in L^1_{\mathbb{C}}(\mathbb{R}) \cap L^2_{\mathbb{C}}(\mathbb{R}).$$

In this case we have $\widehat{g} = \widehat{h}\widehat{f}$, by the convolution–multiplication formula in L^1.

We now suppose that f is in $L^2_{\mathbb{C}}(\mathbb{R})$ (but not necessarily in $L^1_{\mathbb{C}}(\mathbb{R})$). The function

$$f_A(t) := f(t)1_{[-A,+A]}(t)$$

is in $L^1_{\mathbb{C}}(\mathbb{R}) \cap L^2_{\mathbb{C}}(\mathbb{R})$ and $\lim f_A = f$ in $L^2_{\mathbb{C}}(\mathbb{R})$. In particular, $\lim \widehat{f_A} = \widehat{f}$ in $L^2_{\mathbb{C}}(\mathbb{R})$. Introducing the function

$$g_A(t) := \int_{\mathbb{R}} h(t-s)f_A(s)\,ds$$

we have $\widehat{g_A} = \widehat{h}\widehat{f_A}$. Also, $\lim g_A = g$ in $L^2_{\mathbb{C}}(\mathbb{R})$ (use (1.24)), and therefore $\lim \widehat{g_A} = \widehat{g}$ in $L^2_{\mathbb{C}}(\mathbb{R})$. Now, since $\lim \widehat{f_A} = \widehat{f}$ in $L^2_{\mathbb{C}}(\mathbb{R})$ and $\widehat{h}$ is bounded, $\lim \widehat{h}\widehat{f_A} = \widehat{h}\widehat{f}$ in $L^2_{\mathbb{C}}(\mathbb{R})$. Hence the announced result: $\widehat{g} = \widehat{h}\widehat{f}$. $\qquad \square$

The Inversion Formula

So far, we know that the mapping $\varphi : L^2_{\mathbb{C}}(\mathbb{R}) \mapsto L^2_{\mathbb{C}}(\mathbb{R})$ just defined is linear, isometric, and *into*. We shall now show that it is *onto*, and therefore bijective.

Theorem 1.4.5 *Let $\widehat{f}$ be the* FT *of $f \in L^2_{\mathbb{C}}(\mathbb{R})$. Then*

$$\varphi : \widehat{f}(-\nu) \rightarrow f(t),$$

that is,

$$f(t) = \lim_{A \uparrow \infty} \int\limits_{-A}^{+A} \widehat{f}(\nu)e^{2i\pi \nu t}\, d\nu.$$

where the limit is in $L^2_{\mathbb{C}}(\mathbb{R})$, and the equality is almost-everywhere.

We shall prepare the way for the proof with the following:

Lemma 1.4.1 *Let u and v be functions in $L^2_{\mathbb{C}}(\mathbb{R})$. Then*

$$\int\limits_{\mathbb{R}} u(x)\widehat{v}(x)\, dx = \int\limits_{\mathbb{R}} \widehat{u}(x)v(x)\, dx .$$

Proof If Lemma 1.4.1 is true for $u, v \in L^1_{\mathbb{C}}(\mathbb{R}) \cap L^2_{\mathbb{C}}(\mathbb{R})$ then it also holds for $u, v \in L^2_{\mathbb{C}}(\mathbb{R})$. In fact, using the notation $f_A(t) := f(t)1_{[-A,+A]}(t)$, we have

$$\int\limits_{\mathbb{R}} u_A(x)\widehat{(v_A)}(x)\, dx = \int\limits_{\mathbb{R}} \widehat{(u_A)}(x)v_A(x)\, dx,$$

that is, $\langle u_A, \widehat{v_A} \rangle = \langle \widehat{u_A}, v_A \rangle$. Now $u_A, v_A, \widehat{u_A}$ and $\widehat{v_A}$ tend in $L^2_{\mathbb{C}}(\mathbb{R})$ to $u, v, \widehat{u}$ and $\widehat{v}$, respectively, as $A \uparrow \infty$, and therefore, by the continuity property of the inner product, $\langle u, \widehat{v} \rangle = \langle \widehat{u}, v \rangle$.

It remains to prove Lemma 1.4.1 for integrable functions. This is accomplished by Fubini's theorem, observing that $u \times \widehat{v} \in L^1_{\mathbb{C}}(\mathbb{R})$, being the product of two square-integrable functions:

$$\int\limits_{\mathbb{R}} u(x)\widehat{v}(x)\,dx = \int\limits_{\mathbb{R}} u(x)\left\{\int\limits_{\mathbb{R}} v(y)e^{-2i\pi xy}\,dy\right\}dx$$

$$= \int\limits_{\mathbb{R}} v(y)\left\{\int\limits_{\mathbb{R}} u(x)e^{-2i\pi xy}\,dy\right\}dy = \int\limits_{\mathbb{R}} v(y)\widehat{u}(y)\,dy. \qquad \square$$

Proof (of Theorem 1.4.5) Let g be a real function in $L^2_{\mathbb{C}}(\mathbb{R})$ and define $f := \widehat{(g^-)}$, where $g^-(t) := g(-t)$. We have $\widehat{f} = \widehat{g}^*$. In particular, by Lemma 1.4.1:

$$\int\limits_{\mathbb{R}} g(x)\widehat{f}(x)\,dx = \int\limits_{\mathbb{R}} \widehat{g}(x)f(x)\,dx = \int \widehat{g}(x)\widehat{g}(x)^*\,dx.$$

Therefore

$$\|g - \widehat{f}\|^2 = \|g\|^2 - 2\mathrm{Re}\langle g, \widehat{f}\rangle + \|\widehat{f}\|^2$$
$$= \|g\|^2 - 2\|\widehat{g}\|^2 + \|f\|^2.$$

But $\|g\|^2 = \|\widehat{g}\|^2$ and $\|f\|^2 = \|\widehat{g}\|^2$. Therefore $\|g - \widehat{f}\|^2 = 0$, that is to say, $g = \widehat{f}$. In other words, every real (and therefore, every complex) function $g \in L^2_{\mathbb{C}}(\mathbb{R})$ is the Fourier transform of some function of $L^2_{\mathbb{C}}(\mathbb{R})$. In other words, the mapping φ is *onto*. $\qquad \square$

Example 1.4.1 The rectangular pulse $rect_T$ has a Fourier transform that is not integrable, and therefore the inversion formula

$$rect_T(t) = \int\limits_{\mathbb{R}} \widehat{rect_T}(v)e^{2i\pi vt}\,dv$$

does not apply if we interpret the right-hand side as an integral in the usual sense. However, from the above theory, it applies if it is interpreted as the following limit in $L^2_{\mathbb{C}}(\mathbb{R})$

$$\lim_{A\uparrow+\infty} \int\limits_{-A}^{+A} \widehat{rect_T}(v)e^{2i\pi vt}\,dv.$$

The Uncertainty Principle

Fourier analysis has an intrinsic limitation: time resolution is possible only at the expense of frequency resolution, and vice-versa. Indeed, the more spread-out the function, the more concentrated is its Fourier transform, and vice-versa. This imprecise statement is already substantiated by the Doppler theorem

$$|a|^{\frac{1}{2}} f(at) \rightarrow \frac{1}{|a|^{\frac{1}{2}}} \hat{f}(\frac{v}{|a|}).$$

In order to state precisely Heisenberg's uncertainty principle, a definition of the "width" of a function is needed. The following one suits our purpose. Let $w \in L^2_{\mathbb{C}}(\mathbb{R})$ be a nontrivial function. Define the centers of w and $\hat{w}$ respectively by

$$m_w := \frac{1}{E_w} \int_{\mathbb{R}} t\,|w(t)|^2\,dt, \qquad m_{\hat{w}} := \frac{1}{E_w} \int_{\mathbb{R}} v\,|w(v)|^2\,dv$$

where E_w is the energy of w:

$$E_w := \int_{\mathbb{R}} |w(t)|^2\,dt = \int_{\mathbb{R}} |\hat{w}(v)|^2\,dv.$$

The root mean square (RMS) widths of w and $\hat{w}$ are respectively defined by

$$\sigma_w = \left(\frac{1}{E_w} \int_{\mathbb{R}} |t - m_w|^2\,|w(t)|^2\,dt \right)^{\frac{1}{2}}, \qquad \sigma_{\hat{w}} = \left(\frac{1}{E_w} \int_{\mathbb{R}} |v - m_{\hat{w}}|^2\,|\hat{w}(v)|^2\,dv \right)^{\frac{1}{2}}.$$

These quantities are defined as long as m_w and $m_{\hat{w}}$ are well-defined and finite. However, they may be infinite. We assume from now on that

$$\int_{\mathbb{R}} |t|\,|w(t)|^2\,dt < \infty \text{ and } \int_{\mathbb{R}} |v|\,|\hat{w}(v)|^2\,dv < \infty$$

which guarantees the existence of the centers of both w and of its Fourier transform, and the finiteness of their root mean-square widths.

Theorem 1.4.6 *Under the conditions stated above, we have* Heisenberg's inequality:

$$\sigma_w \sigma_{\hat{w}} \geq \frac{1}{4\pi}. \tag{1.25}$$

Proof We assume that the window and its FT are centered at 0, without loss of generality (see Exercise 1.5.23). Denoting by $\|f\|$ the L^2-norm of a function $f \in L^2_{\mathbb{C}}(\mathbb{R})$, we have to show that

$$\|tw\| \times \|v\hat{w}\| \geq \frac{1}{4\pi} \|w\|^2.$$

We first assume that w is a C_c^∞ (infinitely differentiable with compact support). In particular,

$$\widehat{w'}(v) = (2i\pi v)\,\hat{w}(v),$$

and therefore

$$\|v\hat{w}\|^2 = \frac{1}{4\pi^2}\|\widehat{w'}\|^2 = \frac{1}{4\pi^2}\|w'\|^2.$$

Therefore it remains to show that

$$\|tw\| \times \|w'\| \geq \frac{1}{2}\|w\|^2. \tag{*}$$

By Schwarz's inequality

$$\|tw\| \times \|w'\| \geq |\langle tw, w'\rangle| \geq |Re\{\langle tw, w'\rangle\}|.$$

Now,

$$2Re\{\langle tw, w'\rangle\} = \langle tw, w'\rangle + \langle w', tw\rangle = \int_{\mathbb{R}} t(ww'^* + w'w^*)\,dt$$

$$= t|w(t)|^2|_{-\infty}^{+\infty} - \int_{\mathbb{R}} |w(t)|^2\,dt = 0 - \|w\|^2.$$

This gives $(*)$ for $w \in C_c^\infty$. We now show that it remains true in the general case. To see this, we first observe that it suffices to prove $(*)$ in the case where w belongs to the Hilbert space

$$H = \{w \in L_{\mathbb{C}}^2(\mathbb{R}); \ t \to tw(t) \in L_{\mathbb{C}}^2(\mathbb{R}), \ v \to v\hat{w}(v) \in L_{\mathbb{C}}^2(\mathbb{R})\},$$

with the norm $\|w\|_H = \left(\|w\|^2 + \|tw\|^2 + \|v\hat{w}\|^2\right)^{\frac{1}{2}}$ (if w is not in H, Heisenberg's inequality is trivially satisfied). Then we use the fact that the subset of H consisting of the C^∞ functions with compact support is dense in H (Exercise 1.5.24). The result then follows by continuity of the hermitian product. □

Equality in (1.25) takes place if and only if w is proportional to a Gaussian function e^{-ct^2}, where $c > 0$. We do the proof in the case where it is further assumed that $w \in S$ and is real.[3] All the steps in the first part of the proof remain valid since for such functions, $t|w(t)|^2|_{-\infty}^{+\infty} = 0$. Equality is attained in (1.25) if and only if the

---

[3] S is the collection of functions $\varphi : \mathbb{R} \to \mathbb{R}$ thats are infinitely differentiable with rapid decay, in the sense that there for all $n \in \mathbb{N}$ all $k \in \mathbb{N}$, there exists a constant $C_{k,n} < \infty$ such that for all $t \in \mathbb{R}$, $|\varphi^{(k)}(t)| \leq C_{k,n}(1 + |t|)^{-n}$.

functions $t \to t w(t)$ and w' are proportional, say,

$$w'(t) = -c t w(t),$$

and this gives

$$w(t) = A e^{-ct^2},$$

where $c > 0$ because $w \in L_{\mathbb{C}}^2(\mathbb{R})$.

1.4.2 Fourier Series in L_{loc}^2

A T-periodic function f is *locally square-integrable* if $f \in L_{\mathbb{C}}^2([0, T])$, that is,

$$\int_0^T |f(t)|^2 \, dt < \infty.$$

As the Lebesgue measure of $[0, T]$ is finite, $L_{\mathbb{C}}^2([0, T]) \subset L_{\mathbb{C}}^1([0, T])$. In particular, a periodic function in $L_{\mathbb{C}}^2([0, T])$ is also locally integrable.

Theorem 1.4.7 *The sequence* $\{e_n\}_{n \in \mathbb{Z}}$, *where*

$$e_n(t) := \frac{1}{\sqrt{T}} e^{2i\pi \frac{n}{T} t},$$

is a Hilbert basis of $L_{\mathbb{C}}^2([0, T])$.

Proof One first observes that it is an orthonormal system in $L_{\mathbb{C}}^2([0, T])$. In view of Theorem 1.3.16, it remains to show that it is total in $L_{\mathbb{C}}^2([0, T])$. For this, let $f \in L_{\mathbb{C}}^2([0, T])$ and let f_N be its projection on the Hilbert subspace generated by $\{e_n\}_{|n| \le N}$. The coefficient of e_n in this projection is $c_n(f) = \langle f, e_n \rangle_{L_{\mathbb{C}}^2([0,T])}$, and we have that

$$\sum_{n=-N}^{+N} |c_n(f)|^2 + \int_0^T |f(t) - f_N(t)|^2 \, dt = \int_0^T |f(t)|^2 \, dt. \qquad (*)$$

(This is Pythagoras' theorem for projections: $\|P_G(x)\|^2 + \|x - P_G(x)\|^2 = \|x\|^2$.) In particular, $\sum_{n \in \mathbb{Z}} |c_n(f)|^2 < \infty$. It remains to verify condition (b) of Theorem 1.3.16, that is

$$\lim_{N \uparrow \infty} \int_0^T |f(t) - f_N(t)|^2 \, dt = 0.$$

We assume in a first step that f is continuous. For such a function, the formula

$$\varphi(x) := \int_0^T \tilde{f}(x+t) \tilde{f}(t)^* \, dt,$$

where

$$\tilde{f}(t) = \sum_{n \in \mathbb{Z}} f(t+nT) 1_{(0,T]}(t+nT),$$

defines a T-periodic and continuous function φ whose nth Fourier coefficient is

$$
\begin{aligned}
c_n(\varphi) &= \frac{1}{T} \int_0^T \left(\tilde{f}(x+t) \tilde{f}(t)^* \, dt \right) e^{-2i\pi \frac{n}{T} x} \, dx, \\
&= \frac{1}{T} \int_0^T \tilde{f}(t)^* \left\{ \int_0^T \tilde{f}(x+t) e^{-2i\pi \frac{n}{T} x} \, dx \right\} dt \\
&= \frac{1}{T} \int_0^T f(t)^* \left\{ \int_t^{t+T} \tilde{f}(s) e^{-2i\pi \frac{n}{T} s} \, ds \right\} e^{2i\pi \frac{n}{T} t} \, dt = T |c_n(f)|^2.
\end{aligned}
$$

Since $\sum_{n \in \mathbb{Z}} |c_n(f)|^2 < \infty$ and φ is continuous, it follows from the Fourier inversion theorem for locally integrable periodic functions that for *all* $x \in \mathbb{R}$

$$\varphi(x) = \sum_{n \in \mathbb{Z}} |c_n(f)|^2 e^{2i\pi \frac{n}{T} x}.$$

In particular, for $x = 0$

$$\varphi(0) = \int_0^T |f(t)|^2 \, dt = \sum_{n \in \mathbb{Z}} |c_n(f)|^2,$$

and therefore, in view of $(*)$,

$$\lim_{N \uparrow \infty} \int_0^T |f(t) - f_N(t)|^2 \, dt = 0.$$

It remains to pass from the continuous functions to the square-integrable functions. Since the space $\mathcal{C}([0, T])$ of continuous functions from $[0, T]$ into $\mathbb{C}$ is dense in $L^2_{\mathbb{C}}([0, T])$, for any $\varepsilon > 0$, there exists $\varphi \in \mathcal{C}([0, T])$ such that $\|f - \varphi\| \leq \varepsilon/3$. By Bessel's inequality, $\|f_N - \varphi_N\|^2 = \|(f - \varphi)_N\|^2 \leq \|f - \varphi\|^2$, and therefore

$$\|f - f_N\| \leq \|f - \varphi\| + \|\varphi - \varphi_N\| + \|f_N - \varphi_N\|$$
$$\leq \|\varphi - \varphi_N\| + 2\|f - \varphi\| \leq \|\varphi - \varphi_N\| + 2\frac{\varepsilon}{3}.$$

For N sufficiently large, $\|\varphi - \varphi_N\| \leq \frac{\varepsilon}{3}$. Therefore, for N sufficiently large, $\|f - f_N\| \leq \varepsilon$. $\qquad\square$

The Fourier Isometry in L^2_{loc}

Let us consider the Hilbert space $L^2_{\mathbb{C}}([0, T], dt/T)$ of complex functions f such that $\int_0^T |f(t)|^2\, dt < \infty$, with the inner product

$$\langle x, y \rangle_{L^2_{\mathbb{C}}([0,T], \frac{dt}{T})} := \frac{1}{T} \int_0^T x(t) y(t)^*\, dt .$$

Theorem 1.4.8 *Formula*

$$\widehat{f_n} = \frac{1}{T} \int_0^T f(t)\, e^{-2i\pi \frac{n}{T} t}\, dt$$

defines a linear isometry $f \to \{\widehat{f_n}\}_{n \in \mathbb{Z}}$ from $L^2_{\mathbb{C}}([0, T], dt/T)$ onto $\ell^2_{\mathbb{C}}(\mathbb{Z})$, the inverse of which is given by

$$f(t) = \sum_{n \in \mathbb{Z}} \widehat{f_n}\, e^{2i\pi \frac{n}{T} t},$$

where the series on the right-hand side converges in $L^2_{\mathbb{C}}([0, T])$, and the equality is almost-everywhere. This isometry is summarized by the Plancherel–Parseval identity:

$$\sum_{n \in \mathbb{Z}} \widehat{f_n} \widehat{g_n}^* = \frac{1}{T} \int_0^T f(t) g(t)^*\, dt .$$

Proof The result follows from general results on orthonormal bases of Hilbert spaces and Theorem 1.4.7. $\qquad\square$

Example 1.4.2 (Orthonormal systems of shifted functions) Let g be a function of $L^2_{\mathbb{C}}(\mathbb{R})$ and fix $0 < T < \infty$. A necessary and sufficient condition for the family of functions $\{g(\cdot - nT)\}_{n\in\mathbb{Z}}$ to form an orthonormal system of $L^2_{\mathbb{C}}(\mathbb{R})$ is

$$\sum_{n\in\mathbb{Z}}\left|\hat{g}\left(v+\frac{n}{T}\right)\right|^2 = T \quad\text{almost everywhere.}\tag{1.26}$$

Proof The Fourier transform $\hat{g}$ of $g \in L^2_{\mathbb{C}}(\mathbb{R})$ is in $L^2_{\mathbb{C}}(\mathbb{R})$ and, in particular, $|\hat{g}|^2$ is integrable. By Theorem 1.1.20, $\sum_{n\in\mathbb{Z}}|\hat{g}(v+\frac{n}{T})|^2$ is $(1/T)$-periodic and locally integrable, and $T\int_{\mathbb{R}} g(t)\,g(t-nT)^*\,dt$ is its nth Fourier coefficient, as follows from the Plancherel–Parseval formula:

$$T\int_{\mathbb{R}} g(t)g(t-nT)^*\,dt = T\int_{\mathbb{R}}|\hat{g}(v)|^2 e^{-2i\pi vnT}\,dv$$

$$= T\int_{\mathbb{R}}\left\{\sum_{k\in\mathbb{Z}}\left|\hat{g}\left(v+\frac{k}{T}\right)\right|^2\right\}e^{-2i\pi vnT}\,dv.$$

The definition of orthonormality of the system $\{g(\cdot - nT)\}_{n\in\mathbb{Z}}$ is

$$\int_{\mathbb{R}} g(t)\,g(t-nT)^*\,dt = 1_{\{n=0\}}.$$

The proof henceforth follows the argument in the proof of Corollary 1.1.11.   □

L^2-Version of the Shannon–Nyquist Theorem

The L^1 version of the sampling theorem contains a condition bearing on the samples themselves, namely:

$$\sum_{n\in\mathbb{Z}}\left|f\left(\frac{n}{2B}\right)\right| < \infty$$

The simplest way of removing this condition is given by the L^2 version of the Shannon–Nyquist theorem.

Theorem 1.4.9 *Let f be a base-band (B) function in $L^2_{\mathbb{C}}(\mathbb{R})$. Then*

$$\lim_{N\uparrow\infty}\int_{\mathbb{R}}\left|f(t)-\sum_{n=-N}^{+N} b_n\,\mathrm{sinc}(2Bt-n)\right|^2 dt = 0,$$

where

$$b_n = \int_{-B}^{+B} \widehat{f}(v) \, e^{2i\pi v \frac{n}{2B}} \, dv.$$

Proof Denote $L_{\mathbb{C}}^2(\mathbb{R}; B)$ be the Hilbert subspace of $L_{\mathbb{C}}^2(\mathbb{R})$ consisting of the functions therein whose FT has a support contained in $[-B, +B]$. The sequence

$$\left\{ \frac{1}{\sqrt{2B}} h\left(\cdot - \frac{n}{2B} \right) \right\}_{n \in \mathbb{Z}}, \tag{*}$$

where $h(t) \equiv 2B \, \mathrm{sinc}(2Bt)$, is an orthonormal basis of $L_{\mathbb{C}}^2(\mathbb{R}; B)$. Indeed, the functions of this system are in $L_{\mathbb{C}}^2(\mathbb{R}; B)$, and they form an orthonormal system since, by the Plancherel–Parseval formula,

$$\int_{\mathbb{R}} h\left(t - \frac{n}{2B} \right) h\left(t - \frac{k}{2B} \right) dt = \int_{\mathbb{R}} \widehat{h}(v) e^{-2i\pi v \frac{n}{2B}} \left(\widehat{h}(v) e^{-2i\pi v \frac{k}{2B}} \right)^* dv$$

$$= \int_{-B}^{+B} e^{2i\pi v \frac{k-n}{2B}} \, dv = 2B \times 1_{\{n=k\}}.$$

It remains to prove the totality of the orthonormal system $(*)$ (see Theorem 1.3.16). We must show that if $g \in L_{\mathbb{C}}^2(\mathbb{R}; B)$ and

$$\int_{\mathbb{R}} g(t) h\left(t - \frac{n}{2B} \right) dt = 0 \quad \text{for all } n \in \mathbb{Z},$$

then $g \equiv 0$ as a function of $L_{\mathbb{C}}^2(\mathbb{R}; B)$ (or, equivalently, that $g(t) = 0$ almost everywhere).

By the Plancherel–Parseval identity, the last condition is equivalent to

$$\int_{-B}^{+B} \widehat{g}(v) e^{2i\pi v \frac{n}{2B}} \, dv = 0 \quad \text{for all } n \in \mathbb{Z}.$$

This implies $\widehat{g}(v) = 0$ almost everywhere (and consequently $g(t) = 0$ almost everywhere) since the system $\{e^{2i\pi vn/2B}\}_{n \in \mathbb{Z}}$ is total in $L_{\mathbb{C}}^2(\mathbb{R}; B)$ (Theorem 1.4.7).

Expanding $f \in L_{\mathbb{C}}^2(\mathbb{R}; B)$ in the Hilbert basis $(*)$ yields

$$f(t) = \lim_{N \uparrow \infty} \sum_{-N}^{+N} c_n \frac{1}{\sqrt{2B}} h\left(t - \frac{n}{2B} \right),$$

where the limit and the equality are in the L^2 sense, and

$$c_n = \int_{\mathbb{R}} f(t)\, \frac{1}{\sqrt{2B}}\, h\left(t - \frac{n}{2B}\right) dt.$$

By the Plancherel-Parseval identity,

$$c_n = \int_{-B}^{+B} \hat{f}(v)\, \frac{1}{\sqrt{2B}}\, e^{2i\pi v \frac{n}{2B}}\, dv. \qquad\qquad \Box$$

1.5 Exercises

Exercise 1.5.1 (*Symmetries*) What can you say of the FT of an odd (resp., even; resp., real and even) integrable function? (Is it odd, even, real?)

Exercise 1.5.2 (*Convolution of one-sided exponentials*) What is the nth convolution power of $f(t) = e^{-at} 1_{\mathbb{R}_+}(t)$, where $a > 0$? Deduce from this the FT of the function $t \to t^n e^{-at} 1_{\mathbb{R}_+}(t)$.

Exercise 1.5.3 (*Base-band functions*) In Example 1.1.8, prove the following identities:

$$\hat{m}(v) = \frac{\hat{u}(v) + \hat{u}(-v)^*}{2} \text{ and } \hat{n}(v) = \frac{\hat{u}(v) - \hat{u}(-v)^*}{2i}.$$

and

$$\hat{m}(v) = \{\hat{f}(v + v_0) + \hat{f}(v - v_0)\} 1_{[-B,+B]}(v),$$
$$\hat{n}(v) = -i\{\hat{f}(v + v_0) - \hat{f}(v - v_0)\} 1_{[-B,+B]}(v).$$

Exercise 1.5.4 (*A special case of the Fourier inversion formula*) Prove by a direct computation that the Fourier inversion formula holds true for the function $t \to e^{-\alpha t^2 + at}$ where $\alpha \in \mathbb{R}, \alpha > 0, a \in \mathbb{C}$.

Exercise 1.5.5 (*Evaluation of integrals via Fourier transforms*) (1) Which function has the FT $\hat{f}(v) = \frac{2a}{a^2 + 4\pi^2 v^2}$ where $a > 0$? Deduce from this the value of the integral

$$I(a) = \int_{\mathbb{R}} \frac{1}{a^2 + u^2}\, du, \ a > 0.$$

(2) Deduce from the Fourier inversion formula that $\int_{\mathbb{R}} \left(\frac{\sin(t)}{t}\right)^2 dt = \pi$.

Exercise 1.5.6 (*Discontinuities and non-integrability of the* FT) Let $f : \mathbb{R} \to \mathbb{C}$ be an integrable right-continuous function, with a limit from the left at all times. Show that if it is discontinuous at some time t_0, its FT cannot be integrable.

Exercise 1.5.7 (*Autocorrelation function*) Show that the *autocorrelation function* c of a function $f \in L_{\mathbb{C}}^1(\mathbb{R})$ given by formula

$$c(t) := \int_{\mathbb{R}} f(s+t)f^*(s)ds$$

is well-defined, integrable, and that its FT is $|\widehat{f}(v)|^2$.

Exercise 1.5.8 Prove that a T-periodic function cannot be integrable unless it is almost-everywhere null.

Exercise 1.5.9 (*Some typical Fourier series expansions*) Expand in cosine series the following functions:

(1) $f(x) = x^2$, $x \in [-\pi, +\pi]$,
(2) $f(x) = |x|$, $x \in [-\pi, +\pi]$ and
(3) $f(x) = |\sin x|$, $x \in [-\pi, +\pi]$.

Exercise 1.5.10 (*Fourier sum at a discontinuity*) Expand $f(x) = x$, $x \in (-\pi, +\pi)$ in sine series. What happens at the points of discontinuity ($x = -\pi$ or $x = +\pi$)?

Exercise 1.5.11 (*Fourier series and convolution*) Let $h \in L_{\mathbb{R}}^1(\mathbb{R})$ and let $f : \mathbb{R} \to \mathbb{R}$ be a T-periodic locally integrable function. Show that the function

$$y(t) = \int_{\mathbb{R}} h(t-s)f(s)\,ds$$

is a well-defined T-periodic locally integrable function, and give its Fourier coefficients.

Exercise 1.5.12 (*A simple case of convergence of a Fourier series*) Let $\mathbf{c} := \{c_n\}_{n \in \mathbb{Z}}$ be a sequence in $\ell_{\mathbb{C}}^1(\mathbb{Z})$. Show that the function $f : \mathbb{R} \to \mathbb{C}$ defined by

$$f(t) = \sum_{k \in \mathbb{Z}} c_k e^{ikt}$$

is the pointwise limit of its Fourier series.

Exercise 1.5.13 (*A probabilistic statement of the Poisson summation formula*) Let $f \in L_{\mathbb{C}}^1(\mathbb{R})$ and $T > 0$ be such that $\sum_{n \in \mathbb{Z}} |\widehat{f}(\frac{n}{T})| < \infty$. Let U be a random variable uniformly distributed on $[0, T]$. Prove that

$$\sum_{n\in\mathbb{Z}} f(U+nT) = \frac{1}{T}\sum_{n\in\mathbb{Z}} \widehat{f}\left(\frac{n}{T}\right) e^{2i\pi\frac{n}{T}U} \qquad P-a.s.$$

Exercise 1.5.14 (*Pointwise convergence of the Fourier transform*) Let $f \in L^2_{\mathbb{C}}(\mathbb{R})$. Show that for any $B > 0$,

$$\int\limits_{-B}^{+B} \widehat{f}(\nu)e^{2i\pi\nu t}\, d\nu = 2B \int\limits_{\mathbb{R}} f(t+s)\mathrm{sinc}(2Bs)\, ds,$$

and use this to study along the lines of Sect. 1.1.2 the pointwise convergence of the left-hand side as B tends to infinity.

Exercise 1.5.15 (*Impulse response from the frequency response*) Give the impulse response $\mathbf{h} = \{h_n\}_{n\in\mathbb{Z}}$ of the filter with frequency response

$$\tilde{h}(\omega) = \exp(\cos(\omega))e^{i\,\sin(\omega)}.$$

Exercise 1.5.16 (*Laurent expansion*) Give the Laurent series expansion of $H(z) = \frac{1}{(z-\gamma)^r}$, where r is an integer ≥ 1 and $\gamma \in \mathbb{C}$.

Exercise 1.5.17 (*Monotone sequences of Hilbert spaces and projections*) Prove the two following assertions:

(a) Let $\{G_n\}$ be a non-increasing sequence of Hilbert subspaces of H. Then $\bigcap_{n\geq 1} G_n = \varnothing$ if and only if $\lim_{n\uparrow\infty} P_{G_n}(x) = 0$ for all $x \in H$;

(b) Let $\{G_n\}_{n\geq 1}$ be a non-decreasing sequence of Hilbert subspaces of H. Then the closure of $\bigcup_{n\geq 1} G_n$ is H itself if and only if $\lim_{n\uparrow\infty} P_{G_n}(x) = x$ for all $x \in H$.

Exercise 1.5.18 (*A pre-Hilbert space that is not complete*) Let $\mathcal{C}([0,1])$ denote the set of continuous complex-valued functions on $[0,1]$. Show that the formula

$$\langle f, g \rangle = \int\limits_{0}^{1} f(t)\overline{g(t)}\, dt$$

defines an inner product on $\mathcal{C}([0,1])$ considered as a vector space in the usual way. Show that it is not complete with respect to the induced distance.

Exercise 1.5.19 (*Orthogonal complement*) Show that the subset $G = \{f \in L^2_{\mathbb{R}}([0,1]); \ f = 0\,\text{a.e. on}\,[0,\frac{1}{2}]\}$ is a Hilbert subspace of $L^2_{\mathbb{R}}([0,1])$ and find its orthogonal complement.

Exercise 1.5.20 $\ell^1_{\mathbb{C}}(\mathbb{Z}) \subset \ell^2_{\mathbb{C}}(\mathbb{Z})$. Let p and q be positive integers such that $p > q$. Show that $\ell^q_{\mathbb{C}}(\mathbb{Z}) \subset \ell^p_{\mathbb{C}}(\mathbb{Z})$ (strict inclusion).

Exercise 1.5.21 (*Haar system*) (i) Let $\varphi(t) := 1_{(0,1]}(t)$. Show that $\{\varphi(\cdot - n)\}_{n\in\mathbb{Z}}$ is an orthonormal system of $L^2_{\mathbb{R}}(\mathbb{R})$ but not a complete one.
(ii) Let

$$\varphi_{m,n}(t) := 2^{\frac{m}{2}}\varphi(2^m t - n).$$

Show that $\{\varphi_{m,n}\}_{n,m\in\mathbb{Z}}$ is a complete system, but not orthonormal.
(iii) Let $\psi(t) := \varphi(2t) - \varphi(2t - 1)$ and

$$\psi_{m,n}(t) := 2^{\frac{m}{2}}\psi(2^m t - n).$$

Show that $\{\psi_{m,n}\}_{n,m\in\mathbb{Z}}$ is a complete orthonormal system (called the *Haar system*).

Exercise 1.5.22 (*Plancherel–Parseval*) Use Plancherel–Parseval identity to prove that, for positive a and b,

$$\int_{\mathbb{R}} \frac{dt}{(t^2 + a^2)(t^2 + b^2)} = \frac{\pi}{ab(a+b)}.$$

Exercise 1.5.23 (*Mean-square widths*) Check if the centers m_w and $m_{\hat{w}}$ are well-defined, and then compute σ_w, $\sigma_{\hat{w}}$ and the product $\sigma_w\sigma_{\hat{w}}$ in the following cases:

$$w(t) = 1_{[-T,+T]},$$

$$w(t) = e^{-a|t|}, a > 0, and$$

$$w(t) = e^{-at^2}, a > 0.$$

Suppose that the center m_w of the function $w \in L^2$ are well-defined. Show that
(i) the quantity

$$\int_{\mathbb{R}} |t - t_0|^2 |w(t)|^2 dt$$

is minimized by $t_0 = m_w$, and
(ii) for arbitrary $t_0 \in \mathbb{R}$ and arbitrary $v_0 \in \mathbb{R}$,

$$\|(t - t_0)w\| \times \|(v - v_0)\hat{w}\| \geq \frac{1}{4\pi} \|w\|^2.$$

Exercise 1.5.24 (*A sophisticated Hilbert space*) Show that

$$H = \{w \in L^2_{\mathbb{C}}(\mathbb{R}); \ t \to tw(t) \in L^2_{\mathbb{C}}(\mathbb{R}), \ v \to v\hat{w}(v) \in L^2_{\mathbb{C}}(\mathbb{R})\},$$

is a Hilbert space when endowed with the norm

$$\|w\|_H = \left(\|w\|^2_{L^2} + \|tw\|^2_{L^2} + \|v\hat{w}\|^2_{L^2} \right)^{\frac{1}{2}}.$$

Show that the subset of H consisting of the C^∞ functions with compact support is dense in H.

Chapter 2
Fourier Theory of Probability Distributions

Characteristic functions, that is Fourier transforms of probability measures, play a major role in Probability theory, in particular in the Fourier theory of wide-sense stationary stochastic processes, whose starting point is the notion of power spectral measure. It turns out that the existence of such a measure is a direct consequence of Bochner's theorem of characterization of characteristic functions, and that the proof of its unicity is a rephrasing of Paul Lévy's inversion theorem. Another result of Paul Lévy, characterizing convergence in distribution in terms of characteristic functions, intervenes in an essential way in the proof of Bochner's theorem. In fact, characteristic functions are the link between the Fourier theory of deterministic functions and that of stochastic processes. This chapter could have been entitled "Convergence in distribution of random sequences", a classical topic of probability theory. However, we shall need to go slightly beyond this and give the extension of Paul Lévy's convergence theorem to sequences of finite measures (instead of probability distributions) as this is needed in Chap. 5 for the proof of existence of the Bartlett spectral measure.

2.1 Characteristic Functions

2.1.1 Basic Facts

Denote by $M^+(\mathbb{R}^d)$ the collection of *finite measures* on $(\mathbb{R}^d, \mathcal{B}(\mathbb{R}^d))$.

Definition 2.1.1 The Fourier transform of a measure $\mu \in M^+(\mathbb{R}^d)$ is the function $\widehat{\mu} : \mathbb{R}^d \to \mathbb{C}$ defined by

$$\widehat{\mu}(v) = \int_{\mathbb{R}^d} e^{-2i\pi \langle v, x \rangle} \mu(dx),$$

where $\langle v, x \rangle := \sum_{j=1}^{d} v_j x_j$.

© Springer International Publishing Switzerland 2014
P. Brémaud, *Fourier Analysis and Stochastic Processes*, Universitext,
DOI 10.1007/978-3-319-09590-5_2

Theorem 2.1.1 *The Fourier transform of a measure $\mu \in M^+(\mathbb{R}^d)$ is bounded and uniformly continuous.*

Proof The proof is similar to that of Theorem 1.1.1. From the definition, we have that

$$|\widehat{\mu}(v)| \leq \int_{\mathbb{R}^d} |e^{-2i\pi\langle v,x\rangle}|\mu(dx)$$
$$= \int_{\mathbb{R}^d} \mu(dx) = \mu(\mathbb{R}^d),$$

where the last term does not depend on v and is finite. Also, for all $h \in \mathbb{R}^d$,

$$|\widehat{\mu}(v+h) - \widehat{\mu}(v)| \leq \int_{\mathbb{R}^d} \left|e^{-2i\pi\langle v,x+h\rangle} - e^{-2i\pi\langle v,x\rangle}\right|\mu(dx)$$
$$= \int_{\mathbb{R}^d} \left|e^{-2i\pi\langle h,x\rangle} - 1\right|\mu(dx).$$

The last term is independent of v and tends to 0 as $h \to 0$ by dominated convergence (recall that μ is finite). $\qquad\square$

For future reference, we shall quote the following:

Theorem 2.1.2 *Let $\mu \in M^+(\mathbb{R}^d)$ and let $\widehat{f}$ be the Fourier transform of $f \in L^1_{\mathbb{C}}(\mathbb{R}^d)$. Then*

$$\int_{\mathbb{R}^d} \widehat{f}d\mu = \int_{\mathbb{R}^d} f\widehat{\mu}dx.$$

Proof This follows from Fubini's theorem. In fact,

$$\int_{\mathbb{R}^d} \widehat{f}d\mu = \int_{\mathbb{R}^d} \left(\int_{\mathbb{R}^d} f(x)e^{-2i\pi\langle v,x\rangle}dx\right)\mu(dv)$$
$$= \int_{\mathbb{R}^d} f(x)\left(\int_{\mathbb{R}^d} e^{-2i\pi\langle v,x\rangle}\mu(dv)\right)dx.$$

(Interversion of the order of integration is justified by the fact that the function $(x,v) \to |f(x)e^{-2i\pi\langle v,x\rangle}| = |f(x)|$ is integrable with respect to the product measure $dx \times \mu(dv)$. Recall that μ is finite.) $\qquad\square$

Sampling Theorem for the FT of a Finite Measure

The following approach to sampling acknowledges the fact that a function is a combination of complex sinusoids, and therefore starts by obtaining the sampling theorem for this type of elementary functions.

Consider the function

$$f(t) = e^{2i\pi \nu t},$$

where $\nu \in \mathbb{R}$. This function is not integrable and therefore does not fit into the framework of Shannon's sampling theorem. However, the Shannon–Nyquist formula remains essentially true.

Theorem 2.1.3 *For all $t \in \mathbb{R}$ and all $\nu \in (-W, +W)$:*

$$e^{2i\pi \nu t} = \sum_{n\in\mathbb{Z}} e^{2i\pi n 2W\nu} \frac{\sin\left(2\pi W(t - \frac{n}{2W})\right)}{2\pi W(t - \frac{n}{2W})}. \tag{2.1}$$

For all $B < W$, the convergence is uniform in $\nu \in [-B, +B]$.

Proof We first prove that for all $\nu \in \mathbb{R}$ and all $t \in (-W, +W)$

$$e^{2i\pi \nu t} = \sum_{n\in\mathbb{Z}} e^{2i\pi n 2W t} \frac{\sin\left(2\pi W(\nu - \frac{n}{2W})\right)}{2\pi W(\nu - \frac{n}{2W})}, \tag{†}$$

where the series converges uniformly for all $t \in [-B, +B]$ for any $B < W$. The result then follows by exchanging the roles of t and ν.

Let $g : \mathbb{R} \to \mathbb{C}$, whose argument is denoted by t, be the $2W$-periodic function which is equal to $e^{2i\pi \nu t}$ on $(-W, +W]$. The series in (†) is the Fourier series of g. We must therefore show uniform pointwise convergence of this Fourier series to the original function, which was done in Example 1.1.11.   □

In particular, if f is a finite linear combination of complex trigonometric functions, that is

$$f(t) = \sum_{k=1}^{M} \gamma_k e^{2i\pi \nu_k t},$$

where $\gamma_k \in \mathbb{C}$, $\nu_k \in \mathbb{R}$, and if W satisfies the condition

$$W > \sup\{|\nu_k| : 1 \le k \le M\}, \tag{2.2}$$

we have the Shannon–Nyquist reconstruction formula, with $T = \frac{1}{2W}$,

$$f(t) = \sum_{n\in\mathbb{Z}} f(nT) \frac{\sin\left(\frac{\pi}{T}(t - nT)\right)}{\frac{\pi}{T}(t - nT)}. \tag{2.3}$$

Exercise 2.4.1 shows that one really needs the strict inequality in (2.2).

The following extension of the sampling theorem for sinusoids is now straightforward:

Theorem 2.1.4 *Let μ be a finite measure on $[-B, +B]$, where $B \in \mathbb{R}_+ \backslash \{0\}$, and define the function*

$$f(t) = \int_{[-B,+B]} e^{2i\pi vt} \mu(dv).$$

Then for any $T < 1/2B$ and for all $t \in \mathbb{R}$, we have the Shannon–Nyquist reconstruction formula (2.3).

Proof Since μ is finite and the convergence in (2.1) is uniform in $v \in [-B, +B]$,

$$f(t) = \int_{[-B,+B]} \left\{ \sum_{n \in \mathbb{Z}} e^{2i\pi vnT} \frac{\sin\left(\frac{\pi}{T}(t - nT)\right)}{\frac{\pi}{T}(t - nT)} \right\} \mu(dv)$$

$$= \sum_{n \in \mathbb{Z}} \left\{ \int_{[-B,+B]} e^{2i\pi vnT} \frac{\sin\left(\frac{\pi}{T}(t - nT)\right)}{\frac{\pi}{T}(t - nT)} \mu(dv) \right\}$$

$$= \sum_{n \in \mathbb{Z}} \left\{ \int_{[-B,+B]} e^{2i\pi vnT} \mu(dv) \right\} \frac{\sin\left(\frac{\pi}{T}(t - nT)\right)}{\frac{\pi}{T}(t - nT)}. \qquad \square$$

2.1.2 Inversion Formula

The *characteristic function* of a random vector $X = (X_1, \ldots, X_d) \in \mathbb{R}^d$ is the function $\varphi_X : \mathbb{R}^d \to \mathbb{C}$ defined by

$$\varphi_X(u) = E[e^{i\langle u, X \rangle}].$$

In other words, φ_X is the Fourier transform of the probability distribution of X.

It turns out that the characteristic function of a random vector determines uniquely its probability distribution. This result will be obtained as a consequence of *Paul Lévy's inversion formula*:

Theorem 2.1.5 *Let $X \in \mathbb{R}^d$ be a random vector with characteristic function φ. Then for all $1 \leq j \leq d$, all $a_j, b_j \in \mathbb{R}^d$ such that $a_j < b_j$,*

$$\lim_{c\uparrow+\infty} \frac{1}{(2\pi)^d} \int_{-c}^{+c} \cdots \int_{-c}^{+c} \left(\prod_{j=1}^{d} \frac{e^{-iu_ja_j} - e^{-iu_jb_j}}{iu_j} \right) \varphi(u_1, \ldots, u_d) du_1 \ldots du_d$$

$$= E\left[\prod_{j=1}^{d} \left(\frac{1}{2} 1_{\{X_j=a_j \text{ or } b_j\}} + 1_{\{a_j < X_j < b_j\}} \right) \right].$$

Before proving the theorem, we give its main consequence:

Corollary 2.1.1 *The distribution of a random vector of $\mathbb{R}^d$ is uniquely determined by its characteristic function.*

Proof Let X and Y be two vectors of $\mathbb{R}^d$ with the same characteristic function φ. Lévy's inversion formula shows that the distributions of X and Y agree on $\mathcal{A}$, the class of rectangles $A = \prod_{j=1}^{d}(a_j, b_j]$ whose boundary has a null measure with respect to the distributions of both X and Y. Since there are at most a countable number of rectangles whose boundary has positive measure with respect to the distributions of both X and Y, $\mathcal{A}$ generates $\mathcal{B}(\mathbb{R}^d)$. Moreover, $\mathcal{A}$ is a π-system and therefore, by Theorem A.2.1, the two distributions coincide. □

Corollary 2.1.2 *A necessary and sufficient condition for the random variables $X_1, \ldots, X_d$ to be independent is that the characteristic function φ_X of the vector $X = (X_1, \ldots, X_d)$ factorizes as*

$$\varphi_X(u_1, \ldots, u_d) = \prod_{j=1}^{d} \varphi_j(u_j),$$

where for all $1 \le j \le d$, φ_j is a characteristic function. In this case, for all $1 \le j \le d$, $\varphi_j = \varphi_{X_j}$, the characteristic function of X_j.

Proof Necessity. Write

$$\varphi_X(u) = E\left[e^{i\sum_{j=1}^{d} u_j X_j} \right]$$

$$= E\left[\prod_{j=1}^{d} e^{iu_j X_j} \right] = \prod_{j=1}^{d} E\left[e^{iu_j X_j} \right] = \prod_{j=1}^{d} \varphi_{X_j}(u_j),$$

by the product formula for expectations (Theorem A.2.8).

Sufficiency. Let $X' := (X'_1, \ldots, X'_d) \in \mathbb{R}^d$ be a random vector whose *independent* coordinate random variables $X'_1, \ldots, X'_d$ have the respective characteristic functions $\varphi_1, \ldots, \varphi_d$. The characteristic function of X' is $\prod_{j=1}^{d} \varphi_j(u_j)$, and therefore X and X' have the same distribution. In particular $X_1, \ldots, X_d$ are independent random variables with respective characteristic functions $\varphi_1, \ldots, \varphi_d$. □

We now return to the proof of Theorem 2.1.5.

Proof We do the proof in the univariate case for the sake of notational ease. The multivariate case is a straightforward adaptation of it. Let X be a real-valued random variable with cumulative distribution function F and characteristic function φ. We have to show that for any pair of points a, b with $(a < b)$,

$$\lim_{c\uparrow+\infty} \frac{1}{2\pi} \int_{-c}^{+c} \frac{e^{-iua} - e^{-iub}}{iu} \varphi(u)du = E\left[\left(\frac{1}{2}1_{\{X=a \text{ or } b\}} + 1_{\{a<X<b\}}\right)\right]. \qquad (*)$$

For this, write

$$\Phi_c := \frac{1}{2\pi} \int_{-c}^{+c} \frac{e^{-iua} - e^{-iub}}{iu} \varphi(u)du$$

$$= \frac{1}{2\pi} \int_{-c}^{+c} \frac{e^{-iua} - e^{-iub}}{iu} \left(\int_{-\infty}^{+\infty} e^{iux}dF(x)\right) du$$

$$= \frac{1}{2\pi} \int_{-\infty}^{+\infty} \left(\int_{-c}^{+c} \frac{e^{-iua} - e^{-iub}}{iu} e^{iux}du\right) dF(x) = \int_{-\infty}^{+\infty} \Psi_c(x)dF(x),$$

where

$$\Psi_c(x) := \frac{1}{2\pi} \int_{-c}^{+c} \frac{e^{-iua} - e^{-iub}}{iu} e^{-iux}du.$$

The above computations make use of Fubini's theorem. This is allowed since, observing that

$$\left|\frac{e^{-iua} - e^{-iub}}{iu}\right| = \left|\int_a^b e^{-iux}dx\right| \le (b-a),$$

we have

$$\int_{-c}^{+c}\int_{-\infty}^{+\infty} \left|\frac{e^{-iua} - e^{-iub}}{iu} e^{iux}\right| dF(x)du = \int_{-c}^{+c}\int_{-\infty}^{+\infty} \left|\frac{e^{-iua} - e^{-iub}}{iu}\right| dF(x)du$$

$$\le \int_{-c}^{+c}\int_{-\infty}^{+\infty} (b-a)dF(x)du$$

$$= 2c(b-a) < \infty.$$

Since the function $u \to \frac{\cos(au)}{u}$ is antisymmetric, $\int\limits_{-c}^{+c} \frac{\cos(au)}{u} du = 0$, and therefore

$$\Psi_c(x) = \frac{1}{2\pi} \int\limits_{-c}^{+c} \frac{\sin u(x-a) - \sin u(x-b)}{u} du$$

$$= \frac{1}{2\pi} \int\limits_{-c(x-a)}^{+c(x-a)} \frac{\sin u}{u} du - \frac{1}{2\pi} \int\limits_{-c(x-b)}^{+c(x-b)} \frac{\sin u}{u} du.$$

The function $c \to \int_0^c \frac{\sin u}{u} du = \int_{-c}^0 \frac{\sin u}{u}$ is uniformly continuous in c and tends to $\int_0^{+\infty} \frac{\sin u}{u} du = \frac{1}{2}\pi$ as $c \uparrow +\infty$. Therefore the function $(c, x) \to \Psi_c(x)$ is uniformly bounded. Moreover, in view of the above expression for Ψ_c,

$$\lim_{c\uparrow\infty} \Psi_c(x) := \Psi(x) = 0 \quad \text{if } x < a \text{ or } x > b$$

$$= \frac{1}{2} \quad \text{if } x = a \text{ or } x = b$$

$$= 1 \quad \text{if } a < x < b.$$

Therefore, by dominated convergence,

$$\lim_{c\uparrow\infty} \Phi_c = \int\limits_{-\infty}^{+\infty} \lim_{c\uparrow\infty} \Psi_c(x) dF(x)$$

$$= \int\limits_{-\infty}^{+\infty} \Psi(x) dF(x) = E\left[\left(\frac{1}{2} 1_{\{X=a \text{ or } b\}} + 1_{\{a<X<b\}}\right)\right].$$

□

Note that, in the univariate case, denoting by F the cumulative distribution function of the random variable X,

$$E\left[\left(\frac{1}{2} 1_{\{X=a \text{ or } b\}} + 1_{\{a<X<b\}}\right)\right] = \frac{F(b) + F(b-)}{2} - \frac{F(a) + F(a-)}{2},$$

so that formula (∗) takes the perhaps more familiar form

$$\frac{F(b) + F(b-)}{2} - \frac{F(a) + F(a-)}{2} = \lim_{c\uparrow+\infty} \frac{1}{2\pi} \int\limits_{-c}^{+c} \frac{e^{-iua} - e^{-iub}}{iu} \varphi(u) du.$$

Corollary 2.1.3 *If moreover φ is integrable, then X admits a probability density f given by*

$$f(x) = \frac{1}{2\pi} \int_{\mathbb{R}^d} \varphi(u) e^{-i\langle u, x\rangle} du.$$

Proof We do the proof in the univariate case (the extension to the multivariate case follows the same lines of argument). The function f is well-defined, continuous and bounded, and in particular integrable on finite intervals. We have (Fubini)

$$
\begin{aligned}
\int_a^b f(x)dx &= \int_a^b \left(\frac{1}{2\pi} \int_{\mathbb{R}} \varphi(u) e^{-iux} du \right) dx \\
&= \frac{1}{2\pi} \int_{\mathbb{R}} \varphi(u) \left[\int_a^b e^{-iux} dx \right] du \\
&= \lim_{c\uparrow\infty} \frac{1}{2\pi} \int_{-c}^{+c} \varphi(u) \left[\int_a^b e^{-iux} dx \right] du \\
&= \lim_{c\uparrow\infty} \frac{1}{2\pi} \int_{-c}^{+c} \frac{e^{-iua} - e^{-iub}}{iu} \varphi(u)du = F(b) - F(a)
\end{aligned}
$$

for all $a < b$ that are points of continuity of F, from which the result easily follows for any interval $[a, b]$. □

Since a finite probability measure of $M^+(\mathbb{R}^d)$ is a multiple of a probability measure on $\mathbb{R}^d$, which is in turn the probability distribution of some random vector $X \in \mathbb{R}^d$, and since $\hat{\mu}(0)$ enables us to recover the total mass, the uniqueness theorem concerning characteristic functions gives:

Corollary 2.1.4 *A measure of $M^+(\mathbb{R}^d)$ is characterized by its Fourier transform.*

2.1.3 Gaussian Vectors

Gaussian variables and Gaussian vectors play an important role in probabilistic modeling. We decide to include them at this place in the book because they are defined in terms of characteristic functions, and their properties are best studied in these terms.

A Gaussian random variable with mean m and variance $\sigma^2 > 0$ is defined by its probability density

$$f(x) = \frac{1}{\sqrt{2\pi}} e^{-\frac{1}{2}\frac{x^2}{\sigma^2}}.$$

Its characteristic function, which can be derived from the result of Example 1.1.5, is:

$$\varphi_X(u) = \exp\left\{imu - \frac{1}{2}\sigma^2 u^2\right\}. \tag{$\dagger$}$$

Definition 2.1.2 An *extended Gaussian variable* X is any real random variable with a characteristic function of the form ($\dagger$), where $m \in \mathbb{R}$ and $\sigma^2 \in \mathbb{R}_+$. A standard Gaussian variable is one for which $m = 0$ and $\sigma^2 = 1$.

Note that a null variance ($\sigma^2 = 0$) is allowed, and in this case, the random variable X is the constant m. This is precisely the extension we need.

Definition 2.1.3 A random vector $X \in \mathbb{R}^d$ is called a *Gaussian random vector* if for all $\alpha \in \mathbb{R}^d$, the random variable $\langle \alpha, X \rangle$ is an extended Gaussian random variable.

We now connect this definition with another one in terms of characteristic functions.

Theorem 2.1.6 *For a random vector $X \in \mathbb{R}^d$ to be a Gaussian vector, it is necessary and sufficient that its characteristic function ϕ_X be of the form:*

$$\varphi_X(u) = \exp\left\{i\langle u, m \rangle - \frac{1}{2}\langle u, \Gamma u \rangle\right\}. \tag{2.4}$$

where $m \in \mathbb{R}^n$ and where Γ is a symmetric and non-negative definite $d \times d$ matrix.

In this case, we write $X \overset{\mathcal{D}}{\sim} \mathcal{N}(m, \Gamma)$.

Proof Necessity: The characteristic function of a Gaussian vector as defined in Definition 2.1.3 can be expressed as

$$E[e^{i\langle u, X \rangle}] = \varphi_Z(1),$$

where φ_Z is the characteristic function of $Z = \langle u, X \rangle$. The random variable Z being an extended Gaussian variable,

$$\varphi_Z(1) = \exp\left\{im_Z - \frac{1}{2}\sigma_Z^2\right\},$$

where

$$m_Z := E[Z] = \langle u, E[X] \rangle = \langle u, m \rangle$$

and

$$\sigma_Z^2 = E[\langle u, X - m \rangle\langle u, X - m \rangle]$$
$$= \langle u, E[(X - m)(X - m)^T]u \rangle = \langle u, \Gamma u \rangle.$$

Therefore, finally,

$$\varphi_X(u) = \exp\left\{i\langle u, m\rangle - \frac{1}{2}\langle u, \Gamma u\rangle\right\}.$$

Sufficiency: Let X be a random vector with characteristic function given by (2.4). Let $Z = \langle \alpha, X\rangle$, where $\alpha \in \mathbb{R}^d$. The characteristic function of the random variable Z is

$$\begin{aligned}
\varphi_Z(v) &= E[\exp\{ivZ\}] \\
&= E[\exp\{iv\langle \alpha, X\rangle\}] \\
&= \exp\left\{iv(\langle \alpha, m\rangle) - \frac{1}{2}v^2(\langle \alpha, \Gamma\alpha\rangle)\right\}.
\end{aligned}$$

Therefore Z is an extended Gaussian random variable. $\square$

It remains to prove the existence of a random vector with the characteristic function (2.4):

Theorem 2.1.7 *Let Γ be a non-negative definite matrix $d \times d$ matrix and let $m \in \mathbb{R}^d$. There exists a vector $X \in \mathbb{R}^d$ with characteristic functions (2.4).*

Proof Since Γ is non-negative definite, there exists a matrix A of the same dimension and such that $\Gamma = AA^T$. Let $X = m + AZ$ where Z is a vector of independent standard Gaussian variables $Z_1, \ldots, Z_d$. Then X has the required characteristic functions. In fact, the characteristic function of Z is

$$\begin{aligned}
\varphi_Z(u_1, \ldots, u_d) &= E\left[e^{i\sum_{j=1}^d u_j Z_j}\right] = \prod_{j=1}^d E\left[e^{iu_j Z_j}\right] \\
&= \prod_{j=1}^d \varphi_{Z_j}(u_j) = e^{-\frac{1}{2}\sum_{j=1}^d u_j^2} = e^{-\frac{1}{2}\|u\|^2},
\end{aligned}$$

and therefore

$$\begin{aligned}
\varphi_X(u) &= E\left[e^{i\langle u, m+AZ\rangle}\right] = e^{i\langle u, m\rangle}\varphi_Z(u^T A) \\
&= e^{i\langle u, m\rangle}e^{-\frac{1}{2}\|u^T A\|^2} = e^{i\langle u, m\rangle - \frac{1}{2}\langle u, \Gamma u\rangle}.
\end{aligned}$$

                                                                                          $\square$

It is clear from the above proof that the parameters m and Γ in (2.4) are respectively the mean and the covariance matrix of X.

In general, non-correlation does *not* imply independence. However, this is nearly true for Gaussian vectors. We start with a definition that is needed for a precise statement of the corresponding result.

Definition 2.1.4 Two real random vectors X and Y are said to be *jointly Gaussian* if the vector $Z = (X, Y)$ is Gaussian.

Theorem 2.1.8 *Two jointly Gaussian random vectors X and Y of arbitrary dimensions are independent if and only if they are uncorrelated (that is, if $\Gamma_{XY} := E\left[(X - m_X)(Y - m_Y)^T\right] = 0$).*

Proof Necessity: If X and Y are independent, then, by the product formula for expectations,

$$E[(X - m_X)(Y - m_Y)^T] = E[X - m_X]E[Y - m_Y]^T = 0.$$

Sufficiency: The random vector $Z = (X, Y)$ has the mean

$$m_Z = \begin{pmatrix} m_X \\ m_Y \end{pmatrix},$$

and if X and Y are uncorrelated, its covariance matrix is of the form

$$\Gamma_Z = \begin{pmatrix} \Gamma_X & 0 \\ 0 & \Gamma_Y \end{pmatrix}.$$

Vector Z is Gaussian by hypothesis and therefore, with $w = (u, v)$, u and v being real vectors of appropriate dimensions,

$$
\begin{aligned}
E[\exp\{i(u^T X + v^T Y)\}] &= E[\exp\{iw^T Z\}] \\
&= \exp\left\{iw^T m_Z - \frac{1}{2} w^T \Gamma_Z w\right\} \\
&= \exp\left\{i(u^T m_X + v^T m_Y) - \frac{1}{2} u^T \Gamma_X u - \frac{1}{2} v^T \Gamma_Y v\right\} \\
&= E[\exp\{iu^T X\}]E[\exp\{iv^T Y\}],
\end{aligned}
$$

and the conclusion follows from the characteristic function independence criterion

$\square$

Theorem 2.1.9 *Let X be a d-dimensional Gaussian vector with mean m and covariance matrix Γ, and assume that it is non-degenerate, that is:*

$$\langle u, \Gamma u \rangle = u^T \Gamma u = 0 \Rightarrow u = 0.$$

Then X admits the probability distribution function

$$f_X(x) = \frac{1}{(2\pi)^{d/2}(\det \Gamma)^{1/2}} \exp \left\{ -\frac{1}{2} \langle x - m, \Gamma^{-1}(x - m) \rangle \right\}. \qquad (2.5)$$

Proof Since $\Gamma > 0$, there exists a non-singular matrix A of the same dimension as Γ and such that $\Gamma = AA^T$. Define the n-vector $Z = A^{-1}(X - m)$. According to Definition 2.1.3, it is a Gaussian vector, and furthermore $E[Z] = 0$ and

$$\Gamma_Z = A^{-1}\Gamma A^{-T} = A^{-1}AA^T A^{-T} = I,$$

where I is the identity matrix. Therefore

$$E[\exp\{iu^T Z\}] = \exp \left\{ -\tfrac{1}{2} \sum_{i=1}^d u_i^2 \right\}.$$

Since this is the characteristic function of a vector of independent standard (centered, variance 1) Gaussian variables, we can assert that $Z_1, \ldots, Z_d$ are independent standard Gaussian random variables Corollary 2.1.1. In particular, Z admits the probability distribution function

$$f_Z(z) = \prod_{i=1}^d \frac{1}{\sqrt{2\pi}} e^{-\frac{1}{2}z_i^2/2}$$

$$= \frac{1}{(2\pi)^{d/2}} \exp \left\{ -\frac{1}{2} \|z\|^2 \right\}.$$

Now, $X = AZ + m$, and therefore, by the formula of smooth change of variables,

$$f_X(x) = \frac{1}{|\det A|} f_Y(A^{-1}(x - m))$$

$$= \frac{1}{(\det \Gamma)^{1/2}} \frac{1}{(2\pi)^{d/2}} \exp \left\{ -\frac{1}{2} \|A^{-1}(x - m)\|^2 \right\},$$

and this is precisely the announced result since

$$\|A^{-1}(x - m)\|^2 = \langle A^{-1}(x - m), A^{-1}(x - m) \rangle$$
$$= \langle x - m, A^{-T}A^{-1}(x - m) \rangle$$
$$= \langle x - m, \Gamma^{-1}(x - m) \rangle.$$

□

2.2 Convergence in Distribution

2.2.1 Paul Lévy's Theorem

Recall that $C_b(\mathbb{R}^d)$ is the collection of bounded continuous functions $f : \mathbb{R}^d \to \mathbb{R}$.

Definition 2.2.1 (a) The sequence $\{\mu_n\}_{n\geq 1}$ in $M^+(\mathbb{R}^d)$ is said to converge *weakly* to μ if, for all $f \in C_b(\mathbb{R}^d)$, $\lim_{n\uparrow\infty} \int_{\mathbb{R}^d} f d\mu_n = \int_{\mathbb{R}^d} f d\mu$. This is denoted by

$$\mu_n \xrightarrow{w} \mu.$$

(b) The sequence of random vectors $\{X_n\}_{n\geq 1}$ of $\mathbb{R}^d$, with respective probability distributions $\{Q_{X_n}\}_{n\geq 1}$, is said to converge *in distribution* to the random vector $X \in \mathbb{R}^d$ with distribution Q_X if $Q_{X_n} \xrightarrow{w} Q_X$. This is also denoted by

$$X_n \xrightarrow{\mathcal{D}} X.$$

Observe that the vectors X and X_n's need not be defined on the same probability space. Convergence in distribution concerns only probability distributions. As a matter of fact, very often the X_n's are defined on the same probability space but there is no "visible" (that is, defined on the same probability space) limit random vector X. Therefore we sometimes denote convergence in distribution as follows: $X_n \xrightarrow{\mathcal{D}} Q$, where Q is a probability distribution on $\mathbb{R}^d$. If Q is a "famous" probability distribution, for instance a standard Gaussian variable, we then say, that "$\{X_n\}_{n\geq 1}$ converges in distribution to a standard Gaussian distribution". We would then denote this by $X_n \xrightarrow{\mathcal{D}} \mathcal{N}(0, 1)$.

Denote by B^o and B^c respectively the interior and the closure of the set $B \in \mathbb{R}^d$, and by ∂B its boundary $(:= B^c \backslash B^o)$.

Theorem 2.2.1 *Let $\{\mu_n\}_{n\geq 1}$ and μ be probability distributions on $\mathbb{R}^d$. The following conditions are equivalent:*

(i) $\mu_n \xrightarrow{w} \mu$.

(ii) *For any open set $G \subseteq \mathbb{R}^d$, $\liminf_n \mu_n(G) \geq \mu(G)$.*

(iii) *For any closed set $F \subseteq \mathbb{R}^d$, $\limsup_n \mu_n(F) \leq \mu(F)$.*

(iv) *For any measurable set $B \subseteq \mathbb{R}^d$ such that $\mu(\partial B) = 0$, $\lim_n \mu_n(B) = \mu(B)$.*

Proof (i) $\Rightarrow$ (ii). For any open set $G \in \mathbb{R}^d$ there exists a non-decreasing sequence $\{\varphi_k\}_{k\geq 1}$ of non-negative functions of $C_b(\mathbb{R}^d)$ such that $0 \leq \varphi_k \leq 1$ and $\varphi_k \uparrow 1_G$ (for instance, $\varphi_k(x) = 1 - e^{-kd(x,\overline{G})}$). For all $k \geq 1$, $\int 1_G d\mu_n \geq \int \varphi_k d\mu_n$, and therefore

$$\liminf_n \mu_n(G) = \liminf_n \int 1_G d\mu_n \geq \liminf_n \int \varphi_k d\mu_n.$$

Since this is true for all $k \geq 1$,

$$
\liminf_n \mu_n(G) \geq \sup_k \left(\liminf_n \int \varphi_k d\mu_n \right)
$$

$$
= \sup_k \left(\lim_n \int \varphi_k d\mu_n \right)
$$

$$
= \sup_k \int \varphi_k d\mu = \mu(G).
$$

(ii) $\Leftrightarrow$ (iii). Take complements.

(ii) + (iii) $\Rightarrow$ (iv). Indeed, by (ii) and (iii),

$$
\limsup_n \mu_n(B) \leq \limsup_n \mu_n(B^c) \leq \mu(B^c)
$$

and

$$
\liminf_n \mu_n(B) \geq \liminf_n \mu_n(B^o) \geq \mu(B^o).
$$

But, since $\mu(\partial B) = 0$, $\mu(B^o) = \mu(B^c) = \mu(B)$, and therefore (iv) is verified.

(iv) $\Rightarrow$ (i). Let $f \in C_b(\mathbb{R}^d)$. We must show that $\lim_{n \uparrow \infty} \int_{\mathbb{R}^d} f d\mu_n = \int_{\mathbb{R}^d} f d\mu$. It is enough to show this for $f \geq 0$. Let $K < \infty$ be a bound of f. Write, using Fubini,

$$
\int_{\mathbb{R}^d} f(x) d\mu(x) = \int_{\mathbb{R}^d} \left(\int_0^K 1_{\{t \leq f(x)\}} dt \right) d\mu(x)
$$

$$
= \int_0^K \mu(\{x \, ; \, t \leq f(x)\}) dt = \int_0^K \mu(D_t^f) dt,
$$

where $D_t^f := \{x \, ; \, t \leq f(x)\}$. Observe that $\partial D_t^f \subseteq \{x \, ; \, t = f(x)\}$ and that the collection of positive t such that $\mu(\{x \, ; \, t = f(x)\}) > 0$ is at most denumerable (for each positive integer k there are at most k values of t such that $\mu(\{x \, ; \, t = f(x)\}) \geq \frac{1}{k}$). Therefore, by (iv), for almost all t (with respect to the Lebesgue measure),

$$
\lim_n \mu_n(D_t^f) = \mu(D_t^f)
$$

and, by dominated convergence,

$$\lim_n \int f\,d\mu_n = \lim_n \int\limits_0^K \mu_n(D_t^f)\,dt$$

$$= \int\limits_0^K \mu(D_t^f)\,dt = \int f\,d\mu.$$

□

We now state *Lévy's characterization of convergence in distribution*. Its proof will be given later in Theorem 2.3.5.

Theorem 2.2.2 *A necessary and sufficient condition for the sequence* $\{X_n\}_{n\geq 1}$ *of random vectors of* $\mathbb{R}^d$ *to converge in distribution is that the sequence of their characteristic functions* $\{\varphi_n\}_{n\geq 1}$ *converges to some function* φ *that is continuous at* 0. *In such a case,* φ *is the characteristic function of the limit probability distribution.*

The following equivalent formulations of convergence in distribution are then a consequence of Theorems 2.2.2 and 2.2.1.

Corollary 2.2.1 *Let* $\{X_n\}_{n\geq 1}$ *and* X *be random vectors of* $\mathbb{R}^d$ *with respective characteristic functions* $\{\varphi_n\}_{n\geq 1}$ *and* φ. *The following three statements are equivalent.*

(A) $X_n \overset{D}{\to} X$.
(B) *For all continuous and bounded functions* $f : \mathbb{R}^d \to \mathbb{R}$,

$$\lim_{n\uparrow\infty} E[f(X_n)] = E[f(X)].$$

(C) $\lim_{n\uparrow\infty} \varphi_n = \varphi$.

Corollary 2.2.2 *In the univariate case, denote by* F_n *and* F *the cumulative distribution functions of* X_n *and* X *respectively. Call a point* $x \in \mathbb{R}$ *a continuity point of* F *if* $F(x) = F(x_-)$. *Then* $X_n \overset{D}{\to} X$ *if and only*

$$\lim_n F_n(x) = F(x) \quad \text{for all continuity points } x \text{ of } F.$$

Proof Necessity. Let Q_X be the probability distribution of X. If x is a continuity point of F, the boundary of $C := (-\infty, x]$ is $\{x\}$ of null Q_X-measure. Therefore by (iv), $\lim_n Q_{X_n}((-\infty, x]) = Q_X((\infty, x])$, that is $\lim_n F_n(x) = F(x)$.

Sufficiency. Let $f \in C_b(\mathbb{R})$ and let $M < \infty$ be an upper bound of f. For arbitrary $\varepsilon > 0$, there exists a subdivision $-\infty < a = x_0 < x_1 < \cdots < x_k = b < +\infty$ formed by continuity points of F, such that $F(a) < \varepsilon$, $F(b) > 1 - \varepsilon$ and

$|f(x) - f(x_i)| < \varepsilon$ on $[x_{i-1}, x_i]$. By hypothesis,

$$S_n := \sum_{i=1}^{k} f(x_i)(F_n(x_i) - F_n(x_{i-1})) \to S := \sum_{i=1}^{k} f(x_i)(F(x_i) - F(x_{i-1}))$$

Also

$$|E[f(X)] - S| \le \varepsilon + MF(a) + M(1 - F(b)) \le (2M + 1)\varepsilon$$

and

$$|E[f(X_n)] - S_n| \le \varepsilon + MF_n(a) + M(1 - F_n(b)) \to \varepsilon + MF(a) + M(1 - F(b))$$
$$\le (2M + 1)\varepsilon. \tag{2.6}$$

Therefore,

$$\limsup_{n} |E[f(X_n)] - E[f(X)]|$$
$$\le \limsup_{n} |E[f(X_n)] - S_n| + \limsup_{n} |S_n - S| + |E[f(X)] - S|$$
$$\le (4M + 2)\varepsilon.$$

Since ε is arbitrary, $\lim_n |E[f(X_n)] - E[f(X)]| = 0$. □

Theorem 2.2.3 *Let* $\{X_n\}_{n \ge 1}$ *and* $\{Y_n\}_{n \ge 1}$ *be sequences of random vectors of* $\mathbb{R}^d$ *such that* $X_n \overset{D}{\to} X$ *and* $d(X_n, Y_n) \overset{Pr.}{\to} 0$ *where* d *denotes the euclidean distance. Then* $Y_n \overset{D}{\to} X$.

Proof By Corollary 2.2.1, it suffices to show that for all closed sets F, $\limsup_n P(Y_n \in F) \le P(X \in F)$. For all $\varepsilon > 0$, define the closed set $F_\varepsilon = \{x \in \mathbb{R}^d ; d(x, F) \le \varepsilon\}$. We have
$$P(Y_n \in F) \le P(d(X_n, F) \ge \varepsilon) + P(X_n \in F_\varepsilon),$$

and therefore, by Corollary 2.2.1,

$$\limsup_{n} P(Y_n \in F) \le \limsup_{n} P(d(X_n, F) \ge \varepsilon) + \limsup_{n} P(X_n \in F_\varepsilon)$$
$$= \limsup_{n} P(X_n \in F_\varepsilon) \le P(X \in F_\varepsilon).$$

Since $\varepsilon > 0$ is arbitrary and $\lim_{\varepsilon \downarrow 0} P(X \in F_\varepsilon) = P(X \in F)$, $\limsup_n P(Y_n \in F) \le P(X \in F)$. □

Links with Other Types of Convergence

Convergences in probability, almost-sure, and in quadratic mean (in $L^2_{\mathbb{C}}(P)$) are linked to convergence in distribution as the following results show.

Theorem 2.2.4 *If the sequence $\{X_n\}_{n\geq1}$ of random vectors of $\mathbb{R}^d$ converges almost surely to some random vector X, it also converges in distribution to the same vector X.*

Proof By dominated convergence, for all $u \in \mathbb{R}$,

$$\lim_{n\uparrow\infty} E\left[e^{i\langle u,X_n\rangle}\right] = E\left[e^{i\langle u,X\rangle}\right]$$

which implies, by Theorem 2.2.1 that $\{X_n\}_{n\geq1}$ converges in distribution to X. $\square$

In fact, we have the stronger result:

Theorem 2.2.5 *If the sequence $\{X_n\}_{n\geq1}$ of random vectors of $\mathbb{R}^d$ converges in probability to some random vector X, it also converges in distribution to X.*

Proof If this were not the case, one could find a function $f \in C_b(\mathbb{R}^d)$ such that $E[f(X_n)]$ does not converge to $E[f(X)]$. In particular, there would exist a subsequence n_k and some $\varepsilon > 0$ such that $|E[f(X_{n_k})] - E[f(X)]| \geq \varepsilon$ for all k. As $\{X_{n_k}\}_{k\geq1}$ converges in probability to X, one can extract from it a subsequence $\{X_{n_{k_\ell}}\}_{\ell\geq1}$ converging almost-surely to X. In particular, since f is bounded and continuous, $\lim_\ell E[f(X_{n_{k_\ell}})] = E[f(X)]$ by dominated convergence, a contradiction.

$\square$

Theorem 2.2.6 *If the sequence of real random variables $\{Z_n\}_{n\geq1}$ converges in quadratic mean to some real random variable Z, it also converges in probability to the same random variable Z.*

Proof This follows immediately from Chebyshev's inequality

$$P(|Z_n - Z| \geq \varepsilon) \leq \frac{E\left[|Z_n - Z|^2\right]}{\varepsilon^2}.$$

$\square$

Pasting Theorems A.2.17 and 2.2.5, we have:

Theorem 2.2.7 *If the sequence $\{Z_n\}_{n\geq1}$ converges in quadratic mean to some random variable Z, it also converges in distribution to the same random variable Z.*

Example 2.2.1 (*A stability property of the Gaussian distribution*)
Let $\{Z_n\}_{n\geq1}$, where $Z_n = (Z_n^{(1)}, \ldots, Z_n^{(m)})$, be a sequence of Gaussian random vectors of fixed dimension m that converges componentwise in quadratic mean to some vector $Z = (Z^{(1)}, \ldots, Z^{(m)})$. Then the latter vector is Gaussian. In fact, by

continuity of the inner product in $L^2_{\mathbb{R}}(P)$, for all $1 \le i, j \le m$, $\lim_{n \uparrow \infty} E[Z_n^{(i)} Z_n^{(j)}] = E[Z^{(i)} Z^{(j)}]$ and $\lim_{n \uparrow \infty} E[Z_n^{(i)}] = E[Z^{(i)}]$, that is

$$\lim_{n \uparrow \infty} m_{Z_n} = m_Z, \qquad \lim_{n \uparrow \infty} \Gamma_{Z_n} = \Gamma_Z$$

and in particular, for all $u \in \mathbb{R}^m$,

$$\lim_{n \uparrow \infty} E\left[e^{iu^T Z_n}\right] = \lim_{n \uparrow \infty} e^{iu^T \mu_{Z_n} - \frac{i}{2} u^T \Gamma_{Z_n} u}$$
$$= e^{iu^T \mu_Z - \frac{i}{2} u^T \Gamma_Z u}.$$

The sequence $\{u^T Z_n\}_{n \ge 1}$ converges in quadratic mean to $u^T Z$, and therefore it also converges in distribution to $u^T Z$. Therefore, $\lim_{n \uparrow \infty} E\left[e^{iu^T Z_n}\right] = E[e^{iu^T Z}]$, and finally

$$E[e^{iu^T Z}] = e^{iu^T \mu_Z - \frac{i}{2} u^T \Gamma_Z u}$$

for all $u \in \mathbb{R}^m$. This shows that Z is a Gaussian vector.

Therefore, limits in the quadratic mean preserve the Gaussian nature of random vectors. This is the stability property refered to in the title of this example. Note that the Gaussian nature of random vectors is also preserved by linear transformations as we already know.

2.2.2 Bochner's Theorem

The result of this subsection links the classical Fourier theory and the Fourier theory of stochastic processes.

The characteristic function φ of a real random variable X has the following properties:

(A) It is hermitian symmetric, that is $\varphi(-u) = \varphi(u)^*$, and it is uniformly bounded: $|\varphi(u)| \le \varphi(0)$.
(B) It is uniformly continuous on $\mathbb{R}$, and
(C) It is definite non-negative, in the sense that for all integers n, all $u_1, \ldots, u_n \in \mathbb{R}$, and all $z_1, \ldots, z_n \in \mathbb{C}$,

$$\sum_{j=1}^{n} \sum_{k=1}^{n} \varphi(u_j - u_k) z_j z_k^* \ge 0$$

(just observe that the left-hand side equals $E\left[\left|\sum_{j=1}^{n} z_j e^{iu_j X}\right|^2\right]$).

It turns out that Properties A, B and C caracterize characteristic functions (up to a multiplicative constant). This is *Bochner's theorem*:

Theorem 2.2.8 *Let $\varphi : \mathbb{R} \to \mathbb{C}$ be a function satisfying properties A, B and C. Then there exists a constant $0 \leq \beta < \infty$ and a real random variable X such that for all $u \in \mathbb{R}$,*

$$\varphi(u) = \beta E \left[e^{iuX} \right].$$

Proof We henceforth eliminate the trivial case where $\varphi(0) = 0$ (implying, in view of condition A, that φ is the null function). For any continuous function $z : \mathbb{R} \to \mathbb{C}$ and any $A \geq 0$,

$$\int_0^A \int_0^A \varphi(u - v)z(u)z^*(v)du\,dv \geq 0. \tag{$*$}$$

Indeed, as the integrand is continuous, the integral is the limit as $n \uparrow \infty$ of

$$\frac{A^2}{4^n} \sum_{j=1}^{2^n} \sum_{k=1}^{2^n} \varphi\left(\frac{A(j-k)}{2^n}\right) z\left(\frac{Aj}{2^n}\right) z\left(\frac{Ak}{2^n}\right)^*,$$

a non-negative quantity, by condition C. From $(*)$ with $z = e^{-ixu}$, we have that

$$g(x, A) := \frac{1}{2\pi A} \int_0^A \int_0^A \varphi(u - v)e^{-ix(u-v)}du\,dv \geq 0.$$

Changing variables, we obtain for $g(x, A)$ the alternative expression

$$g(x, A) := \frac{1}{2\pi} \int_{-A}^A \left(1 - \frac{|u|}{A}\right) \varphi(u)e^{-iux}du$$

$$= \frac{1}{2\pi} \int_{-\infty}^{+\infty} h\left(\frac{u}{A}\right) \varphi(u)e^{-iux}du,$$

where

$$h(u) = (1 - |u|)1_{\{|u| \leq 1\}}.$$

Let $M > 0$. We have

$$
\int\limits_{-\infty}^{+\infty} h\left(\frac{x}{2M}\right) g(x, A)dx = \frac{1}{2\pi} \int\limits_{-\infty}^{+\infty} h\left(\frac{u}{A}\right) \varphi(u) \left(\int\limits_{-\infty}^{+\infty} h\left(\frac{x}{2M}\right) e^{-iux}dx \right) du
$$

$$
= \frac{1}{\pi} M \int\limits_{-\infty}^{+\infty} h\left(\frac{u}{A}\right) \varphi(u) \left(\frac{\sin Mu}{Mu}\right)^2 du.
$$

Therefore

$$
\int\limits_{-\infty}^{+\infty} h\left(\frac{x}{2M}\right) g(x, A)dx \leq \frac{1}{\pi} M \int\limits_{-\infty}^{+\infty} h\left(\frac{u}{A}\right) |\varphi(u)| \left(\frac{\sin Mu}{Mu}\right)^2 du
$$

$$
\leq \frac{1}{\pi} \varphi(0) \int\limits_{-\infty}^{+\infty} \left(\frac{\sin u}{u}\right)^2 du = \varphi(0).
$$

By monotone convergence,

$$
\lim_{M\uparrow\infty} \int\limits_{-\infty}^{+\infty} h\left(\frac{x}{2M}\right) g(x, A)dx = \int\limits_{-\infty}^{+\infty} g(x, A)dx,
$$

and therefore

$$
\int\limits_{-\infty}^{+\infty} g(x, A)dx \leq \varphi(0).
$$

The function $x \to g(x, A)$ is therefore integrable and it is the Fourier transform of the integrable and continuous function $u \to h\left(\frac{u}{A}\right) \varphi(u)$. Therefore, by the Fourier inversion formula:

$$
h\left(\frac{u}{A}\right) \varphi(u) = \int\limits_{-\infty}^{+\infty} g(x, A)e^{iux}dx.
$$

In particular, with $u = 0$, $\int_{-\infty}^{+\infty} g(x, A)dx = \varphi(0)$. Therefore, $f(x, A) := \frac{g(x,A)}{\varphi(0)}$ is the probability density of some real random variable with characteristic function $h\left(\frac{u}{A}\right) \frac{\varphi(u)}{\varphi(0)}$. But

$$
\lim_{A\uparrow\infty} h\left(\frac{u}{A}\right) \frac{\varphi(u)}{\varphi(0)} = \frac{\varphi(u)}{\varphi(0)}.
$$

This limit of a sequence of characteristic functions is continuous at 0, and therefore it is a characteristic function (Paul Lévy's criterion, Theorem 2.2.2).   □

2.3 Weak Convergence of Finite Measures

2.3.1 Helly's Theorem

We are now left with the task of proving Paul Lévy's theorem. We shall need the following slight extension of Riesz's theorem (Theorem A.1.31), Part (ii) of the following:

Theorem 2.3.1 *(i) Let $\mu \in M^+(\mathbb{R}^d)$. The linear form $L : C_0(\mathbb{R}^d) \to \mathbb{R}$ defined by $L(f) := \int_{\mathbb{R}^d} f\,d\mu$ is positive ($L(f) \geq 0$ whenever $f \geq 0$) and continuous, and its norm is $\mu(\mathbb{R}^d)$.*
(ii) Let $L : C_0(\mathbb{R}^d) \to \mathbb{R}$ be a positive continuous linear form. There exists a unique measure $\mu \in M^+(\mathbb{R}^d)$ such that for all $f \in C_0(\mathbb{R}^d)$,

$$L(f) = \int_{\mathbb{R}^d} f\,d\mu.$$

Proof Part (i) is left as an exercise Exercise 2.4.13. We turn to the proof of (ii). The restriction of L to C_c is a positive Radon linear form, and therefore, according to Riesz's theorem A.1.31, there exists a locally finite μ on $(\mathbb{R}^d, \mathcal{B}(\mathbb{R}^d))$ such that for all $f \in C_c(\mathbb{R}^d)$, $L(f) = \int_{\mathbb{R}^d} f\,d\mu$.

We show that μ is a *finite* (not just locally finite) measure. If not, there would exist a sequence $\{K_m\}_{m \geq 1}$ of compact subsets of $\mathbb{R}^d$ such that $\mu(K_m) \geq 3^m$ for all $m \geq 1$. Let then $\{\varphi_m\}_{m \geq 1}$ be a sequence of non-negative functions in $C_c(\mathbb{R}^d)$ with values in $[0, 1]$ and such that for all $m \geq 1$, $\varphi_m(x) = 1$ for all $x \in K_m$. In particular, the function $\varphi := \sum_{m \geq 1} 2^{-m}\varphi_m$ is in $C_0(\mathbb{R}^d)$ and

$$
\begin{aligned}
L(\varphi) &\geq L\left(\sum_{m=1}^{k} 2^{-m}\varphi_m\right) = \sum_{m=1}^{k} 2^{-m} L(\varphi_m) \\
&= \sum_{m=1}^{k} 2^{-m} \int_{\mathbb{R}^d} \varphi_m\,d\mu \geq \sum_{m=1}^{k} 2^{-m}\mu(K_m) \geq \left(\frac{3}{2}\right)^k.
\end{aligned}
$$

Letting $k \uparrow \infty$ leads to $L(\varphi) = \infty$, a contradiction.

We show that L is continuous. Suppose it is not. We could then find a sequence $\{\varphi_m\}_{m \geq 1}$ of functions in $C_c(\mathbb{R}^d)$ such that $|\varphi_m| \leq 1$ and $L(\varphi_m) \geq 3^m$. The function

$\varphi := \sum_{m \geq 1} 2^{-m} \varphi_m$ is in $C_0(\mathbb{R}^d)$ and

$$L(\varphi) \geq L(\sum_{m=1}^{k} 2^{-m} \varphi_m) = \sum_{m=1}^{k} 2^{-m} L(\varphi_m)$$

$$= \sum_{m=1}^{k} 2^{-m} \int_{\mathbb{R}^d} \varphi_m d\mu \geq \left(\frac{3}{2}\right)^k,$$

Letting $k \uparrow \infty$ again leads to $L(\varphi) = \infty$, a contradiction.

It remains to show that $L(f) = \int_{\mathbb{R}^d} f d\mu$ for all $f \in C_0(\mathbb{R}^d)$. For this we consider a sequence $\{f_m\}_{m \geq 1}$ of functions in $C_c(\mathbb{R}^d)$ converging uniformly to $f \in C_0(\mathbb{R}^d)$. We have, since L is continuous, $\lim_{m \uparrow \infty} L(f_m) = L(f)$ and, since μ is finite, $\lim_{m \uparrow \infty} \int f_m d\mu = \int f d\mu$ by dominated convergence. Therefore, since $L(f_m) = \int f_m d\mu$ for all $m \geq 1$, $L(f) = \int f d\mu$. $\qquad \square$

We now introduce the notion of *vague convergence*.

Definition 2.3.1 The sequence $\{\mu_n\}_{n \geq 1}$ in $M^+(\mathbb{R}^d)$ is said to converge *vaguely* to μ if, for all $f \in C_0(\mathbb{R}^d)$, $\lim_{n \uparrow \infty} \int_{\mathbb{R}^d} f d\mu_n = \int_{\mathbb{R}^d} f d\mu$.

Theorem 2.3.2 *The sequence $\{\mu_n\}_{n \geq 1}$ in $M^+(\mathbb{R}^d)$ converges vaguely if and only if*

(a) $\sup_n \mu_n(\mathbb{R}^d) < \infty$, *and*
(b) *There exists a dense subset $\mathcal{E}$ in $C_0(\mathbb{R}_d)$ such that for all $f \in \mathcal{E}$, there exists* $\lim_{n \uparrow \infty} \int_{\mathbb{R}^d} f d\mu_n$.

Proof Necessity. If the sequence converges vaguely, it obviously satisfies (b). As for (a), it is a consequence of the Banach–Steinhaus theorem.[1] Indeed, $\mu_n(\mathbb{R}^d)$ is the norm of L_n, where L_n is the continuous linear form $f \to \int_{\mathbb{R}^d} f d\mu_n$ from the Banach space $C_0(\mathbb{R}^d)$ (with the sup norm) to $\mathbb{R}$, and for all $f \in C_0(\mathbb{R}^d)$, $\sup_n \left| \int_{\mathbb{R}^d} f d\mu_n \right| < \infty$.

Sufficiency. Suppose the sequence satisfies (a) and (b). Let $f \in C_0(\mathbb{R}^d)$. For all $\varphi \in \mathcal{E}$,

$$\left| \int f d\mu_m - \int f d\mu_n \right|$$

$$\leq \left| \int \varphi d\mu_m - \int \varphi d\mu_n \right| + \left| \int f d\mu_m - \int \varphi d\mu_m \right| + \left| \int f d\mu_n - \int \varphi d\mu_n \right|$$

$$\leq \left| \int \varphi d\mu_m - \int \varphi d\mu_n \right| + \sup_{x \in \mathbb{R}^d} |f(x) - \varphi(x)| \times \sup_n \mu_n(\mathbb{R}^d).$$

---

[1] Let E be a Banach space and F be a normed vector space. Let $\{L_i\}_{i \in I}$ be a family of continuous linear mappings from E to F such that, $\sup_{i \in I} \|L_i(x)\| < \infty$ for all $x \in E$. Then $\sup_{i \in I} \|L_i\| < \infty$. See for instance Rudin (1986), Theorem 5.8.

Since $\sup_{x\in\mathbb{R}^d}|f(x)-\varphi(x)|$ can be made arbitrarily small by a proper choice of φ, this shows that the sequence $\{\int f d\mu_n\}_{n\geq1}$ is a Cauchy sequence. It therefore converges to some $L(f)$, and L so defined is a positive linear form on $C_0(\mathbb{R}^d)$. Therefore, there exists $\mu\in M^+(\mathbb{R}^d)$ such that $L(f)=\int_{\mathbb{R}^d}f d\mu$ and $\{\mu_n\}_{n\geq1}$ converges vaguely to μ. □

We now give *Helly's theorem*.

Theorem 2.3.3 *Fom any bounded sequence of $M^+(\mathbb{R}^d)$, one can extract a vaguely convergent subsequence.*

Proof Let $\{\mu_n\}_{n\geq1}$ be a bounded sequence of $M^+(\mathbb{R}^d)$. Let $\{f_n\}_{n\geq1}$ be a dense sequence of $C_0(\mathbb{R}^d)$.

Since the sequence $\{\int f_1 d\mu_n\}_{n\geq1}$ is bounded, one can extract from it a convergent subsequence $\{\int f_1 d\mu_{1,n}\}_{n\geq1}$. Since the sequence $\{\int f_2 d\mu_{1,n}\}_{n\geq1}$ is bounded, one can extract from it a convergent subsequence $\{\int f_2 d\mu_{2,n}\}_{n\geq1}$. This diagonal selection process is continued. At step k, since the sequence $\{\int f_{k+1} d\mu_{k,n}\}_{n\geq1}$ is bounded, one can extract from it a convergent subsequence $\{\int f_{k+1} d\mu_{k+1,n}\}_{n\geq1}$. The sequence $\{\nu_k\}_{k\geq1}$ where $\nu_k=\mu_{k,k}$ (the "diagonal" sequence) is extracted from the original sequence and for all f_n, the sequence $\{\int f_n d\nu_k\}_{k\geq1}$ converges. The conclusion follows from Theorem 2.3.2. □

2.3.2 Proof of Paul Lévy's Theorem

For the next definition, recall that $C_b(\mathbb{R}^d)$ denotes the collection of uniformly bounded and continuous functions from $\mathbb{R}^d$ to $\mathbb{R}$.

Theorem 2.3.4 *The sequence $\{\mu_n\}_{n\geq1}$ in $M^+(\mathbb{R}^d)$ converges weakly to μ if and only if*

(i) It converges vaguely to μ, and
(ii) $\lim_{n\uparrow\infty}\mu_n(\mathbb{R}^d)=\mu(\mathbb{R}^d)$.

Proof The necessity of (i) immediately follows from the observation that $C_0(\mathbb{R}^d)\subset C_b(\mathbb{R}^d)$. The necessity of (ii) follows from the fact that the function that is the constant 1 is in $C_b(\mathbb{R}^d)$ and therefore $\int 1 d\mu_n=\mu_n(\mathbb{R}^d)$ tends to $\int 1 d\mu=\mu(\mathbb{R}^d)$ as $n\uparrow\infty$.

Sufficiency. Suppose that (i) and (ii) are satisfied. To prove weak convergence, it suffices to prove that $\lim_{n\uparrow\infty}\int_{\mathbb{R}^d}f d\mu_n=\int_{\mathbb{R}^d}f d\mu$ for any non-negative function $f\in C_b(\mathbb{R}^d)$.

Since the measure μ is of finite total mass, for any $\varepsilon>0$ one can find a compact set $K_\varepsilon=K$ such that $\mu(\overline{K})\leq\varepsilon$. Choose a continuous function with compact support φ with values in $[0,1]$ and such that $\varphi\geq 1_K$. Since $|f-f\varphi)|\leq\|f\|(1-\varphi)$

(where $\|f\| = \sup_{x \in \mathbb{R}^d} |f(x)|$),

$$
\limsup_{n \uparrow \infty} \left| \int f d\mu_n - \int f\varphi d\mu_n \right| \leq \limsup_{n \uparrow \infty} \|f\| \int (1 - \varphi) d\mu_n
$$

$$
= \|f\| \left(\lim_{n \uparrow \infty} \int d\mu_n - \lim_{n \uparrow \infty} \int \varphi d\mu_n \right)
$$

$$
= \|f\| \int (1 - \varphi) d\mu \leq \varepsilon \|f\|.
$$

Similarly, $\left| \int f d\mu - \int f\varphi d\mu \right| \leq \varepsilon \|f\|$. Therefore, for all $\varepsilon > 0$,

$$
\limsup_{n \uparrow \infty} \left| \int f d\mu_n - \int f d\mu \right| \leq 2\varepsilon \|f\|,
$$

and this completes the proof.   □

We will be ready for the generalization of Paul Lévy's criterion of convergence in distribution after a few preliminaries.

Definition 2.3.2 A family $\{\alpha_t\}_{t>0}$ of functions $\alpha_t : \mathbb{R}^d \to \mathbb{C}$ in $L_{\mathbb{C}}^1(\mathbb{R}^d)$ is called an *approximation of the Dirac distribution* in $\mathbb{R}^d$ if it satisfies the following three conditions:

(i) $\int_{\mathbb{R}^d} \alpha_t(x) dx = 1$,
(ii) $\sup_{t>0} \int_{\mathbb{R}^d} |\alpha_t(x)| dx := M < \infty$, and
(iii) For any compact neighborhood V of $0 \in \mathbb{R}^d$, $\lim_{t \downarrow 0} \int_V |\alpha_t(x)| dx = 0$.

Lemma 2.3.1 *Let $\{\alpha_t\}_{t>0}$ be an approximation of the Dirac distribution in $\mathbb{R}^d$. Let $f : \mathbb{R}^d \to \mathbb{C}$ be a bounded function continuous at all points of a compact $K \subset \mathbb{R}^d$. Then $\lim_{t \downarrow 0} f * \alpha_t$ uniformly in K.*

Proof We will show later that

$$
\lim_{y \to 0} \sup_{x \in K} |f(x - y) - f(x)| \to 0. \tag{*}
$$

V being a compact neighborhood of 0, we have that

$$
\sup_{x \in K} |f(x) - (f * \alpha_t)(x)| \leq M \sup_{y \in V} \sup_{x \in K} |f(x - y) - f(x)|
$$

$$
+ 2| \sup_{x \in \mathbb{R}^d} |f(x)| \int_V |\alpha_t(y)| dy.
$$

This quantity can be made smaller than an arbitrary $\varepsilon > 0$ by choosing V such that the first term be $< \frac{1}{2}\varepsilon$ (uniform continuity of f on compact sets) and the second term can then be made $< \frac{1}{2}\varepsilon$ by letting $t \downarrow$ (condition (iii) of Definition 2.3.2).

Proof of (∗). Let $\varepsilon > 0$ be given. For all $x \in K$, there exists an open and symmetric neihborhood V_x of 0 such that for all $y \in V_x$, $f(x-y)_f(x)| \leq \frac{1}{2}\varepsilon$. Also, one can find an open and symmetric neighborhood W_x of 0 such that $W_x + W_x \subset V_x$. The union of open sets $\cup_{x \in K}\{x + W_x\}$ obviously covers K, and since the latter is a compact set, one can extract a finite covering of K: $\cup_{j=1}^{m}(x_j + W_{x_j})$. Define $W = \cap_{j=1}^{m} W_{x_j}$, open neighborhood of 0.

Let $y \in W$. Any $x \in K$ belongs to some $x_j + W_{x_j}$, and for such j, and

$$|f(x-y) - f(x)| \leq |f(x_j) - f(x_j - (x_j - x))|$$
$$+ |f(x_j) - f(x_j - (x_j - x + y))|.$$

But $x_j - x \in W_{x_j}$ and $x_j - x + y \in W_{x_j} + W \subset V_{x_j}$. Therefore

$$|f(x-y) - f(x)| \leq \frac{1}{2}\varepsilon + \frac{1}{2}\varepsilon = \varepsilon.$$

$\square$

Theorem 2.3.5 *Let $\{\mu_n\}_{n\geq 1}$ be a sequence of $M^+(\mathbb{R}^d)$ such that for all $v \in \mathbb{R}^d$, there exists $\lim_{n\uparrow\infty} \widehat{\mu}_n(v) = \varphi(v)$ for some function φ that is continuous at 0. Then $\{\mu_n\}_{n\geq 1}$ converges weakly to a finite measure μ whose Fourier transform is φ.*

Proof The sequence $\{\mu_n\}_{n\geq 1}$ is bounded (that is $\sup_n \mu_n(\mathbb{R}^d) < \infty$) since $\mu_n(\mathbb{R}^d) = \widehat{\mu}_n(1)$ has a limit as $n \uparrow \infty$. In particular

$$\widehat{\mu}_n(v) \leq \mu_n(\mathbb{R}^d) \leq \sup_n \mu_n(\mathbb{R}^d) < \infty. \tag{†}$$

If $f \in L^1_{\mathbb{R}}(\mathbb{R}^d)$, then by Theorem 2.1.2, $\int \widehat{f} d\mu_n = \int f \widehat{\mu}_n dx$. By dominated convergence (using (†)), $\lim_{n\uparrow\infty} \int f \widehat{\mu}_n dx = \int f \varphi dx$. Therefore

$$\lim_{n\uparrow\infty} \int \widehat{f} d\mu_n = \int f \varphi dx.$$

One can replace in the above equality $\widehat{f}$ by any function in $\mathcal{D}(\mathbb{R}^d)$, since such function is always the Fourier transform of some integrable function (Corollary 1.1.5).

Therefore, by Theorem 2.3.2, $\{\mu_n\}_{n\geq 1}$ converges vaguely to some finite measure μ.

We now show that it converges *weakly* to μ. Let f be an integrable function of integral 1 such that $f(x) = f(-x)$ and $\widehat{f} \in \mathcal{D}(\mathbb{R}^d)$. For $t > 0$, define $f_t(x) := t^{-d} f(t^{-1}x))$. Using Theorem 2.1.2, we have

$$\int \widehat{f}(tx)\mu_n(dx) = \int f_t(x)\widehat{\mu}_n(x)dx = (f_t * \widehat{\mu}_n)(0).$$

By dominated convergence,

$$\lim_{n\uparrow\infty} \int \widehat{f}(tx)\mu_n(dx) = (f_t * \varphi)(0),$$

and by vague convergence,

$$\lim_{n\uparrow\infty} \int \widehat{f}(tx)\mu_n(dx) = \int \widehat{f}(tx)\mu(dx).$$

Therefore, for all $t > 0$, $\int \widehat{f}(tx)\mu(dx) = (f_t * \varphi)(0)$.

Since the function φ is bounded and continuous at the origin, by Lemma 2.3.1, $\lim_{t\downarrow 0}(f_t * \varphi)(0) = \varphi(0)$. Also, by dominated convergence $\lim_{t\downarrow 0}\int \widehat{f}(tx)\mu(dx) = \mu(\mathbb{R}^d)$. Therefore,

$$\mu(\mathbb{R}^d) = \varphi(0) = \lim_{n\uparrow\infty}\mu_n(\mathbb{R}^d).$$

Therefore, by Theorem 2.3.4, $\{\mu_n\}_{n\geq 1}$ converges *weakly* to μ. Since the function $x \to e^{-2i\pi\langle v,x\rangle}$ is continuous and bounded,

$$\widehat{\mu}(v) = \int e^{-2i\pi\langle v,x\rangle}\mu(dx) = \lim_{n\uparrow\infty}\int e^{-2i\pi\langle v,x\rangle}\mu_n(dx) = \varphi(v).$$

$\square$

2.4 Exercises

Exercise 2.4.1 (*Sampling a sinusoid*) Check with a single sinusoid that you really need the strict inequality in (2.2).

Exercise 2.4.2 (*Characteristic function of the residual time*) Let $G : \mathbb{R}_+ \to \mathbb{C}$ be the primitive function of the locally integrable function $g : \mathbb{R}_+ \to \mathbb{C}$, that is, for all $x \geq 0$,

$$G(x) = G(0) + \int_0^x g(u)du.$$

(a) Let X be a non-negative random variable with finite mean μ and such that $E[|G(X)|] < \infty$. Show that

$$E[G(X)] = G(0) + \int_0^\infty g(x)P(X \geq x)dx.$$

(b) Let X be as in (a), and let $\overline{X}$ be a non-negative random variable with the probability density $\mu^{-1} P(X \geq x)$. Compute the characteristic function of $\overline{X}$.

Exercise 2.4.3 (*Characteristic functions of lattice distributions*) Let X be a real random variable whose characteristic function ψ is such that $|\psi(t_0)| = 1$ for some $t_0 \neq 0$. Show that there exists some $a \in \mathbb{R}$ such that $\sum_{n \in \mathbb{Z}} P(X = a + n\frac{2\pi}{t_0}) = 1$.

Exercise 2.4.4 (*Probability of the positive quadrant*) Let (X, Y) be a 2-dimensional Gaussian vector with probability density

$$f(x, y) = \frac{1}{2\pi (1 - \rho^2)^{1/2}} \exp\left\{-\frac{1}{2(1 - \rho^2)} \left(x^2 - 2\rho x y + y^2\right)\right\}$$

where $|\rho| < 1$.

(a) Show that X and $(Y - \rho X) / (1 - \rho^2)^{1/2}$ are independent Gaussian random variable with mean 0 and variance 1.
(b) Prove that

$$P(X > 0, Y > 0) = \frac{1}{4} + \frac{1}{2\pi} \sin^{-1}(\rho).$$

Exercise 2.4.5 (*Discrete uniform variable and convergence in distribution*) Let Z_n be a random variable uniformly distributed on $\{\frac{1}{2^n}, 2\frac{1}{2^n}, \ldots, 2^n \frac{1}{2^n} = 1\}$. Does it converge in distribution?

Exercise 2.4.6 (*Convergence in distribution for integer-valued variables*) Show that if the random variables X_n's and X take integer values, $X_n \xrightarrow{D} X$ if and only if for all $k \geq 0$,

$$\lim_{n \uparrow \infty} P(X_n = k) = P(X = k).$$

Exercise 2.4.7 (*Poisson's law of rare events in the plane*) Let $Z_1, \ldots, Z_M$ be M bidimensional IID random vectors uniformly distributed on the square $[0, A] \times [0, A] = \Gamma_A$. Define for any measurable set $C \subseteq \Gamma_A$, $N(C)$ to be the number of random vectors Z_i that fall in C. Let $C_1, \ldots, C_K$ be measurable *disjoint* subsets of Γ_A.

(i) Give the characteristic function of the vectors $(N(C_1), \ldots, N(C_K))$.
(ii) We now let M be a function of $A \in \mathbb{R}_+$ such that $\frac{M(A)}{A^2} = \lambda > 0$. Show that, as $A \uparrow \infty$, $(N(C_1), \ldots, N(C_K))$ converges in distribution. Identify the limit distribution.

Exercise 2.4.8 (*Cauchy random variable*) A random variable X with PDF

$$f(x) = \frac{1}{\pi(1+x^2)}$$

is called a *Cauchy* random variable. Compute the Fourier transform of $u \to e^{-|u|}$ and deduce from the result that the CF of the Cauchy random variable is $\psi(u) = e^{-|u|}$.

Exercise 2.4.9 (*Cauchy random variables and convergences*) Let $\{X_n\}_{n\geq 1}$ be a sequence of IID Cauchy random variables.

(A.) What is the limit in distribution of $\frac{X_1+\cdots+X_n}{n}$?

(B.) Does $\frac{X_1+\cdots+X_n}{n^2}$ converge in distribution?

(C.) Does $\frac{X_1+\cdots+X_n}{n}$ converge almost surely to a deterministic constant?

Exercise 2.4.10 (*A simple counter example*) Let Z be a random variable with a symmetric distribution (that is, Z and $-Z$ have the same distribution). Define the sequence $\{Z_n\}_{n\geq 1}$ as follows: $Z_n = Z$ if n is odd, $Z_n = -Z$ if n is even. In particular, $\{Z_n\}_{n\geq 1}$ converges in *distribution* to Z. Show that if Z is nondegenerate, then $\{Z_n\}_{n\geq 1}$ does NOT converge to Z *in probability*.

Exercise 2.4.11 (*Infimum of IID uniform variables*) Let $\{Y_n\}_{n\geq 1}$ be a sequence of IID random variables uniformly distributed on $[0, 1]$. Show that

$$X_n = n\min(Y_1,\ldots,Y_n) \xrightarrow{\mathcal{D}} \mathcal{E}(1),$$

(the exponential distribution with mean 1).

Exercise 2.4.12 (*A stability property of the Gaussian distribution*) Prove the statement (ii) of Example 2.2.1.

Exercise 2.4.13 (*Riesz's theorem*) Prove Part (i) of Theorem 2.3.1.

Exercise 2.4.14 (*Dirac measures and vague convergence*)

(i) Show that the sequence of Dirac measures on $(\mathbb{R}^d, \mathcal{B}(\mathbb{R}^d))$, $\{\varepsilon_{a_n}\}_{n\geq 1}$ converges vaguely to ε_a if and only if $\lim_{n\uparrow\infty} a_n = a$.

(ii) Show that ε_a converges vaguely to the null measure when $a \to \infty$.

(iii) Show that the measure $\mu_a = \frac{1}{2B}1_{[-B,+B]}\ell$ on $(\mathbb{R}, \mathcal{B}(\mathbb{R}))$ converges vaguely to the null measure as $B \to \infty$.

(iv) Let $f \in L^1_{\mathbb{C}}(\mathbb{R}^d)$ be non-negative and such that $\int_{\mathbb{R}^d} f(x)dx = 1$. Let $f_t(x) := t^{-d}f(t^{-1}x)$. Show that, as $n \uparrow \infty$, the measures on $\mathbb{R}^d$ with densities $f_{\frac{1}{n}}$ and f_n converge vaguely to the Dirac measure ε_0 and the null measure respectively.

Exercise 2.4.15 (*Polya's theorem*) Prove the following result:

A non-negative symmetric continuous function $\varphi : \mathbb{R} \to \mathbb{R}$ that is convex on $[0, +\infty)$ and such that $\varphi(0) = 1$ and $\lim_{|u|\uparrow\infty}\varphi(u) = 0$ is a characteristic function.

Hint: first show that this is true if φ is in addition piecewise linear.

Use the result to prove that the fact that two real random variables whose characteristic functions coincide on an interval of $\mathbb{R}$ do not necessarily have the same distribution.

Chapter 3
Fourier Analysis of Stochastic Processes

This chapter is devoted to continuous-time wide-sense stationary stochastic processes and continuous parameter random fields. First, we introduce the notion of power spectral measure, that is, in a broad sense, the Fourier transform of the autocovariance function of such processes. (As we mentioned before, the existence and the unicity of the spectral measure are immediate consequences of Bochner's theorem and of Paul Lévy's inversion formula of the previous chapter.) Then, we look at the trajectories of the stochastic process themselves and give their Fourier decomposition. The classical Fourier theory of the first chapter does not apply there since the trajectories are in general not in L^1 or L^2. The corresponding result—Cramér–Khinchin's decomposition—, is in terms of an integral of a special type, the Doob–Wiener integral of a function with respect to a stochastic process with uncorrelated increments, such as for instance the Wiener process, a particular case of Gaussian processes.

3.1 Review of Stochastic Processes

3.1.1 Distribution of a Stochastic Process

The main results of the theory of stochastic processes needed in this chapter and the following will be reviewed. A *stochastic process* is a family of random variables $\{X(t)\}_{t \in \mathbb{R}}$ defined on a given probability space $(\Omega, \mathcal{F}, P)$. (A more general definition will be given soon.)

Example 3.1.1 (Random sinusoid) Let A be some real non-negative random variable, let $\nu_0 \in \mathbb{R}$ be a positive constant, and let Φ be a random variable with values in $[0, 2\pi]$. Formula

$$X(t) = A \sin(2\pi\nu_0 t + \Phi)$$

defines a stochastic process. For each sample $\omega \in \Omega$, the function $t \mapsto X(t, \omega)$ is a sinusoid with frequency ν_0, random amplitude $A(\omega)$ and random phase $\Phi(\omega)$.

© Springer International Publishing Switzerland 2014
P. Brémaud, *Fourier Analysis and Stochastic Processes*, Universitext,
DOI 10.1007/978-3-319-09590-5_3

Example 3.1.2 (Counting processes) Let $\{T_n\}_{n\in\mathbb{Z}}$ be a sequence of real—possibly infinite—random variables such that, almost-surely,

(i) $T_0 \leq 0 < T_1$,
(ii) $T_n < T_{n+1}$ whenever $|T_n| < \infty$, and
(iii) $\lim_{|n|\uparrow\infty} |T_n| = +\infty$.

The T_n's are called *event times* or *points* and the sequence itself is called a *point process*. Condition (ii) implies that there is almost surely no multiple points at finite distance (simplicity), whereas condition (iii) guarantees that the number of points in a bounded interval is almost surely finite (local finiteness). The random variable

$$N((a, b]) = \sum_{n\in\mathbb{Z}} 1_{(a,b]}(T_n)$$

counts the event times occuring in the time interval $(a, b] \subset \mathbb{R}$. Let $N(t)$ be the random variable defined for all $t \in \mathbb{R}$ by $N(0) = 0$ and $N((a, b]) = N(b) - N(a)$. For each ω, the function $t \to N(t, \omega)$ is right-continuous, non-decreasing, has limits on the left, and jumps one unit upwards at each event time $T_n(\omega)$. The family of random variables $\{N(t)\}_{t\in\mathbb{R}}$ is a stochastic process, called the *counting process* of the point process $\{T_n\}_{n\in\mathbb{Z}}$.

Example 3.1.3 (Step processes) Let $\{T_n\}_{n\in\mathbb{Z}}$ be as in the previous example, and let $\{X_n\}_{n\in\mathbb{Z}}$ be another sequence of random variables, with values in some measurable space $(E, \mathcal{E})$. Define a stochastic process $\{X(t)\}_{t\in\mathbb{R}}$, called a *step process*, by the formula

$$X(t) = X_{N(t)},$$

that is, $X(t) = X_n$ if $T_n \leq t < T_{n+1}$. (The random variable X_n can be viewed as a mark associated with the event time T_n, this mark remaining visible during the whole interval $[T_n, T_{n+1})$.) The function $t \to X(t, \omega)$ is, for each ω, right-continuous, non-decreasing, and has limits on the left.

Since we want to include random functions with spatial arguments ($t \in \mathbb{R}^m$ or $t \in \mathbb{Z}^m$), or even with arguments that are sets (as in Example 3.2.3 introducing the Brownian pseudo-measure), and since the values taken by these functions may be largely arbitrary, we shall adopt a more general definition of a stochastic process.

Definition 3.1.1 Let $\mathbf{T}$ be an arbitrary index set. A *stochastic process* is a family $\{X(t)\}_{t\in\mathbf{T}}$ of random variables defined on the same probability space $(\Omega, \mathcal{F}, P)$ and taking their values values in a measurable space $(E, \mathcal{E})$.

It is called a *real* (resp., *complex*) stochastic process if it takes real (resp., complex) values. It is called a *continuous-time* stochastic process when the index set is $\mathbb{R}$ or $\mathbb{R}_+$, and a *discrete-time* stochastic process when it is $\mathbb{N}$ or $\mathbb{Z}$. If the index set is $\mathbb{R}^m$ (*resp.*, $\mathbb{Z}^m$) for some $m \geq 2$, it is called a continuous-space (resp., discrete-space) *random field.* For each $\omega \in \Omega$, the function $t \in \mathbf{T} \to X(t, \omega) \in E$ is called the *ω-trajectory*, or *ω-sample path*, of the stochastic process. When the index set is $\mathbb{N}$ or $\mathbb{Z}$, we shall usually prefer to use the notation n instead of t for the time index, and write X_n for $X(n)$.

For theoretical as well as computational purposes, the probabilistic behaviour of a stochastic process is determined by its finite-dimensional distribution.

Definition 3.1.2 By definition, the *finite-dimensional (fidi) distribution* of a stochastic process $\{X(t)\}_{t\in\mathbf{T}}$ is the collection of probability distributions of the random vectors

$$(X(t_1), \ldots, X(t_k))$$

for all $k \geq 1$ and all $t_1, \ldots, t_k \in \mathbf{T}$.

Of course, in applications, one prefers to use as models stochastic processes whose finite-dimensional distribution receives a compact expression and which are easy to handle mathematically, such as Gaussian processes, Markov processes and so on. Such processes will be introduced in due time after we have dealt with the necessary general concepts.

The probability distribution of the vector $(X(t_1), \ldots, X(t_k))$ is a probability measure $Q_{(t_1,\ldots,t_k)}$ on $(E^k, \mathcal{E}^k)$. It satisfies the following obvious properties, called the *compatibility conditions*:

C_1. For all $(t_1, \ldots, t_k) \in \mathbf{T}^k$, and any permutation σ on $\{1, 2, \ldots, k\}$,

$$Q_{(t_{\sigma(1)},\ldots,t_{\sigma(k)})} = Q_{t_1,\ldots,t_k} \circ \tilde{\sigma}^{-1},$$

where $\tilde{\sigma}(x_1, \ldots, x_k) := (x_{\sigma(1)}, \ldots, x_{\sigma(k)})$, and

C_2. For all $(t_1, \ldots, t_k, t_{k+1}) \in \mathbf{T}^{k+1}$, all $A \in \mathcal{E}^k$

$$Q_{(t_1,\ldots,t_k)}(A) = Q_{(t_1,\ldots,t_{k+1})}(A \times E).$$

The list of examples to follow concerns Poisson processes and Markov chains, in discrete or continuous time, and constitute a quick review of their basic properties directly useful for our purpose.

Example 3.1.4 (Discrete-time HMC, I) Let $\{X_n\}_{n\in\mathbb{N}}$ be a discrete-time stochastic process with countable state space E. If for all integers $n \geq 0$ and all states $i_0, i_1, \ldots, i_{n-1}, i, j$,

$$P(X_{n+1} = j \mid X_n = i, X_{n-1} = i_{n-1}, \ldots, X_0 = i_0) = P(X_{n+1} = j \mid X_n = i),$$

the above stochastic process is called a *Markov chain*, and a *homogeneous* Markov chain (HMC) if, in addition, the right-hand side is independent of n. In this case, the matrix $\mathbf{P} = \{p_{ij}\}_{i,j\in E}$, where

$$p_{ij} = P(X_1 = j \mid X_0 = i),$$

is called the *transition matrix* of the HMC. Since the entries are probabilities, and since a transition from any state i must be to some state, it follows that for all states i, j, $p_{ij} \geq 0$ and $\sum_{k \in E} p_{ik} = 1$ (lines sum up to 1). A matrix $\mathbf{P}$ indexed by E and satisfying the above properties is called a *stochastic matrix*.

One sometimes associates with a transition matrix its *transition graph*. It is an oriented graph whose set of vertices is the state space E and with an oriented edge from i to j if and only if $p_{ij} > 0$. This oriented edge then receives the label p_{ij}.

Notation: any collection $x = \{x_i\}_{i \in E}$ will be considered as a *column* vector, so that its transpose x^T is then a row vector.

The distribution at time n of the chain is the vector $\nu_n = \{\nu_n(i)\}_{i \in E}$, where

$$\nu_n(i) = P(X_n = i).$$

From the Bayes rule of exclusive and exhaustive causes, $\nu_{n+1}(j) = \sum_{i \in E} \nu_n(i) p_{ij}$, that is, in matrix form, $\nu_{n+1}^T = \nu_n^T \mathbf{P}$. Iteration of this equality yields

$$\nu_n^T = \nu_0^T \mathbf{P}^n. \tag{3.1}$$

The matrix $\mathbf{P}^m$ is called the *m-step transition matrix* because its general term is

$$p_{ij}(m) = P(X_{n+m} = j \mid X_n = i).$$

In fact, the Bayes sequential rule and the Markov property give for the right-hand side of the latter equality

$$\sum_{i_1, \dots, i_{m-1} \in E} p_{ii_1} p_{i_1 i_2} \cdots p_{i_{m-1} j},$$

which is the general term of the m-th power of $\mathbf{P}$.

The probability distribution ν_0 of the *initial state* is called the *initial distribution*. From the Bayes sequential rule, the homogeneous Markov property and the definition of the transition matrix,

$$P(X_0 = i_0, X_1 = i_1, \dots, X_k = i_k) = \nu_0(i_0) p_{i_0 i_1} \cdots p_{i_{k-1} i_k}. \tag{3.2}$$

We therefore see that in this case, the finite-dimensional distribution of the chain is uniquely determined by the initial distribution and the transition matrix.

Example 3.1.5 (Continuous-time HMC, I.) Let E be a countable state space. A stochastic process $\{X(t)\}_{t \in \mathbb{R}_+}$ with values in E is called a *continuous-time Markov chain* if for all $i, j, i_1, \dots, i_k \in E$, all $t, s \geq 0$, all $s_1, \dots, s_k \geq 0$ with $s_\ell < s$ $(1 \leq \ell \leq k)$,

$$P(X(t+s) = j \mid X(s) = i, X(s_1) = i_1, \dots, X(s_k) = i_k)$$
$$= P(X(t+s) = j \mid X(s) = i).$$

If the right-hand side is independent of s, this continuous-time Markov chain is called *homogeneous*. We then say: a *continuous-time* HMC. Define in this case the matrix

$$\mathbf{P}(t) := \{p_{ij}(t)\}_{i,j\in E},$$

where

$$p_{ij}(t) := P(X(t+s) = j \mid X(s) = i).$$

The family $\{\mathbf{P}(t)\}_{t\in\mathbb{R}_+}$ is called the *transition semi-group* of the continuous-time HMC. A simple application of Bayes's rule of exclusive and exhaustive causes leads to the *Chapman–Kolmogorov equation*

$$p_{ij}(t+s) = \sum_{k\in E} p_{ik}(t)p_{kj}(s),$$

that is, in compact form,

$$\mathbf{P}(t+s) = \mathbf{P}(t)\mathbf{P}(s).$$

Also, clearly, $\mathbf{P}(0) = I$, where I is the identity matrix.

The *distribution at time t* of the chain is the vector $\nu(t) = \{\nu_i(t)\}_{i\in E}$ where $\nu_i(t) = P(X(t) = i)$. It is obtained from the initial distribution by the formula

$$\nu(t)^T = \nu(0)^T \mathbf{P}(t). \tag{3.3}$$

More generally, for all $t_1, \ldots, t_k$ such that $0 \leq t_1 \leq t_2 \leq \cdots \leq t_k$, and for all states $i_0, i_1, \ldots, i_k$,

$$P\left(X(t_1) = i_1, \ldots, X(t_k) = i_k\right) = \sum_{i_0\in E} \nu_0(i_0)\, p_{i_0 i_1}(t_1)\, p_{i_1 i_2}(t_2 - t_1) \cdots$$

$$p_{i_{k-1} i_k}(t_k - t_{k-1}).$$

The above formulas are easy applications of the elementary Bayes calculus. The last one shows in particular that the finite-dimensional distribution of a continuous-time HMC is entirely determined by its initial distribution and its transition semi-group.

Example 3.1.6 (Homogeneous Poisson process) The point process $\{T_n\}_{n\in\mathbb{Z}}$ of Example 3.1.2 is called a homogeneous Poisson process (HPP) with *intensity* $\lambda > 0$ if its counting process has the following properties:

(α) For all $k \in \mathbb{N}_+$, all mutually disjoint intervals $I_j = (a_j, b_j]$ $(1 \leq j \leq k)$, the random variables $N(I_j)$ $(1 \leq j \leq k)$ are independent.

(β) For any interval $(a, b] \subset \mathbb{R}_+$, $N((a, b])$ is a Poisson random variable with mean $\lambda(b - a)$.

Therefore, for all $k \geq 0$,

$$P(N((a, b]) = k) = e^{-\lambda(b-a)} \frac{[\lambda(b-a)]^k}{k!},$$

and in particular, $E[N((a, b])] = \lambda(b-a)$. In this sense, λ is the average density of points.

The counting process $\{N(t)\}_{t \in \mathbb{R}_+}$ is a continuous-time homogeneous Markov chain (Exercise 3.5.1).

We recall the following result. The inter-event sequence $\{S_n\}_{n \geq 1}$ defined by $S_1 := T_1$ and for $n \geq 2$, by $S_n = T_n - T_{n-1}$, is IID (independent and identically distributed) with exponential distribution of parameter λ:

$$P(S_n \leq t) = 1 - e^{-\lambda t}.$$

In particular
$$E[S_n] = \lambda^{-1},$$

that is, the average number of events per unit of time equals the inverse of the average interevent time.

Next example is a specific instance of Example 3.1.3 showing how to construct a stochastic process from a point process and a sequence of random variables.

Example 3.1.7 (The uniform Markov chain, I) Let $\{\widehat{X}_n\}_{n \in \mathbb{N}}$ be a discrete-time HMC with countable state space E and transition matrix $\mathbf{K} = \{k_{ij}\}_{i,j \in E}$ and let N be an HPP on $\mathbb{R}_+$ with intensity $\lambda > 0$ and associated time sequence $\{T_n\}_{n \in \mathbb{N}_+}$. Suppose that $\{\widehat{X}_n\}_{n \in \mathbb{N}}$ and N are independent. The stochastic process with values in E defined by

$$X(t) = \widehat{X}_{N(t)}$$

is called a *uniform Markov chain*. The Poisson process N is called the *clock*, and the chain $\{\widehat{X}_n\}_{n \in \mathbb{N}}$ is called the *subordinated chain*. Observe that $X(T_n) = \widehat{X}_n$ for all $n \in \mathbb{N}$. Observe also that the discontinuity times of the uniform chain are all event times of N but that not all event times of N are discontinuity times, since it may well occur that $\widehat{X}_{n-1} = \widehat{X}_n$ (a transition of type $i \to i$ of the subordinated chain).

The process $\{X(t)\}_{t \in \mathbb{R}_+}$ is a continuous-time HMC (Exercise 3.5.2). Its transition semigroup is given by

$$\mathbf{P}(t) = \sum_{n=0}^{\infty} e^{-\lambda t} \frac{(\lambda t)^n}{n!} \mathbf{K}^n. \tag{3.4}$$

In fact,

$$p_{ij}(t) = P(X(t) = j | X(0) = i) = P(\widehat{X}_{N(t)} = j | X(0) = i)$$

$$= \sum_{n=0}^{\infty} P(\widehat{X}_n = j, N(t) = n | X(0) = i)$$

$$= \sum_{n=0}^{\infty} P(\widehat{X}_n = j | X(0) = i) P(N(t) = n) = \sum_{n=0}^{\infty} e^{-\lambda t} \frac{(\lambda t)^n}{n!} k_{ij}(n).$$

Kolmogorov's Theorem

In the above examples, we have assumed the existence of a process that is a homogeneous Markov chain. Does there exist such a process? In other terms, can one construct a probability space $(\Omega, \mathcal{F}, P)$ and, for instance, a family $\{X_n\}_{n \geq 0}$ of E-valued random variables on it, such that the resulting process is a homogeneous Markov chain with given stochastic matrix P as transition matrix and prescribed initial distribution ν_0? or equivalently with finite-dimensional distribution given by (3.2)? This is answered by Kolmogorov's theorem guaranteeing the existence of a stochastic process with prescribed finite-dimensional distribution. In order to state Kolmogorov's theorem, we need the notion of a *canonical space*.

Let $E^{\mathbf{T}}$ be the collection of all functions $x : \mathbf{T} \to E$. An element $x \in E^{\mathbf{T}}$ is therefore a function from $\mathbf{T}$ to E:

$$x = (x(t), t \in \mathbf{T}).$$

Let $\mathcal{E}^{\otimes \mathbf{T}}$ be the smallest sigma-field containing all the sets of the form

$$\{x \in E^{\mathbf{T}} ; x(t) \in C\}$$

where t ranges over $\mathbf{T}$ and C ranges over $\mathcal{E}$. The measurable space $(E^{\mathbf{T}}, \mathcal{E}^{\otimes \mathbf{T}})$ so defined is called the *canonical (measurable) space* of stochastic processes indexed by $\mathbf{T}$ with values in $(E, \mathcal{E})$ (we say: "with values in E" if the choice of the sigma-field on this space is clear in the given context). Denote by π_t the *coordinate application* at $t \in \mathbf{T}$, that is the mapping from $E^{\mathbf{T}}$ to E defined by

$$\pi_t(x) = x(t).$$

It associates with the function x its value at t. This π_t is a random variable since when $C \in \mathcal{B}(\mathbb{R})$, the set $\{x ; \pi_t(x) \in C\} = \{x ; x(t) \in C\}$ belongs to $\mathcal{E}^{\otimes \mathbf{T}}$, by definition of the latter. The family $\{\pi_t\}_{t \in \mathbf{T}}$ is called the *coordinate process* on the (canonical) measurable space $(E^{\mathbf{T}}, \mathcal{E}^{\otimes \mathbf{T}})$.

Recall the definition of a *Polish space*.[1] It is a topological space whose topology is metrizable (generated by some distance function), complete (for this distance) and separable (there exists a countable dense subset). Assume that E is a Polish space, and let $\mathcal{E}$ be its Borel sigma-field (the sigma-field generated by the open sets).

---

[1] For all applications in this book, it will be just some $\mathbb{R}^m$ with the Euclidean topology.

Theorem 3.1.1 *Let* $\mathcal{Q} = \{Q_{(t_1,\ldots,t_k)}; k \geq 1, (t_1,\ldots,t_k) \in \mathbf{T}^k\}$ *be a family of probability distributions (more precisely:* $Q_{(t_1,\ldots,t_k)}$ *is a probability distribution on* $(E^k, \mathcal{E}^k)$*) satisfying the compatibility conditions C1 and C2. Then there exists one and only one probability* $\mathcal{P}$ *on the canonical measurable space* $(E^{\mathbf{T}}, \mathcal{E}^{\otimes \mathbf{T}})$ *such that the coordinate process* $\{\pi_t\}_{t \in \mathbf{T}}$ *admits the finite distribution* $\mathcal{Q}$.

This result is the *Kolmogorov existence and uniqueness theorem*. We will admit it without proof.[2]

Example 3.1.8 (IID sequences) Take $E = \mathbb{R}$, $\mathbf{T} = \mathbb{Z}$, and let the *fidi* distributions $\mathcal{Q}_{t_1,\ldots,t_k}$ be of the form

$$\mathcal{Q}_{(t_1,\ldots,t_k)} = Q_{t_1} \times \cdots \times Q_{t_k}$$

where for each $t \in \mathbf{T}$, Q_t is a probability distribution on $(E, \mathcal{E})$. This collection of finite dimensional distributions obviously satisfies the compatibility conditions, and the resulting coordinate process is an independent random sequence indexed by the relative integers. It is an IID (that is: independent and identically distributed) sequence if $Q_t = Q$ for all $t \in \mathbf{T}$.

Example 3.1.9 (Discrete-time HMC, II) For memory (see the discussion before Kolmogorov's theorem): Let $\mathbf{P}$ be a stochastic matrix on the countable state space E and let ν_0 be a probability distribution on E. There exists a process $\{X_n\}_{n \geq 0}$ that is a homogeneous Markov chain with transition matrix $\mathbf{P}$ and initial distribution ν_0.

Example 3.1.10 (Continuous-time HMC, II) A result similar to that of the previous example holds true for continuous-time homogeneous Markov chains: there exists such chain with given initial distribution and transition semi-group.

Transport in a Canonical Space

Let be given on the probability space $(\Omega, \mathcal{F}, P)$ a stochastic process $\{X(t)\}_{t \in \mathbf{T}}$ with values in a Polish space E. We shall need an avatar of this stochastic process that lives on a canonical space, and has the same finite-dimensional distribution. For this, define the mapping $h : (\Omega, \mathcal{F}) \to (E^{\mathbf{T}}, \mathcal{E}^{\otimes \mathbf{T}})$ by

$$h(\omega) = (X(t, \omega), t \in \mathbf{T}).$$

This mapping is measurable. To show this, it is enough to verify that $h^{-1}(C) \in \mathcal{F}$ for all $C \in \mathcal{C}$, where $\mathcal{C}$ is a collection of subsets of $E^{\mathbf{T}}$ that generates $\mathcal{E}^{\otimes \mathbf{T}}$, for instance, the collection of sets of the form $C = \{x; x(t) \in A\}$ for some $t \in \mathbf{T}$ and $A \in \mathcal{E}$. But $h^{-1}(\{x; x(t) \in A\}) = (\{\omega; X(t, \omega) \in A\}) \in \mathcal{F}$ since $X(t)$ is a random variable. Denote by $\mathcal{P}_X$ the image of P by h. The fidi distribution with respect to $\mathcal{P}_X$ of the coordinate process of the canonical measurable space is the same as that of the original stochastic process.

---

[2] For the case $E = \mathbb{R}^m$, which is all we need in the present book, see Theorem 36.1 of (Billingsley 1995).

Definition 3.1.3 The probability $\mathcal{P}_X$ on $(E^{\mathbf{T}}, \mathcal{E}^{\otimes \mathbf{T}})$ is called the *distribution* of $\{X(t)\}_{t \in \mathbf{T}}$.

An immediate consequence of the uniqueness part of Kolmogorov's theorem is:

Theorem 3.1.2 *Two stochastic processes with the same fidi distribution have the same distribution.*

Independence of Stochastic Processes

Let $\{X(t)\}_{t \in \mathbf{T}}$ be a stochastic process. We define the sigma-field

$$\mathcal{F}^X = \sigma(X(t); t \in \mathbf{T})$$

recording all the events relative to this process. (By definition, it is the smallest sigma-field on Ω that makes all mappings $\omega \to X(t, \omega)$, $t \in \mathbf{T}$, measurable; or, equivalently, the smallest sigma-field that contains all the sets $\{\omega ; X(t, \omega) \in C\}$, $t \in \mathbf{T}, C \in \mathcal{E}$.)

Definition 3.1.4 Two stochastic processes $\{X(t)\}_{t \in \mathbf{T}}$ and $\{Y(t)\}_{t \in \mathbf{T}'}$ defined on the same probability space, with values in $(E, \mathcal{E})$ and $(E', \mathcal{E}')$ respectively, are called *independent* if the sigma-fields $\mathcal{F}^X$ and $\mathcal{F}^Y$ are independent.

The verification of independence is simplified by the following result.

Theorem 3.1.3 *Two stochastic processes $\{X(t)\}_{t \in \mathbf{T}}$ and $\{Y(t)\}_{t \in \mathbf{T}'}$ defined on the same probability space, with values in $(E, \mathcal{E})$ and $(E', \mathcal{E}')$ repectively, are independent if for all $t_1, \ldots, t_k \in \mathbf{T}$, and all $s_1, \ldots, s_\ell \in \mathbf{T}'$ the vectors $(X(t_1), \ldots, X(t_k))$ and $(Y(s_1), \ldots, Y(s_\ell))$ are independent.*

Proof The collection of events of the type $\{X(t_1) \in C_1, \ldots, X(t_k) \in C_k\}$, where the C_i's are in $\mathcal{E}$, is a collection of sets closed by finite intersections and generating $\mathcal{F}^X$, with a similar observation for $\mathcal{F}^Y$. The result follows from these observations and Theorem A.2.7. □

Stationarity

In this subsection, the index set is assumed to be one of the following: $\mathbb{N}^d$, $\mathbb{Z}^d$, $(\mathbb{R}_+)^d$ or $\mathbb{R}^d$, for some integer $d \geq 1$.

Definition 3.1.5 A stochastic process $\{X(t)\}_{t \in \mathbf{T}}$ is called (strictly) *stationary* if for all $k \geq 1$, all $(t_1, \ldots, t_k) \in \mathbf{T}^k$, the probability distribution of the random vector

$$(X(t_1 + a), \ldots, X(t_k + a))$$

is independent of $a \in \mathbf{T}$ such that $t_1 + a, \ldots, t_k + a \in \mathbf{T}$.

Example 3.1.11 (Discrete-time HMC, III) Consider the HMC of Example 3.1.4. A probability distribution π satisfying the *global balance equation*

$$\pi^T = \pi^T \mathbf{P}$$

is called a *stationary distribution* of the transition matrix $\mathbf{P}$, or of the corresponding HMC. The global balance equation says that for all states i,

$$\pi(i) = \sum_{j \in E} \pi(j) p_{ji}.$$

Iteration of the global balance equation gives $\pi^T = \pi^T \mathbf{P}^n$ for all $n \geq 0$, and therefore, in view of (3.1), if the initial distribution $\nu = \pi$, then $\nu_n = \pi$ for all $n \geq 0$. Thus, if a chain is started with a stationary distribution, it keeps the same distribution forever. But there is more, because then,

$$P(X_n = i_0, X_{n+1} = i_1, \ldots, X_{n+k} = i_k) = P(X_n = i_0) p_{i_0 i_1} \cdots p_{i_{k-1} i_k}$$
$$= \pi(i_0) p_{i_0 i_1} \cdots p_{i_{k-1} i_k}$$

does not depend on n. In this sense, the chain is *stationary*. In summary: A HMC whose initial distribution is a stationary distribution is stationary.

Example 3.1.12 (Continuous-time HMC, III) Consider the continuous-time HMC of Example 3.1.5. The probability distribution π on E is called a *stationary distribution of the continuous-time* HMC, or of its transition semi-group, if for all $t \geq 0$,

$$\pi^T \mathbf{P}(t) = \pi^T.$$

From (3.3), we see that if the initial distribution of the chain is a stationary distribution, then the distribution at any time $t \geq 0$ is π, and moreover, the chain is stationary, since for all $k \geq 1$, all $0 \leq t_1 < \ldots < t_k$, and all $i_1, \ldots, i_k \in E$, the quantity

$$P(X(t_1 + a) = i_1, \ldots, X(t_k + a) = i_k) = \pi(i_1) p_{i_1, i_2}(t_2 - t_1) \cdots p_{i_{k-1}, i_k}(t_k - t_{k-1})$$

does not depend on $a \geq 0$. Therefore, a continuous-time HMC having for initial distribution a stationary distribution of the transition semi-group is stationary.

Example 3.1.13 (The uniform Markov chain, II) Let $\{X(t)\}_{t \in \mathbb{R}_+}$ be the uniform Markov chain of Example 3.1.7. If π is a stationary distribution of the subordinated chain, $\pi^T \mathbf{K}^n = \pi^T$, and therefore in view of (3.4), $\pi^T \mathbf{P}(t) = \pi^T$. Conversely, if π is a stationary distribution of the continuous-time HMC, then, by (3.4),

$$\pi^T = \sum_{n=0}^{\infty} e^{-\lambda t} \frac{(\lambda t)^n}{n!} \pi^T \mathbf{K}^n,$$

and letting $t \downarrow 0$, we obtain $\pi^T = \pi^T \mathbf{K}$.

Example 3.1.14 (Discrete-time HMC, IV) Most Markov chains arising in signal processing have a finite state space. The situation with respect to the existence and uniqueness of a stationary distribution is then very simple. We first need the notion of irreducibility. State j is said to be *accessible* from state i if there exists $M \geq 0$ such that $p_{ij}(M) > 0$. In particular, a state i is always accessible from itself, since $p_{ii}(0) = 1$. States i and j are said to *communicate* if i is accessible from j *and* j is accessible from i, and this is denoted by $i \leftrightarrow j$. For $M \geq 1$, $p_{ij}(M) = \sum_{i_1,\ldots,i_{M-1}} p_{ii_1} \cdots p_{i_{M-1}j}$, and therefore $p_{ij}(M) > 0$ if and only if there exists at least one path $i, i_1, \ldots, i_{M-1}, j$ from i to j such that

$$p_{ii_1} p_{i_1 i_2} \cdots p_{i_{M-1}j} > 0,$$

or, equivalently, if there is an oriented path from i to j in the transition graph G. Clearly, the communication relation ($\leftrightarrow$) is an equivalence relation. It generates a partition of the state space E into disjoint equivalence classes called *communication classes*. If there exists only one communication class, then the chain, its transition matrix and its transition graph are said to be *irreducible*.

We can now recall the following classical: An irreducible HMC with a finite state space has a unique stationary probability distribution.[3]

Wide-Sense Stationarity

A notion weaker than strict stationarity concerns second-order processes with values in $E = \mathbb{C}$, whose definition is the following:

Definition 3.1.6 A complex stochastic process $\{X(t)\}_{t \in \mathbf{T}}$ satisfying the condition

$$E[|X(t)|^2] < \infty$$

for all $t \in \mathbf{T}$ is called a *second-order* stochastic process.

In other words, for all $t \in \mathbf{T}$, $X(t) \in L^2_{\mathbb{C}}(P)$. Recall that $L^2_{\mathbb{C}}(P)$ is a Hilbert space with scalar field $\mathbb{C}$ and associated inner product

$$\langle X, Y \rangle = E\left[XY^*\right].$$

The subset of $L^2_{\mathbb{C}}(P)$ consisting of the *real* random variables thereof is a Hilbert space with scalar field $\mathbb{R}$ denoted by $L^2_{\mathbb{R}}(P)$. Recall Schwarz's inequality: For all $X, Y \in L^2_{\mathbb{C}}(P)$

$$E\left[|X||Y|\right] \leq E\left[|X|^2\right]^{\frac{1}{2}} E\left[|Y|^2\right]^{\frac{1}{2}}. \tag{3.5}$$

---

[3] See for instance (Brémaud 1999), *Markov chains*, Theorem 3.3.

The *mean* function $m : \mathbf{T} \to \mathbb{C}$ and the *covariance* function $\Gamma : \mathbf{T}^2 \to \mathbb{C}$ of a second order process are well-defined by

$$m(t) = E[X(t)]$$

and

$$\Gamma(t, s) = \operatorname{cov}(X(t), X(s)) = E[X(t)X(s)^*] - m(t)m(s)^*.$$

When the mean function is the null function, the stochastic process is said to be *centered*.

Theorem 3.1.4 *Let $\{X(t)\}_{t \in \mathbf{T}}$ be a second-order stochastic process with mean function m and covariance function Γ. Then*

$$E\left[|X(t) - m(t)|\right] \le \Gamma(t, t)^{\frac{1}{2}}$$

and

$$|\Gamma(t, s)| \le \Gamma(t, t)^{\frac{1}{2}} \Gamma(s, s)^{\frac{1}{2}}.$$

Proof Apply Schwarz's inequality (3.5) with $X := X(t) - m(t)$ and $Y := 1$ for the first inequality, and $X := X(t) - m(t)$ and $Y := X(s) - m(s)$ for the second one. $\square$

For a stationary second–order stochastic process, for all $s, t \in \mathbf{T}$

$$m(t) \equiv m \tag{3.6}$$

where $m \in \mathbb{C}$ and

$$\Gamma(t, s) = C(t - s), \tag{3.7}$$

for some function $C : \mathbf{T} \to \mathbb{C}$, also called the *covariance function* of the process. The complex number m is called the *mean* of the process.

Definition 3.1.7 A complex stochastic process $\{X(t)\}_{t \in \mathbf{T}}$ is called *wide-sense stationary* if conditions (3.6) and (3.7) are satisfied for all $s, t \in \mathbf{T}$.

Note that $C(0)$ is the variance σ_X^2 of any of the random variables $X(t)$.

Exercise 3.5.3 asks to find a simple example of a discrete-time stochastic proces that is wide-sense stationary process that is not strictly stationary.

An immediate corollary of Theorem 3.1.4 is the following.

Corollary 3.1.1 *Let $\{X(t)\}_{t \in \mathbf{T}}$ be a wide-sense stationary stochastic process with mean m and covariance function C. Then*

$$E\left[|X(t) - m|\right] \le C(0)^{\frac{1}{2}} \text{ and } |C(\tau)| \le C(0).$$

3.1.2 Measurability and Sample Path Properties

It is often convenient to view a stochastic process as a *mapping* $X : \mathbf{T} \times \Omega \to E$, defined by

$$(t, \omega) \to X(t, \omega). \qquad\qquad (*)$$

Definition 3.1.8 The stochastic process $\{X(t)\}_{t \in \mathbb{R}}$ is said to be *measurable* iff the mapping $(*)$ is measurable with respect to the sigma-fields $\mathcal{B}(\mathbb{R}) \otimes \mathcal{F}$ and $\mathcal{E}$.

In particular, for any $\omega \in \Omega$ the mapping $t \to X(t, \omega)$ is measurable with respect to the sigma-fields $\mathcal{B}(\mathbb{R})$ and $\mathcal{E}$. Also, if $E = \mathbb{R}$ and if $X(t)$ is non-negative, one can define the Lebesgue integrals $\int_{\mathbb{R}} X(t, \omega) \, dt$ for each $\omega \in \Omega$, and also apply Tonelli's theorem to obtain

$$E\left[\int_{\mathbb{R}} X(t) dt \right] = \int_{\mathbb{R}} E\left[X(t) \right] dt.$$

By Fubini's theorem, the last equality also holds true for measurable stochastic processes of arbitrary sign satisfying

$$\int_{\mathbb{R}} E\left[|X(t)| \right] dt < \infty.$$

Example 3.1.15 A real-valued right-continuous (*resp.* left-continuous) stochastic process is measurable.

Proof For all $n \geq 0$ and all $t \in \mathbb{R}$, define $X^{(n)}(t)$ by

$$X^{(n)}(t) = X((k+1)t/2^{-n}) \text{ if } t \in [(k)2^{-n}, (k+1)2^{-n}), \ k \in \mathbb{Z}.$$

The stochastic process $\{X^{(n)}(t)\}_{t \in \mathbb{R}}$ is clearly measurable. If $\{X(t)\}_{t \in \mathbb{R}}$ is right-continuous, $X(t, \omega)$ is the limit of $X^{(n)}(t, \omega)$ for all $(t, \omega) \in \mathbb{R} \times \Omega$, and therefore the mapping $(t, \omega) \to X(t, \omega)$ is measurable with respect to $\mathcal{B}(\mathbb{R}) \otimes \mathcal{F}$ and $\mathcal{E}$ as a pointwise limit of the measurable mappings $(t, \omega) \to X^{(n)}(t, \omega)$. The left-continuous case is treated in a similar way. $\qquad\square$

Theorem 3.1.5 *Let $\{X(t)\}_{t \in \mathbb{R}}$ be a second-order complex-valued measurable stochastic process with mean function m and covariance function Γ. Let $f : \mathbb{R} \to \mathbb{C}$ be such that*

$$\int_{\mathbb{R}} |f(t)| |m(t)| \, dt < \infty.$$

Then the integral $\int_{\mathbb{R}} f(t)X(t)\,dt$ is almost-surely well-defined and

$$E\left[\int_{\mathbb{R}} f(t)X(t)\,dt\right] = \int_{\mathbb{R}} f(t)m(t)\,dt.$$

Suppose in addition that f satisfies the condition

$$\int_{\mathbb{R}} |f(t)||\Gamma(t,t)|^{\frac{1}{2}}\,dt < \infty$$

and let $g : \mathbb{R} \to \mathbb{C}$ be a function with the same properties as f. Then $\int_{\mathbb{R}} f(t)X(t)\,dt$ is square-integrable and

$$cov\left(\int_{\mathbb{R}} f(t)X(t)\,dt, \int_{\mathbb{R}} g(t)X(t)\,dt\right) = \int_{\mathbb{R}}\int_{\mathbb{R}} f(t)g^*(s)\Gamma(t,s)\,dt\,ds.$$

Proof By Tonelli's theorem

$$E\left[\int_{\mathbb{R}} |f(t)||X(t)|\,dt\right] = \int_{\mathbb{R}} |f(t)||m(t)|\,dt < \infty$$

and therefore, almost surely $\int_{\mathbb{R}} |f(t)||X(t)|\,dt < \infty$, so that almost surely the integral $\int_{\mathbb{R}} f(t)X(t)\,dt$ is well-defined and finite. Also (Fubini)

$$E\left[\int_{\mathbb{R}} f(t)X(t)\,dt\right] = \int_{\mathbb{R}} E\left[f(t)X(t)\right]\,dt = \int_{\mathbb{R}} f(t)E\left[X(t)\right]\,dt.$$

We now suppose (without loss of generality) that the process is centered. By Tonelli's theorem

$$E\left[\left(\int_{\mathbb{R}} |f(t)||X(t)|\,dt\right)\left(\int_{\mathbb{R}} |g(t)||X(t)|\,dt\right)\right]$$
$$= \int_{\mathbb{R}}\int_{\mathbb{R}} |f(t)||g(s)|E\left[|X(t)||X(s)|\right]\,dt\,ds.$$

But (Schwarz) $E\left[|X(t)||X(s)|\right] \le \Gamma(t,t)|^{\frac{1}{2}}\Gamma(s,s)|^{\frac{1}{2}}$, and therefore the right-hand side of the last equality is bounded by

$$\left(\int_{\mathbb{R}} |f(t)|\Gamma(t,t)|^{\frac{1}{2}}\, dt\right)\left(\int_{\mathbb{R}} |g(s)|\Gamma(s,s)|^{\frac{1}{2}}\, ds\right) < \infty.$$

We can therefore apply Fubini's theorem to obtain

$$E\left[\left(\int_{\mathbb{R}} f(t)X(t)\, dt\right)\left(\int_{\mathbb{R}} g(t)X(t)\, dt\right)\right] = \int_{\mathbb{R}}\int_{\mathbb{R}} f(t)g^*(s)E\left[X(t)X(s)\right]\, dt\, ds.$$

$\square$

A stochastic process can be regarded as a *random function*. For fixed $\omega \in \Omega$, we can then discuss the continuity properties of the sample paths of the process. (The state space E is assumed to be a topological space endowed with its Borel sigma-field $\mathcal{E}$, that is, the sigma-field generated by the open sets.) For example, with $\mathbf{T} = \mathbb{R}$, if for all $\omega \in \Omega$ the ω-sample path is right-continuous, we call this stochastic process right-continuous. It is called P–a.s. right-continuous if the ω-sample paths are right-continuous for all $\omega \in \Omega$, except perhaps for $\omega \in N$, where N is a P-negligible set. One defines similarly (P–a.s.) left-continuity, (P–a.s.) continuity, etc.

Note that Kolmogorov's existence and uniqueness theorem of a (canonical) stochastic process with given finite-dimensional distribution says nothing about the properties of the sample paths such as, for instance, continuity or differentiability. It is therefore useful to find conditions bearing only on the finite-dimensional distributions and guaranteeing that the sample paths have the desired properties. This is in general not feasible but, in certain cases, a reasonable modification (that is, a modification that preserves the finite-dimensional distributions and therefore the distribution) possesses the desired properties.[4]

Definition 3.1.9 Two stochastic processes $\{X(t)\}_{t\in\mathbf{T}}$ and $\{Y(t)\}_{t\in\mathbf{T}}$ defined on the same probability space $(\Omega, \mathcal{F}, P)$ are *modifications* or *versions* of one another iff

$$P\left(\{\omega;\ X(t) \ne Y(t)(\omega)\}\right) = 0 \text{ for all } t \in \mathbf{T}.$$

They are said to be P-*undistinguishable* if

$$P\left(\{\omega;\ X(t,\omega) = Y(t,\omega) \text{ for all } t \in \mathbf{T}\}\right) = 1,$$

that is, if they have identical trajectories except on a $P - null$ set.

Clearly, two undistinguishable processes are modifications of one another.

---

[4] The classic text of H. Cramér and M. R. Leadbetter, *Stationary and Related Stochastic Processes*, Wiley, New York, 1967, contains results in this vein, p. 63 and following.

3.2 Gaussian Processes

3.2.1 Gaussian Subspaces

Let **T** be an arbitrary index.

Definition 3.2.1 The *real*-valued stochastic process $\{X(t)\}_{t\in\mathbf{T}}$ is called a *Gaussian process* if for all $n \geq 1$ and for all $t_1, \ldots, t_n \in \mathbf{T}$, the random vector $(X(t_1), \ldots, X(t_n))$ is Gaussian.

In particular, its characteristic function is given by the formula

$$E\left[\exp\left\{i\sum_{j=1}^{n}u_jX(t_j)\right\}\right] = \exp\left\{i\sum_{j=1}^{n}u_jm(t_j) - \frac{1}{2}\sum_{j=1}^{n}\sum_{k=1}^{n}u_ju_k\Gamma(t_j,t_k)\right\},$$
(3.8)

where $u_1, \ldots, u_n \in \mathbb{R}$ and where m and Γ are the mean and covariance functions respectively.

Theorem 3.2.1 *Let* $\Gamma : \mathbf{T}^2 \to \mathbb{R}$ *be a non-negative definite function, that is, such that for all* $t_1, \ldots, t_k \in \mathbf{T}$, *all* $u_1, \ldots, u_k \in \mathbb{R}'$

$$\sum_{i=1}^{k}\sum_{j=1}^{k}u_iu_j\Gamma(t_i,t_j) \geq 0.$$

Then, there exists a centered Gaussian process with covariance function Γ.

Proof By Theorem 2.1.7, there exists a centered Gaussian vector with covariance matrix $\{\Gamma(t_i,t_j)\}_{1\leq i,j\leq k}$. Let $Q_{t_1,\ldots,t_k}$ be the probability distribution of this vector. The family $\{Q_{t_1,\ldots,t_k}; t_1, \ldots, t_k \in \mathbf{T}\}$ is obviously compatible, and therefore a centered Gaussian process with covariance function Γ exists (Kolmogorov's existence Theorem 3.1.1). □

Theorem 3.2.2 *For a Gaussian process with index set* $\mathbf{T} = \mathbb{R}$ *or* $\mathbb{Z}$ *to be stationary, it is necessary and sufficient that* $m(t) = m$ *and* $\Gamma(t,s) = C(t-s)$ *for all* $s, t \in \mathbf{T}$.

Proof The necessity is obvious, whereas the sufficiency is proven by replacing the t_ℓ's in (3.8) by $t_\ell+h$ to obtain the characteristic function of $(X(t_1+h), \ldots, X(t_n+h))$, namely,

$$\exp\left\{i\sum_{j=1}^{n}u_jm - \frac{1}{2}\sum_{j=1}^{n}\sum_{k=1}^{n}u_ju_kC(t_j-t_k)\right\},$$

and then observing that this quantity is independent of h. □

Example 3.2.1 (Clipped Gaussian process, I) Let $\{X(t)\}_{t\in\mathbb{R}}$ be a centered stationary Gaussian process with covariance function $C_X(\tau)$. Define the *clipped* (or *hard-limited*) process

$$Y(t) = \text{sign } X(t),$$

with the convention sign $X(t) = 0$ if $X(t) = 0$ (note however that this occurs with null probability if $C_X(0) = \sigma_X^2 > 0$, which is henceforth assumed). Clearly this stochastic process is centered. Moreover, it is unchanged when $\{X(t)\}_{t\in\mathbb{R}}$ is multiplied by a positive constant. In particular, we may assume that the variance $C_X(0)$ equals 1, so that the covariance matrix of the vector $(X(0), X(\tau))^T$ is

$$\Gamma(\tau) = \begin{pmatrix} 1 & \rho_X(\tau) \\ \rho_X(\tau) & 1 \end{pmatrix},$$

where $\rho_X(\tau)$ is the correlation coefficient of $X(0)$ and $X(\tau)$. We assume that $\Gamma(\tau)$ is invertible, that is, $|\rho_X(\tau)| < 1$.

We then have the *Van Vleck–Middleton formula*:

$$C_Y(\tau) = \frac{2}{\pi} \sin^{-1}\left(\frac{C_X(\tau)}{C_X(0)}\right).$$

Proof Since for each t the random variable $Y(t)$ takes the values ± 1 and 0, the latter with null probability, we can express the autocovariance function of the clipped process as

$$C_Y(\tau) = 2\{P(X(0) > 0, X(\tau) > 0) - P(X(0) > 0, X(\tau) < 0)\},$$

where it was noted that

$$P(X(0) < 0, X(\tau) < 0) = P(X(0) > 0, X(\tau) > 0)$$

and that

$$P(X(0) < 0, X(\tau) > 0) = P(X(0) > 0, X(\tau) < 0).$$

The result then follows from that of Exercise 2.2.1 with $\rho = \rho_X(\tau)$. $\qquad\square$

Definition 3.2.2 Let $\{X_i\}_{i\in I}$ be an arbitrary collection of complex (*resp.*, real) random variables in $L_{\mathbb{C}}^2(P)$ (*resp.*, $L_{\mathbb{R}}^2(P)$). The Hilbert subspace of $L_{\mathbb{C}}^2(P)$ (*resp.*, $L_{\mathbb{R}}^2(P)$) consisting of the closure of the vector space of finite linear complex (*resp.*, real) combinations of elements of $\{X_i\}_{i\in I}$, is called the complex (*resp.*, real) *Hilbert subspace generated by* $\{X_i\}_{i\in I}$, and is denoted by $H_{\mathbb{C}}(X_i, i \in I)$ (*resp.*, $H_{\mathbb{R}}(X_i, i \in I)$).

More explicitly, in the complex case for instance: the Hilbert subspace $H_{\mathbb{C}}$ $(X_i, i \in I) \subseteq L^2_{\mathbb{C}}(P)$ consists of all complex square-integrable random variables that are limits in the quadratic mean (that is, limits in $L^2_{\mathbb{C}}(P)$) of some sequence of finite complex linear combinations of elements in the set $\{X_i\}_{i \in I}$.

Definition 3.2.3 A collection $\{X_i\}_{i \in I}$ of real random variables defined on the same probability space, where I is an arbitrary index set, is called a *Gaussian family* if for all finite set of indices $i_1, \ldots, i_k \in I$, the random vector $(X_{i_1}, \ldots, X_{i_k})$ is Gaussian. A Hilbert subspace G of the real Hilbert space $L^2_{\mathbb{R}}(P)$ is called a *Gaussian (Hilbert) subspace* if it is a Gaussian family.

Theorem 3.2.3 *Let $\{X_i\}_{i \in I}$ be a Gaussian family of random variables of $L^2_{\mathbb{R}}(P)$. Then the Hilbert subspace $H_{\mathbb{R}}(X_i, i \in I)$ generated by $\{X_i\}_{i \in I}$ is a Gaussian subspace of $L^2_{\mathbb{R}}(P)$.*

Proof By definition, the Hilbert subspace $H_{\mathbb{R}}(X_i, i \in I)$ consists of all the random variables in $L^2_{\mathbb{R}}(P)$ that are limits in quadratic mean of finite linear combinations of elements of the family $\{X_i\}_{i \in I}$. The result follows from that in Example 2.2.1.   □

3.2.2 Brownian Motion

Definition 3.2.4 By definition, a *standard Brownian motion*, or *standard Wiener process*, is a continuous centered Gaussian process $\{W(t)\}_{t \in \mathbb{R}}$ with independent increments, such that $W(0) = 0$, and such that for any interval $[a, b] \subset \mathbb{R}$, the variance of $W(b) - W(a)$ is equal to $b - a$.

In particular, the vector $(W(t_1), \ldots, W(t_k))$ with $0 < t_1 < \ldots < t_k$ admits the probability density function

$$\frac{1}{(\sqrt{2\pi})^k \sqrt{t_1 (t_2 - t_1) \cdots (t_k - t_{k-1})}} e^{-\frac{1}{2}\left(\frac{x_1^2}{t_1} + \frac{(x_1+x_2)^2}{t_2-t_1} + \cdots + \frac{(x_1+\cdots+x_k)^2}{t_k-t_{k-1}}\right)}.$$

Note for future reference that for $s, t \in \mathbb{R}_+$,

$$E[W(t)W(s)] = t \wedge s. \tag{3.9}$$

In fact, for $0 \leq s \leq t$,

$$\begin{aligned} E[W(t)W(s)] &= E[(W(t) - W(s))W(s)] + E[W(s)^2] \\ &= E[(W(t) - W(s))(W(s) - W(0))] + E[(W(s) - W(0))^2] \\ &= 0 + s = t \wedge s. \end{aligned}$$

The existence of a process with such finite distributions on a canonical space is guaranteed by Kolmogorov's theorem (the verification of compatibility of the above collection of probability distributions is immediate). But this theorem says nothing about the continuity of the trajectories. However:

Theorem 3.2.4 *Consider a stochastic process such as the one in Definition 3.2.4, except that continuity of the trajectories is not assumed. There exists a version of it that has continuous paths.*

Therefore, from now on, we shall assume without loss of generality that the trajectories of the Wiener process are continuous.

Definition 3.2.4 of the Wiener process does not tell much about the qualitative behavior of this process. Although the trajectories of the Brownian motion are continuous functions, their behaviour is rather chaotic. First of all we observe that, for fixed $t_0 > 0$, the random variable

$$\frac{W(t_0 + h) - W(t_0)}{h} \sim \mathcal{N}\left(0, h^{-1}\right)$$

and therefore it cannot converge in distribution as $h \downarrow 0$ since the limit of its characteristic function is the null function, which is not a characteristic function. In particular, it does not converge almost-surely to any random variable. Therefore, for any $t_0 > 0$,

$$P\left(t \to W(t) \text{ is not differentiable at } t_0\right) = 1.$$

But the situation is even more dramatic:

Theorem 3.2.5 *Almost all the paths of the Wiener process are nowhere differentiable.*

We shall not prove this result here, but state one of its consequences.

Corollary 3.2.1 *Almost all the paths of the Wiener process are of unbounded variation on finite intervals.*

Proof This is because any function of bounded variation is differentiable almost-everywhere[5] (with respect to the Lebesgue measure). □

3.2.3 The Wiener–Doob Integral

The *Doob stochastic integral*, a special case of which is the *Wiener stochastic integral*

$$\int_{\mathbb{R}} f(t)\, dW(t) \qquad (*)$$

---

[5] Rudin (1986), Theorem 8.19.

that we proceed to define for a certain class of measurable functions f is not of
the usual types. For instance, it cannot be defined pathwise as a Stieltjes-Lebesgue
integral since the trajectories of the Brownian motion are of unbounded variation
(Corollary 3.2.1). This integral cannot either be interpreted as $\int_{\mathbb{R}} f(t) \dot{W}(t) \, dt$ (where
the dot denotes derivation), since the the Brownian motion sample paths are not
differentiable (Theorem 3.2.5).

The integral in $(*)$ will therefore be defined in a radically different way. In fact,
the *Doob stochastic integral* will be defined more generally, with respect to a process
with centered and uncorrelated increments.

Definition 3.2.5 Let $\{Z(t)\}_{t\in\mathbb{R}}$ be a complex stochastic process such that for all
intervals $[t_1, t_2] \subset \mathbb{R}$ the increments $Z(t_2) - Z(t_1)$ are in $L_{\mathbb{C}}^2(P)$, centered and such
that for some locally finite measure μ on $(\mathbb{R}, \mathcal{B}(\mathbb{R}))$:

$$E[(Z(t_2) - Z(t_1))(Z(t_4) - Z(t_3))^*] = \mu((t_1, t_2] \cap (t_3, t_4])$$

for all $[t_1, t_2] \subset \mathbb{R}$ and all $[t_3, t_4] \subset \mathbb{R}$. Such stochastic process $\{Z(t)\}_{t\in\mathbb{R}}$ is called a
stochastic process with *centered* and *uncorrelated increments*, and μ is its *structural
measure*.

We shall observe in particular that if the intervals $(t_1, t_2]$ and $(t_3, t_4]$ are disjoint,
$Z(t_2) - Z(t_1)$ and $Z(t_4) - Z(t_3)$ are orthogonal elements of $L_{\mathbb{C}}^2(P)$.

Example 3.2.2 (Wiener process) The Wiener process $\{W(t)\}_{t\in\mathbb{R}}$ is a process with
centered and uncorrelated increments whose structural measure is the Lebesgue
measure.

The *Doob integral* is defined for all integrands $f \in L_{\mathbb{C}}^2(\mu)$ in the following
manner. First of all, we define this integral for all $f \in \mathcal{L}$, the vector subspace
of $L_{\mathbb{C}}^2(\mu)$ formed by the finite complex linear combinations of interval indicator
functions

$$f(t) = \sum_{i=1}^{N} \alpha_i 1_{(a_i, b_i]}(t).$$

For such functions, *by definition*,

$$\int_{\mathbb{R}} f(t) \, dZ(t) := \sum_{i=1}^{N} \alpha_i (Z(b_i) - Z(a_i)).$$

Observe that this random variable belongs to the Hilbert subspace $H_{\mathbb{C}}(Z)$ of $L_{\mathbb{C}}^2(P)$
generated by $\{Z(t)\}_{t\in\mathbb{R}}$. One easily verifies that the linear mapping

$$\varphi : f \in \mathcal{L} \to \int_{\mathbb{R}} f(t) \, dZ(t) \in L_{\mathbb{C}}^2(P)$$

is an isometry, that is,

$$\int_{\mathbb{R}} |f(t)|^2 \, \mu(dt) = E\left[\left|\int_{\mathbb{R}} f(t) \, dZ(t)\right|^2\right].$$

Since $\mathcal{L}$ is a dense subset of $L_{\mathbb{C}}^2(\mu)$ (Corollary A.3.1), φ can be uniquely extended to an isometric linear mapping of $L_{\mathbb{C}}^2(\mu)$ into $H_{\mathbb{C}}(Z)$ (see Sect. 1.3.1). We continue to call this extension φ and then define, for all $f \in L_{\mathbb{C}}^2(\mu)$, the Doob integral of f with respect to $\{Z(t)\}_{t \in \mathbb{R}}$ by

$$\int_{\mathbb{R}} f(t) \, dZ(t) := \varphi(f).$$

The fact that φ is an isometry is expressed by the *Doob isometry formula*

$$E\left[\left(\int_{\mathbb{R}} f(t) \, dZ(t)\right)\left(\int_{\mathbb{R}} g(t) \, dZ(t)\right)^*\right] = \int_{\mathbb{R}} f(t) g^*(t) \, \mu(dt), \qquad (3.10)$$

where f and g are in $L_{\mathbb{C}}^2(\mu)$. Note also that for all $f \in L_{\mathbb{C}}^2(\mu)$:

$$E\left[\int_{\mathbb{R}} f(t) \, dZ(t)\right] = 0, \qquad (3.11)$$

since the Doob integral is the limit in $L_{\mathbb{C}}^2(\mu)$ of random variables of the type $\sum_{i=1}^{N} \alpha_i (Z(b_i) - Z(a_i))$ that have mean 0 (use the continuity of the inner product in $L_{\mathbb{C}}^2(P)$).

Definition 3.2.6 Let $\{W(t)\}_{t \in \mathbb{R}}$ be a standard Wiener process, and let for all $t \in \mathbb{R}$

$$X(t) = (2\alpha)^{\frac{1}{2}} \int_{-\infty}^{t} e^{-\alpha(t-s)} \sigma \, dW(s),$$

where $\alpha > 0$ and $\sigma > 0$ and. The process $\{X(t)\}_{t \in \mathbb{R}}$ defined in this way is called the *Ornstein–Uhlenbeck process*.

Since for all $t \in \mathbb{R}$, $X(t)$ belongs to $H_{\mathbb{R}}(W)$, it is a Gaussian process (Theorem 3.2.3). It is centered, with covariance function

$$\Gamma(t, s) = e^{-\alpha|t-s|},$$

as follows directly from the isometry formula (3.10).

Doob's Integral with Respect to Random Pseudo-measures

Doob's integral can be straightforwardly extended to the spatial case. We shall first define a random pseudo-measure.[6]

Definition 3.2.7 A collection $\{Z(A)\}_{A \in \mathcal{B}_b(\mathbb{R}^m)}$ (where $\mathcal{B}_b(\mathbb{R}^m)$ is the collection of *bounded* Borel sets of $\mathbb{R}^m$) of centered random variables in $L^2_\mathbb{C}(P)$ such that

$$E[Z(A)Z(B)^*] = \mu(A \cap B) \qquad (3.12)$$

for all $A, B \in \mathcal{B}_b(\mathbb{R}^m)$, where μ is a locally finite measure on $(\mathbb{R}^m, \mathcal{B}(\mathbb{R}^m))$, is called a *second-order random pseudo-measure* on $\mathbb{R}^m$ with structural measure μ.

It is called second-order because, in particular, for $A \in \mathcal{B}_b(\mathbb{R}^m)$, $E[|Z(A)|^2] = \mu(A) < \infty$. It has uncorrelated spatial increments in the sense that, if A and B are disjoint bounded Borel sets, then $Z(A)$ and $Z(B)$ are uncorrelated, in view of (3.12), since in this case $\mu(A \cap B) = \mu(\varnothing) = 0$.

Example 3.2.3 (The Brownian pseudo-measure) Let μ be a locally finite measure on $\mathbb{R}^d$ and let $\mathbf{T} = \{A \in \mathcal{B}_b(\mathbb{R}^d)\}$. Define $\Gamma : \mathbf{T}^2 \to \mathbb{R}$ by

$$\Gamma(A, B) = \mu(A \cap B).$$

This is a non-negative definite function since

$$\sum_{i=1}^{k} \sum_{j=1}^{k} u_i u_j \Gamma(A_i, A_j) = \sum_{i=1}^{k} \sum_{j=1}^{k} u_i u_j \mu(A_i \cap A_j)$$

$$= \int_{\mathbb{R}^d} \left(\sum_{i=1}^{k} 1_{A_i} \right)^2 d\mu \geq 0$$

Therefore, by Theorem 3.2.1, there exists a centered Gaussian process $\{Z(A)\}_{A \in \mathcal{B}_b(\mathbb{R}^d)}$ with this covariance function. It is called a *Brownian pseudo-measure with structural measure* μ.

Example 3.2.4 The processes with uncorrelated increments $\{Z(t)\}_{t \in \mathbb{R}}$ of Definition 3.2.5 are associated to a random field $\{Z(A)\}_{A \in \mathcal{B}_b(\mathbb{R})}$ by the formula $Z(A) = \int_\mathbb{R} 1_A(t) dZ(t)$. That this defines a centered random field with uncorrelated spatial

---

[6] Sometimes called a "random field". However, in this text, we have reserved the appelation "random field" for stochastic processes with a spatial argument. For us, a random pseudo-measure is in fact a particular case of a stochastic process with an index set that is a collection of sets. See Example 3.2.3.

increments and structural measure μ in the sense of Definition 3.2.7 follows from the isometry formulas (3.10) and (3.11) with $f = 1_A$ and $g = 1_B$.

The term "pseudo-measure" refers to the following property: Almost surely

$$Z(A \cup B) = Z(A) + Z(B)$$

(Exercise 3.5.12). This additivity property cannot always be extended to an infinite countable sum (sigma-additivity) as the forthcoming counterexample shows.

Example 3.2.5 In the previous example, let $Z(t) = W(t)$ (the standard Wiener process). If the resulting pseudo-measure were almost-surely a random signed measure, then the process $\{Z((0, t])\}_{t \geq 0} = \{W(t)\}_{t \geq 0}$ would be a process of locally bounded variation, which is false (Theorem 3.2.5).

Let $\{Z(A)\}_{A \in \mathcal{B}_b(\mathbb{R}^m)}$ be a pseudo-measure on $\mathbb{R}^m$ with (locally finite) structural measure μ. For all $f \in L^2_{\mathbb{C}}(\mu)$ one can construct a random variable in $L^2_{\mathbb{C}}(P)$ denoted by $\int_{\mathbb{R}^m} f(t) Z(dt)$ such that for all f and g in $L^2_{\mathbb{C}}(\mu)$

$$E\left[\left(\int_{\mathbb{R}^m} f(t) Z(dt)\right)\left(\int_{\mathbb{R}^m} g(t) Z(dt)\right)^*\right] = \int_{\mathbb{R}^m} f(t) g^*(t) \mu(dt),$$

and

$$E\left[\int_{\mathbb{R}^m} f(t) Z(dt)\right] = 0,$$

The construction is similar to that on the real line (only this time we start from the definition of the integral for functions of the type $f = \sum_{i=1}^N \alpha_i 1_{A_i}$ where $A_i \in \mathcal{B}_b(\mathbb{R}^m)$: $\int_{\mathbb{R}^m} f(t) Z(dt) = \sum_{i=1}^N \alpha_i Z(A_i)$).

The integral $\int_{\mathbb{R}^m} f(t) Z(dt)$ is the *Doob integral* of f with respect to $\{Z(A)\}_{A \in \mathcal{B}_b(\mathbb{R}^m)}$.

3.3 The Power Spectral Measure

3.3.1 The Covariance Function

In what follows, the index set **T** is $\mathbb{N}$, $\mathbb{Z}$, $\mathbb{R}_+$, $\mathbb{R}$, or some $\mathbb{R}^m$. We are considering stochastic processes that are wide-sense stationary according to Definition 3.1.7.

Example 3.3.1 (Harmonic process) Let $\{U_k\}_{k \geq 1}$ be centered random variables of $L^2_{\mathbb{C}}(P)$ that are mutually uncorrelated. Let $\{\Phi_k\}_{k \geq 1}$ be completely random phases, that is, real random variables uniformly distributed on $[0, 2\pi]$. Suppose moreover that the U variables are independent of the Φ variables. Finally, suppose that $\sum_{k=1}^{\infty} E[|U_k|^2] < \infty$. For all $t \in \mathbb{R}$, the series in the right hand side of

$$X(t) = \sum_{k=1}^{\infty} U_k \cos(2\pi\nu_k t + \Phi_k),$$

where the ν_k's are arbitrary real numbers (frequencies), is convergent in $L^2_{\mathbb{C}}(P)$ and defines a centered WSS stochastic process with covariance function

$$C(\tau) = \sum_{k=1}^{\infty} \frac{1}{2} E[|U_k|^2] \cos(2\pi\nu_k \tau).$$

(This stochastic process is called a *harmonic process*.)

Proof We first do the proof for a finite number N of terms, that is with $X(t) = \sum_{k=1}^{N} U_k \cos(2\pi\nu_k t + \Phi_k)$. We then have

$$E[X(t)] = E\left[\sum_{k=1}^{N} U_k \cos(2\pi\nu_k t + \Phi_k)\right]$$

$$= \sum_{k=1}^{N} E[U_k \cos(2\pi\nu_k t + \Phi_k)] = \sum_{k=1}^{N} E[U_k]E[\cos(2\pi\nu_k t + \Phi_k)] = 0$$

and

$$E[X(t+\tau)X(t)^*] = E\left[\sum_{k=1}^{N}\sum_{\ell=1}^{N} U_k U_\ell^* \cos(2\pi\nu_k(t+\tau) + \Phi_k) \cos(2\pi\nu_\ell t + \Phi_\ell)\right]$$

$$= \sum_{k=1}^{N}\sum_{\ell=1}^{N} E[U_k U_\ell^* \cos(2\pi\nu_k(t+\tau) + \Phi_k) \cos(2\pi\nu_\ell t + \Phi_\ell)]$$

$$= \sum_{k=1}^{N}\sum_{\ell=1}^{N} E[U_k U_\ell^*]E[\cos(2\pi\nu_k(t+\tau) + \Phi_k) \cos(2\pi\nu_\ell t + \Phi_\ell)]$$

$$= \sum_{k=1}^{N} E[|U_k|^2]E[\cos(2\pi\nu_k(t+\tau) + \Phi_k) \cos(2\pi\nu_k t + \Phi_k)]$$

$$= \sum_{k=1}^{N} E[|U_k|^2]E\left[\frac{1}{2}(\cos(2\pi\nu_k(2t+\tau) + 2\Phi_k) + \cos(2\pi\nu_k\tau))\right].$$

The announced result then follows since

$$E[\cos(2\pi\nu_k(2t+\tau)+2\Phi_k)] = \frac{1}{2\pi}\int_0^{2\pi}\cos(2\pi\nu_k(2t+\tau)+2\varphi)\,d\varphi = 0.$$

The extension of this result to an infinite sum of complex exponentials is a consequence of the result of Exercise 1.4.21. □

Recall the definition of the *correlation coefficient* ρ between two, say real, square integrable random variables X and Y, with respective means and variances m_X and m_Y, and σ_X^2 and σ_Y^2, that are both not almost-surely null:

$$\rho = \frac{\text{cov}\,(X,Y)}{\sigma_X\sigma_Y}.$$

The random variable $aX+b$ that minimizes the function $F(a,b) := E\left[(Y-aX-b)^2\right]$ is

$$\widehat{Y} = m_Y + \frac{\text{cov}\,(X,Y)}{\sigma_X^2}(X-m_X)$$

and moreover

$$E\left[(\widehat{Y}-Y)^2\right] = \left(1-\rho^2\right)\sigma_Y^2$$

(Exercise 3.5.6). The variable $\widehat{Y}$ is called the *best linear-quadratic estimate* of Y given X, or the *linear regression* of Y on X.

For a WSS stochastic process with covariance function C, the function

$$\rho(\tau) = \frac{C(\tau)}{C(0)}$$

is called the *autocorrelation function*. It is in fact, for any t, the correlation coefficient between $X(t)$ and $X(t+\tau)$. In particular, the best linear-quadratic estimate of $X(t+\tau)$ given $X(t)$ is

$$\widehat{X}(t+\tau|t) = m + \rho(\tau)(X(t)-m).$$

The estimation error is then, according to the above,

$$E\left[(\widehat{X}(t+\tau|t)-X(t+\tau))^2\right] = \sigma_X^2\left(1-\rho(\tau)^2\right).$$

This shows that if the support of the covariance function is concentrated around $\tau=0$, the process tends to be "unpredictable". We shall come back to this when we introduce the notion of white noise.

We henceforth assume that the covariance function is continuous. For this to be true, it suffices that the covariance function be continuous at the origin. This is in turn equivalent to continuity in the quadratic mean of the stochastic process, that is: For all $t \in \mathbb{R}$,

$$\lim_{h \to 0} E\left[|X(t+h) - X(t)|^2\right] = 0.$$

In fact, the covariance function is then uniformly continuous on $\mathbb{R}$.

Proof We can assume that the process is centered. Then:

$$
\begin{aligned}
E\left[|X(t+h) - X(t)|^2\right] &= E\left[|X(t+h)|^2\right] + E\left[|X(t)|^2\right] \\
&\quad - E\left[X(t)X(t+h)^*\right] - E\left[X(t)^*X(t+h)\right] \\
&= 2C_X(0) - C_X(h) - C_X(h)^*
\end{aligned}
$$

and therefore, continuity in quadratic mean follows from the continuity at the origin of the autocovariance function. For the uniform continuity property, write

$$
\begin{aligned}
|C_X(\tau+h) - C_X(\tau)| &= \left|E\left[X(\tau+h)X(0)^*\right] - E\left[X(\tau)X(0)^*\right]\right| \\
&= \left|E\left[(X(\tau+h) - X(\tau))X(0)^*\right]\right| \\
&\leq E\left[|X(0)|^2\right]^{\frac{1}{2}} \times E\left[|X(\tau+h) - X(\tau)|^2\right]^{\frac{1}{2}} \\
&= E\left[|X(0)|^2\right]^{\frac{1}{2}} \times E\left[|X(h) - X(0)|^2\right]^{\frac{1}{2}}.
\end{aligned}
$$

$\square$

3.3.2 Fourier Representation of the Covariance

We begin with simple examples, and then give the general result.

Example 3.3.2 (Absolutely continuous spectrum) Consider a WSS stochastic process with *integrable and continuous* covariance function C, in which case the Fourier transform f of the latter is well-defined by

$$f(\nu) = \int_{\mathbb{R}} e^{-2i\pi\nu\tau} C(\tau)\, d\tau.$$

It is called the *power spectral density* (PSD). It turns out, as we shall soon see when we consider the general case, that it is non-negative and integrable. Since it is integrable, the Fourier inversion formula

$$C(\tau) = \int_{\mathbb{R}} e^{2i\pi\nu\tau} f(\nu)\, d\nu, \tag{3.13}$$

holds true, and it holds true for all $t \in \mathbb{R}$ since C is continuous (Theorem 1.1.8). Also f is the unique integrable function such that (3.13) holds. In the context of WSS stochastic processes, (3.13) is called the *Bochner formula*. Letting $\tau = 0$ in this formula, we obtain, since $C(0) = \mathrm{Var}(X(t)) := \sigma^2$,

$$\sigma^2 = \int_{\mathbb{R}} f(\nu)\, d\nu.$$

Example 3.3.3 (Ornstein–Uhlenbeck process) Consider the Ornstein–Uhlenbeck process of Definition 3.2.6. It is a centered Gaussian process with covariance function

$$\Gamma(t, s) = C(t - s) = e^{-\alpha|t-s|}.$$

The function C is integrable and therefore the power spectral density is the Fourier transform of the covariance function:

$$f(\nu) = \int_{\mathbb{R}} e^{-2i\pi\nu\tau} e^{-\alpha|\tau|}\, d\tau = \frac{2\alpha}{\alpha^2 + 4\pi^2\nu^2}.$$

Not all WSS stochastic processes admit a power spectral density. For instance:

Example 3.3.4 (Line spectrum) Consider a wide-sense stationary process with a covariance function of the form

$$C(\tau) = \sum_{k\in\mathbb{Z}} P_k e^{2i\pi\nu_k\tau},$$

where

$$P_k \geq 0 \text{ and } \sum_{k\in\mathbb{Z}} P_k < \infty,$$

(for instance, the harmonic process of Example 3.3.1). Clearly, C is not integrable, and in fact there does not exist a power spectral density. In particular, a representation of the covariance function such as (3.13) is not available, at least if the function f is interpreted in the ordinary sense. However, there is a formula such as (3.13) if we consent to define the PSD in this case to be the *pseudo-function*

$$f(\nu) = \sum_{k\in\mathbb{Z}} P_k\, \delta(\nu - \nu_k),$$

where $\delta(\nu - a)$ is the delayed Dirac pseudo-function informally defined by

$$\int_{\mathbb{R}} \varphi(\nu)\,\delta(\nu - a)\,d\nu = \varphi(a).$$

Indeed, with such a convention,

$$\int_{\mathbb{R}} f(\nu)e^{2i\pi\nu\tau}\,f(\nu)\,d\nu = \sum_{k\in\mathbb{Z}} P_k \int_{\mathbb{R}} e^{2i\pi\nu\tau}\delta(\nu - \nu_k)\,d\nu = \sum_{k\in\mathbb{Z}} P_k e^{2i\pi\nu_k\tau}$$

We can (and perhaps should) however avoid recourse to Dirac pseudo-functions, and the general result to follow will tell us what to do.

As we just saw, it may happen that the covariance function is not integrable and/or that there does not exist a line spectrum.

Theorem 3.3.1 *Let* $\{X(t)\}_{t\in\mathbb{R}}$ *be a* WSS *stochastic process continuous in the quadratic mean, with covariance function* C. *Then, there exists a* unique *measure* μ *on* $\mathbb{R}$ *such that*

$$C(\tau) = \int_{\mathbb{R}} e^{2i\pi\nu\tau}\mu(d\nu). \tag{3.14}$$

In particular, μ is a *finite* measure:

$$\mu(\mathbb{R}) = C(0) = \mathrm{Var}(X(0)) < \infty. \tag{3.15}$$

Proof The covariance function of a WSS stochastic process that is continuous in the quadratic mean shares the following properties with the characteristic function of a real random variable:

(a) It is hermitian symmetric, and $|C(\tau)| \le C(0)$ (Schwarz's inequality).
(b) It is uniformly continuous.
(c) It is definite non-negative, in the sense that for all integers n, all $\tau_1, \ldots, \tau_n \in \mathbb{R}$, and all $z_1, \ldots, z_n \in \mathbb{C}$,

$$\sum_{j=1}^{n}\sum_{k=1}^{n} C(\tau_j - \tau_k)z_j z_k^* \ge 0$$

(just observe that the left-hand side is equal to $E\left[\left|\sum_{j=1}^{n} z_j X(t_j)\right|^2\right]$). Therefore, by Theorem 2.2.8, the covariance function C is (up to a multiplicative constant) a characteristic function. This is exactly what (3.14) says, since μ thereof is a finite measure, that is, up to a multilicative constant, a probability distribution.

Uniqueness of the power spectral measure follows from the fact that a finite measure is characterized by its Fourier transform (Corollary 2.1.4).   □

The following simple result is stated for future reference.

Theorem 3.3.2 *Let* $\{X(t)\}_{t\in\mathbb{R}}$ *be a* real WSS *stochastic process with continuous covariance function. Then its spectral measure is symmetric.*

Proof (Exercise 3.5.15)   □

The case of an absolutely continuous spectrum corresponds to the situation where μ admits a density with respect to Lebesgue measure: $\mu(d\nu) = f(\nu)\,d\nu$. We then say that the WSS stochastic process in question admits the *power spectral density* (PSD) f. If such a power spectral density exists, it has the properties mentioned in Example 3.3.2 without proof: it is non-negative and it is integrable.

The case of a line spectrum corresponds to a spectral measure that is a weighted sum of Dirac measures:

$$\mu(d\nu) = \sum_{k\in\mathbb{Z}} P_k\,\varepsilon_{\nu_k}(d\nu),$$

where the P_k's are non-negative and have a finite sum, as mentioned in Example 3.3.4.

3.3.3 Filtering and White Noise

Let now $\{X(t)\}_{t\in\mathbb{R}}$ be a WSS stochastic process with the continuous covariance function C_X. We examine the effect of filtering on this process. The output process is

$$Y(t) = \int_{\mathbb{R}} h(t-s)X(s)\mathrm{d}s. \tag{3.16}$$

The integral (3.16) is well-defined if the "impulse response" h is integrable. This follows from Theorem 3.1.5 according to which the integral

$$\int_{\mathbb{R}} f(s)X(s,\omega)\,ds$$

is well-defined for P-almost all ω when f is integrable (in the special case of WSS stochastic processes, $m(t) = m$ and $\Gamma(t,t) = C(0) + |m|^2$, and therefore the conditions on f and g thereof reduce to integrability of these functions). Refering to the same theorem, we have

$$E[\int_{\mathbb{R}} f(t)X(t)\,dt] = \int_{\mathbb{R}} f(t)E[X(t)]\,dt = m\int_{\mathbb{R}} f(t)\,dt. \tag{3.17}$$

Let now $f, g : \mathbb{R} \to \mathbb{C}$ and be integrable functions. As a special case of Theorem 3.1.5, we have

$$\operatorname{cov}\left(\int_{\mathbb{R}} f(t)X(t)\,dt\,,\ \int_{\mathbb{R}} g(s)X(s)\,ds\right) = \int_{\mathbb{R}}\int_{\mathbb{R}} f(t)g^*(s)C(t-s)\,dt\,ds. \quad (3.18)$$

We shall see that, in addition,

$$\operatorname{cov}\left(\int_{\mathbb{R}} f(t)X(t)\,dt\,,\ \int_{\mathbb{R}} g(s)X(s)\,ds\right) = \int_{\mathbb{R}}\int_{\mathbb{R}} \widehat{f}(-\nu)\widehat{g}^*(-\nu)\mu(d\nu). \quad (3.19)$$

Proof We assume without loss of generality that $m = 0$. From Bochner's representation of the covariance function, we obtain for the last double integral in (3.18)

$$\int_{\mathbb{R}}\int_{\mathbb{R}} f(t)g^*(s)\left(\int_{\mathbb{R}} e^{+2j\pi\nu(t-s)}\,\mu(d\nu)\right)dt\,ds =$$

$$\int_{\mathbb{R}}\left(\int_{\mathbb{R}} f(t)e^{+2j\pi\nu t}\,dt\right)\left(\int_{\mathbb{R}} g(s)e^{+2j\pi\nu s}\,ds\right)^{*}\mu(d\nu).$$

Here again we have to justify the change of order of integration using Fubini's theorem. For this, it suffices to show that the function $(t, s, \nu) \to \left|f(t)g^*(s)e^{+2j\pi\nu(t-s)}\right| = |f(t)|\,|g(s)|\,1_{\mathbb{R}}(\nu)$ is integrable with respect to the product measure $\ell \times \ell \times \mu$. This is true indeed, the integral being equal to $(\int_{\mathbb{R}} |f(t)|\,dt) \times (\int_{\mathbb{R}} |g(t)|\,dt) \times \mu(\mathbb{R})$. □

In view of the above results, the right-hand side of formula (3.16) is well-defined. Moreover

Theorem 3.3.3 *When the input process $\{X(t)\}_{t\in\mathbb{R}}$ is a WSS stochastic process with power spectral measure μ_X, the output $\{Y(t)\}_{t\in\mathbb{R}}$ of a stable convolutional filter of transmittance $\widehat{h}$ is a WSS stochastic process with the power spectral measure*

$$\mu_Y(d\nu) = |\widehat{h}(\nu)|^2\mu_X(d\nu).$$

This formula will be refered to as the *fundamental filtering formula* in continuous time.

Proof We just have to apply formulas (3.17) and (3.19) with the functions

$$f(u) = h(t - u), \qquad g(v) = h(s - v),$$

to obtain

$$E[Y(t)] = m\int_{\mathbb{R}} h(t)\,dt,$$

and

$$E[(Y(t) - m)(Y(s) - m)^*] = \int_{\mathbb{R}} |\widehat{h}(\nu)|^2 e^{+2j\pi\nu(t-s)} \mu(d\nu).$$

$\square$

Example 3.3.5 (Two special cases) In particular, if the input process admits a PSD f_X, the output process admits also a PSD given by

$$f_Y(\nu) = |\widehat{h}(\nu)|^2 f_X(\nu) \, d\nu$$

When the input process has a line spectrum, the power spectral measure of the output process takes the form

$$\mu_Y(d\nu) = \sum_{k=1}^{\infty} P_k |\widehat{h}(\nu_k)|^2 \varepsilon_{\nu_k}(d\nu).$$

By analogy with Optics, one calls *white noise* any centered WSS stochastic process $\{B(t)\}_{t \in \mathbb{R}}$ with constant power spectral density $f_B(\nu) = C$. Such a definition presents a theoretical difficulty, because

$$\int_{-\infty}^{+\infty} f_B(\nu) \, d\nu = +\infty,$$

which contradicts the finite power property of wide-sense stationary processes.

First Approach

From a pragmatic point of view, one could define a white noise to be a centered WSS stochastic process whose PSD is constant over a "large", yet bounded, range of frequencies $[-A, +A]$. The calculations below show what happens as A tends to infinity. Let therefore $\{X(t)\}_{t \in \mathbb{R}}$ be a centered WSS stochastic process with PSD

$$f(\nu) = \frac{N_0}{2} 1_{[-A,+A]}(\nu).$$

(The strange notation $N_0/2$ for what will be interpreted as the power of the white noise comes from Physics and is a standard one in communications theory when dealing with white noise channels.) Let φ_1, $\varphi_2 : \mathbb{R} \to \mathbb{C}$ be two functions in $L^1_{\mathbb{C}}(\mathbb{R}) \cap L^2_{\mathbb{C}}(\mathbb{R})$ with Fourier transforms $\widehat{\varphi}_1$ and $\widehat{\varphi}_2$ respectively. Then

$$\lim_{A\uparrow\infty} E\left[\left(\int_{\mathbb{R}} \varphi_1(t)X(t)\,dt\right)\left(\int_{\mathbb{R}} \varphi_2(t)X(t)\,dt\right)^*\right] = \frac{N_0}{2}\int_{\mathbb{R}} \varphi_1(t)\varphi_2^*(t)\,dt$$

$$= \frac{N_0}{2}\int_{\mathbb{R}} \widehat{\varphi}_1(\nu)\widehat{\varphi}_2^*(\nu)\,d\nu.$$

Proof We have

$$E\left[\left(\int_{\mathbb{R}} \varphi_1(t)X(t)\,dt\right)\left(\int_{\mathbb{R}} \varphi_2(t)X(t)\,dt\right)^*\right] = \int_{\mathbb{R}}\int_{\mathbb{R}} \varphi_1(u)\varphi_2(v)^* C_X(u-v)\,du\,dv.$$

The latter quantity is equal to

$$\frac{N_0}{2}\int_{-\infty}^{+\infty} \varphi_1(u)\varphi_2(v)^* \left(\int_{-A}^{+A} e^{2i\pi\nu(u-v)}\,dv\right) du\,dv$$

$$= \frac{N_0}{2}\int_{-A}^{+A} \left(\int_{-\infty}^{+\infty} \varphi_1(u)e^{2i\pi\nu u}\,du\right)\left(\int_{-\infty}^{+\infty} \varphi_2(v)^* e^{-2i\pi\nu v}\,dv\right) d\nu$$

$$= \frac{N_0}{2}\int_{-A}^{+A} \widehat{\varphi}_1(-\nu)\widehat{\varphi}_2(-\nu)^*\,d\nu,$$

and the limit of this quantity as $A\uparrow\infty$ is:

$$\frac{N_0}{2}\int_{-\infty}^{+\infty} \widehat{\varphi}_1(\nu)\widehat{\varphi}_2^*(\nu)\,d\nu = \frac{N_0}{2}\int_{-\infty}^{+\infty} \varphi_1(t)\varphi_2(t)^*\,dt$$

where the last equality is the Plancherel–Parseval identity.   □

Let now $h : \mathbb{R} \to \mathbb{C}$ be in $L_{\mathbb{C}}^1(\mathbb{R}) \cap L_{\mathbb{C}}^2(\mathbb{R})$, and define

$$Y(t) = \int_{\mathbb{R}} h(t-s)X(s)\,ds.$$

Applying the above result with $\varphi_1(u) = h(t-u)$ and $\varphi_2(v) = h(t+\tau-v)$, we find that the covariance function C_Y of this WSS stochastic process is such that

$$\lim_{A\uparrow\infty} C_Y(\tau) = \int_{\mathbb{R}} e^{2i\pi\nu\tau}|\widehat{h}(\nu)|^2 \frac{N_0}{2}\,d\nu.$$

The limit is finite since $\widehat{h} \in L^2_{\mathbb{C}}(\mathbb{R})$ and is a covariance function corresponding to a *bona fide* (that is integrable) PDF $f_Y(\nu) = |\widehat{h}(\nu)|^2 \frac{N_0}{2}$. With $f(\nu) = \frac{N_0}{2}$, we formally retrieve the usual filtering formula,

$$f_Y(\nu) = |\widehat{h}(\nu)|^2 f(\nu).$$

White Noise via the Doob–Wiener Integral

We take another point of view for white noise, more formal. We shall now work right away "at the limit". We do not attempt to define the white noise $\{B(t)\}_{t\in\mathbb{R}}$ directly (for good reasons since it does not exist as a *bona fide* WSS stochastic process as we noted earlier). Instead we define directly the symbolic integral $\int_{\mathbb{R}} f(t)B(t)\, dt$ for integrands f to be made precise below, by

$$\int_{\mathbb{R}} f(t)B(t)\, dt := \sqrt{\frac{N_0}{2}} \int_{\mathbb{R}} f(t)\, dZ(t), \qquad (3.20)$$

where $\{Z(t)\}_{t\in\mathbb{R}}$ is a centered stochastic process with uncorrelated increments with unit variance. We say that $\{B(t)\}_{t\in\mathbb{R}}$ is a *white noise* and that $\left\{\sqrt{\frac{N_0}{2}}\, Z(t)\right\}_{t\in\mathbb{R}}$ is an *integrated white noise*. For all $f, g \in L^2_{\mathbb{C}}(\mathbb{R})$, we have that

$$E\left[\int_{\mathbb{R}} f(t)\, B(t)\, dt \right] = 0,$$

and from Doob's isometry formulas,

$$E\left[\left(\int_{\mathbb{R}} f(t)\, B(t)\, dt \right) \left(\int_{\mathbb{R}} g(t)\, B(t)\, dt \right)^* \right] = \frac{N_0}{2} \int_{\mathbb{R}} f(t)g(t)^*\, dt\,,$$

which can be formally rewritten, using the Dirac symbolism:

$$\int_{\mathbb{R}} f(t)g(s)^*\, E\left[B(t)B^*(s) \right] dt\, ds = \int_{\mathbb{R}} f(t)g(s)^* \frac{N_0}{2}\delta(t-s)\, dt\, ds.$$

Hence "the covariance function of the white noise $\{B(t)\}_{t\in\mathbb{R}}$ is a Dirac pseudo-function: $C_B(\tau) = \frac{N_0}{2}\delta(\tau)$".

When $\{Z(t)\}_{t\in\mathbb{R}} \equiv \{W(t)\}_{t\in\mathbb{R}}$, a standard Brownian motion, $\{B(t)\}_{t\in\mathbb{R}}$ is called a *Gaussian white noise* (GWN). In this case, the Wiener–Doob integral is certainly not a Stieltjes–Lebesgue integral since the trajectories of the Wiener process are of unbounded variation on any finite interval.

Let $\{B(t)\}_{t\in\mathbb{R}}$ be a white noise with PSD $N_0/2$. Let $h : \mathbb{R} \to \mathbb{C}$ be in $L^1_\mathbb{C} \cap L^2_\mathbb{C}$ and define the output of a filter with impulse response h when the white noise $\{B(t)\}_{t\in\mathbb{R}}$ is the input, by

$$Y(t) = \int_\mathbb{R} h(t-s)B(t)\,ds.$$

By the isometry formula for the Wiener–Doob integral,

$$E[Y(t)Y(s)^*] = \frac{N_0}{2}\int_\mathbb{R} h(t-s-u)h^*(u)\,du,$$

and therefore (Plancherel–Parseval equality)

$$C_Y(\tau) = \int_\mathbb{R} e^{2i\pi\nu\tau}|\widehat{h}(\nu)|^2 \frac{N_0}{2}\,d\nu.$$

The stochastic process $\{Y(t)\}_{t\in\mathbb{R}}$ is therefore centered and WSS, with PSD

$$f_Y(\nu) = |\widehat{h}(\nu)|^2 f_B(\nu),$$

where

$$f_B(\nu) := \frac{N_0}{2}.$$

We therefore once more recover formally the fundamental equation of linear filtering of WSS continuous-time stochastic processes

The Approximate Derivative Approach

There is a third approach to white noise. As we saw before, the derivative of the Brownian motion does not exist. But one can approximate it by the "finitesimal" derivative

$$B_h(t) = \frac{W(t+h) - W(t)}{h}.$$

For fixed $h > 0$ this defines a proper WSS stochastic process centered, with covariance function

$$C_h(\tau) = \frac{(h - |\tau|)^+}{h^2}$$

and power spectral density

$$f_h(\nu) = \left(\frac{\sin \pi\nu h}{\pi\nu h}\right)^2.$$

Note that, as $h \downarrow 0$, the power spectral density tends to the constant function 1, the power spectral density of the "white noise". At the same time, the covariance function "tends to the Dirac function" and the energy $C_h(0) = \frac{1}{h}$ tends to infinity. This is another feature of white noise: inpredictability. Indeed, for $\tau \geq h$, the value $B_h(t + \tau)$ cannot be predicted fron the value $B_h(t)$, since both are independent random variables.

In order to connect with the second approach, Exercise 3.5.18 asks you to show that for all $f \in L^2_{\mathbb{C}}(\mathbb{R}_+) \cap L^1_{\mathbb{C}}(\mathbb{R}_+)$,

$$\lim_{h\downarrow 0} \int_{\mathbb{R}_+} f(t) B_h(t)\, dt = \int_{\mathbb{R}_+} f(t)\, dW(t)$$

in the quadratic mean.

3.3.4 Long-range Dependence and Self-similarity

The "classical" models of WSS stochastic processes have a covariance function which is geometrically bounded, that is:

$$C(|\tau|) \leq K e^{-\alpha|\tau|},$$

where $K > 0$ and $\alpha > 0$. A *long memory* WSS stochastic process is one for which exponential decay is replaced by polynomial decay. More precisely,

$$C(|\tau|) \sim K|\tau|^{2d-1} \quad \text{as } \tau \to \infty, \tag{3.21}$$

where $K > 0$ and $d < \frac{1}{2}$. One sometimes makes the distinction between *intermediate memory* and *long memory* according to whether $d < 0$, in which case

$$\int_{\mathbb{R}} C(\tau)\, d\tau < \infty,$$

or $0 < d < \frac{1}{2}$, in which case

$$\int_{\mathbb{R}} C(\tau)\,d\tau = \infty.$$

The terminology is not yet standardized,[7] and sometimes one talks about *long-range dependence* rather than *long-memory*. Here, a long-range dependent stochastic process will be one whose covariance function is not summable. In particular a long memory stochastic process is long-range dependent.

Long-range dependence can be approached from the spectral point of view. Indeed, suppose that a WSS stochastic process admits a power spectral density f whose behaviour at the origin is the following:

$$\lim_{\nu \to 0} f(\nu)\nu^\beta = r > 0, \tag{3.22}$$

for some $\beta \in (0, 1)$. Then writing

$$C(\tau)\tau^{1-\beta} = \int_{-\infty}^{+\infty} f(\nu)\tau^{1-\beta}e^{+2i\pi\nu\tau}\,d\nu = \int_{-\infty}^{+\infty} \tau^{1-\beta}\frac{1}{\tau}f\left(\frac{\nu}{\tau}\right)\,d\nu$$

and observing that

$$\lim_{\tau\uparrow\infty} \tau^{1-\beta}\frac{1}{\tau}f\left(\frac{\nu}{\tau}\right) = \frac{r}{\nu^\beta}$$

we have that (with mild additional conditions on f guaranteeing that Lebesgue's dominated convergence theorem is applicable)

$$\lim_{\tau\uparrow\infty} C(\tau)\tau^{1-\beta} = r\int_0^\infty \frac{e^{+2i\pi\nu}}{\nu^\beta}\,d\nu.$$

Therefore, at infinity, the covariance function behaves as $\frac{1}{\tau^{1-\beta}}$ where $1-\beta \in (0,1)$: the process is long-range dependent.

Starting with a non-negative integrable function f with property (3.22) we can construct a WSS stochastic process $\{X(t)\}_{t\in\mathbb{R}}$ with the PSD f by the formula

$$X(t) = \int_0^\infty \sqrt{f(\nu)}e^{+2i\pi\nu t}\,dZ(\nu),$$

[7] For a more detailed introduction to long range dependence, see Gennady Samorodnitsky, *Long Range Dependence*, Foundations and Trends in Stochastic Systems, Vol. 1, 3, 2006, 163–257.

where $\{Z(\nu)\}_{\nu \in \mathbb{R}}$ is a centered stochastic process with uncorrelated increments and the Lebesgue measure for structural measure (for instance a Wiener process) and where the integral is a Doob integral.

We now introduce the notion of self-similarity, which is related to long-range dependence as we shall see in the next subsection.

Self-similarity and the Fractal Brownian Motion

The Wiener process $\{W(t)\}_{t \in \mathbb{R}_+}$ has the following property. If c is a positive constant, the process $\{W_c(t)\}_{t \in \mathbb{R}_+} := \{c^{-\frac{1}{2}} W(ct)\}_{t \in \mathbb{R}_+}$ is also a Wiener process. It is indeed a centered Gaussian process with independent increments, null at the origin of times, and for $0 < a < b$,

$$E\left[|W_c(b) - W_c(a)|^2 \right] = c^{-1} E\left[|W(cb) - W(ca)|^2 \right] = c^{-1}(cb - ca) = b - a.$$

This is a particular instance of a self-similar stochastic process, as defined below:

Definition 3.3.1 A real-valued stochastic process $\{Y(t)\}_{t \in \mathbb{R}_+}$ is called *self-similar* with *(Hurst) self-similarity parameter* H if for any $c > 0$,

$$\{Y(t)\}_{t \in \mathbb{R}_+} \overset{\mathcal{D}}{\sim} \{c^{-H} Y(ct)\}_{t \in \mathbb{R}_+}.$$

(The Wiener process is therefore self-similar with similarity parameter $H = \frac{1}{2}$.)

It follows from the definition that $Y(t) \overset{\mathcal{D}}{\sim} t^H Y(1)$, and therefore, if $P(Y(1) \neq 0) > 0$:

If $H < 0$, $Y(t) \to 0$ in distribution as $t \to \infty$ and $Y(t) \to \infty$ in distribution as $t \to 0$.

If $H > 0$, $Y(t) \to \infty$ in distribution as $t \to 0$ and $Y(t) \to 0$ in distribution as $t \to \infty$.

If $H = 0$, $Y(t)$ has a distribution independent of t.

In particular, when $H \neq 0$, a self-similar process cannot be stationary (strictly or in the wide sense).

We shall be interested in self-similar processes that have *stationary increments*. We must restrict attention to non-negative self-similarity parameters, because of the following negative result[8]: for any strictly negative value of the self-similarity parameter, a self-similar stochastic process with independent increments is not measurable (except of course for the trivial case where the process is identically null).

Theorem 3.3.4 Let $\{Y(t)\}_{t \in \mathbb{R}_+}$ be a self-similar stochastic process with stationary increments and self-similarity parameter $H > 0$ (in particular $Y(0) = 0$). Its covariance function is given by

---

[8] W. Vervaat (1987), "Properties of general self-similar processes" *Bull. Int. Statist. Inst.*, 52, 4, 199–216.

$$\Gamma(s,t) := \text{cov}\,(Y(s), Y(t)) = \frac{1}{2}\sigma^2 \left[t^{2H} - |t - s|^{2H} + s^{2H}\right]$$

where $\sigma^2 = E\left[(Y(t+1) - Y(t))^2\right] = E\left[Y(1)^2\right]$.

Proof Assume without loss of generality that the process is centered. Let $0 \leq s \leq t$. Then

$$E\left[(Y(t) - Y(s))^2\right] = E\left[(Y(t-s) - Y(0))^2\right]$$
$$= E\left[(Y(t-s))^2\right] = \sigma^2(t-s)^{2H}$$

and

$$2E\,[Y(t)Y(s)] = E\left[Y(t)^2\right] + E\left[Y(s)^2\right] - E\left[(Y(t) - Y(s))^2\right],$$

hence the result.   □

The *fractal Brownian motion*[9] is a Gaussian process that in a sense generalizes the Wiener process.

Definition 3.3.2 A fractal Brownian motion on $\mathbb{R}_+$ with Hurst parameter $H \in (0, 1)$ is a centered Gaussian process $\{B_H(t)\}_{t \in \mathbb{R}_+}$ with continuous paths, such that $B_H(0) = 0$, and with covariance function

$$E[B_H(t)B_H(s)] = \frac{1}{2}\left(|t|^{2H} + |s|^{2H} - |t - s|^{2H}\right). \tag{3.23}$$

The existence of such process follows from Theorem 3.2.1 as soon as we show that the right-hand side of (3.23) is a non-negative definite function. This can be done directly, although we choose another path. We shall prove the existence of the fractal Brownian motion by constructing it as a Doob integral with respect to a Wiener process. More precisely, define for $0 < H < 1$, $w_H(t,s) = 0$ for $t \leq s$,

$$w_H(t,s) = (t-s)^{H-\frac{1}{2}} \text{ for } 0 \leq s \leq t$$

and

$$w_H(t,s) = (t-s)^{H-\frac{1}{2}} - (-s)^{H-\frac{1}{2}} \text{ for } s < 0.$$

Observe that for any $c > 0$

$$w_H(ct,s) = c^{H-\frac{1}{2}}w_H(t, sc^{-1}).$$

---

[9] Benoit Mandelbrot and John W. Van Ness, *Fractional Brownian Motions, Fractional Noises and Applications*, SIAM review, Vol. 10, 4, 1968.

Define

$$B_H(t) := \int_{\mathbb{R}} w_H(t, s)\, dW(s)$$

The Doob integral of the right-hand side is, more explicitly,

$$A - B := \int_0^t (t - s)^{H - \frac{1}{2}}\, dW(s) - \int_{-\infty}^0 \left((t - s)^{H - \frac{1}{2}} - (-s)^{H - \frac{1}{2}}\right) dW(s). \quad (3.24)$$

It is well defined and with the change of variable $u = c^{-1}s$ it becomes

$$c^{H - \frac{1}{2}} \int_{\mathbb{R}} w_H(t, u)\, dW(cu)$$

Using the self-similarity of the Wiener process, the process defined by the last display has the same distribution as the process defined by

$$c^{H - \frac{1}{2}} c^{\frac{1}{2}} \int_{\mathbb{R}} w_H(t, u)\, dW(u).$$

Therefore $\{B_H(t)\}_{t \in \mathbb{R}_+}$ is self-similar with similarity parameter H.

It is tempting to rewrite (3.24) as $Z(t) - Z(0)$, where

$$Z(t) = \int_{-\infty}^t (t - s)^{H - \frac{1}{2}}\, dW(s).$$

However this last integral is not well-defined as a Doob integral since for all $H > 0$, the function $s \to (t - s)^{H - \frac{1}{2}} 1_{\{s \le t\}}$ is not in $L^2_{\mathbb{R}}(\mathbb{R})$.

3.4 Fourier Analysis of the Trajectories

3.4.1 The Cramér–Khinchin Decomposition

Almost-surely, a trajectory of a stationary stochastic process is neither in $L^1_{\mathbb{C}}(\mathbb{R})$ nor in $L^2_{\mathbb{C}}(\mathbb{R})$ unless it is identically null. The formal argument will not be given here, but the examples show this convincingly. Therefore such trajectory does not have a Fourier transform in the usual senses. There exists however, in some particular sense, a kind of Fourier spectral decomposition of the trajectories of a WSS stochastic process, as we shall now see.

Theorem 3.4.1 *Let* $\{X(t)\}_{t\in\mathbb{R}}$ *be a centered* WSS *stochastic process, continuous in quadratic mean, and let* μ *be its power spectral measure. There exists a* unique *(more precision below the theorem) centered stochastic process* $\{x(\nu)\}_{\nu\in\mathbb{R}}$ *with uncorrelated increments and with structural measure* μ*, such that for all* $t \in \mathbb{R}$*,* P*–a.s.,*

$$X(t) = \int_{\mathbb{R}} e^{2i\pi\nu t}\, dx(\nu), \qquad (3.25)$$

where the integral of the right-hand side is a Doob integral.

The decomposition (3.25) is unique in the following sense: If there exists another centered stochastic process $\{\tilde{x}(\nu)\}_{\nu\in\mathbb{R}}$ with uncorrelated increments, and with finite structural measure $\tilde{\mu}$, such that for all $t \in \mathbb{R}$, we have P–a.s., $X(t) = \int_{\mathbb{R}} e^{2i\pi\nu t}\, d\tilde{x}(\nu)$, then for all $a, b \in \mathbb{R}$, $a \leq b$, $\tilde{x}(b) - \tilde{x}(a) = x(b) - x(a)$, P-a.s.

We will now and then say: "$dx(\nu)$ is the *(Cramér–Khinchin) spectral decomposition*" of the WSS stochastic process.

Proof 1. Denote by $\mathcal{H}(X)$ the vector subspace of $L^2_{\mathbb{C}}(P)$ formed by the finite complex linear combinations of the type

$$Z = \sum_{k=1}^{K} \lambda_k X(t_k),$$

and by φ the mapping of $\mathcal{H}(X)$ into $L^2_{\mathbb{C}}(\mu)$ defined by

$$\varphi : Z \to \sum_{k=1}^{K} \lambda_k e^{2i\pi\nu t_k}.$$

We verify that it is a linear isometry of $\mathcal{H}(X)$ into $L^2_{\mathbb{C}}(\mu)$. In fact,

$$E\left[\left|\sum_{k=1}^{K} \lambda_k X(t_k)\right|^2\right] = \sum_{k=1}^{K}\sum_{\ell=1}^{K} \lambda_k \lambda_\ell^* E\left[X(t_k)X(t_\ell)^*\right]$$

$$= \sum_{k=1}^{K}\sum_{\ell=1}^{K} \lambda_k \lambda_\ell^* C(t_k - t_\ell),$$

and using Bochner's theorem, this quantity is equal to

$$\sum_{k=1}^{K} \sum_{\ell=1}^{K} \lambda_k \lambda_\ell^* \int_R e^{2i\pi\nu(t_k - t_\ell)} \mu(d\nu) = \int_R \left(\sum_{k=1}^{K} \sum_{\ell=1}^{K} \lambda_k \lambda_\ell^* e^{2i\pi\nu(t_k - t_\ell)} \right) \mu(d\nu)$$

$$= \int_R \left| \sum_{k=1}^{K} \lambda_k e^{2i\pi\nu t_k} \right|^2 \mu(d\nu).$$

2. This isometric linear mapping can be uniquely extended to an isometric linear mapping (that we shall continue to call φ) from $\mathcal{H}(X)$, the closure of $H(X)$, into $L_{\mathbb{C}}^2(\mu)$ (Theorem 1.3.7). As the combinations $\sum_{k=1}^{K} \lambda_k e^{2i\pi\nu t_k}$ are dense in $L_{\mathbb{C}}^2(\mu)$ when μ is a finite measure (Corollary A.3.1), φ is *onto*. Therefore, it is a linear isometric bijection between $H(X)$ and $L_{\mathbb{C}}^2(\mu)$.

3. We shall define $x(\nu_0)$ to be the random variable in $H(X)$ that corresponds in this isometry to the function $1_{(-\infty,\nu_0]}(\nu)$ of $L_{\mathbb{C}}^2(\mu)$. First, we observe that

$$E[x(\nu_2) - x(\nu_1)] = 0$$

since $H(X)$ is the closure in $L_{\mathbb{C}}^2(P)$ of a family of centered random variables. Also, by isometry,

$$E[(x(\nu_2) - x(\nu_1))(x(\nu_4) - x(\nu_3))^*] = \int_{\mathbb{R}} 1_{(\nu_1,\nu_2]}(\nu) 1_{(\nu_3,\nu_4]}(\nu) \, \mu(d\nu)$$

$$= \mu((\nu_1, \nu_2] \cap (\nu_3, \nu_4]).$$

We can therefore define the Doob integral $\int_{\mathbb{R}} f(\nu) \, dx(\nu)$ for all $f \in L_{\mathbb{C}}^2(\mu)$.

4. Let now

$$Z_n(t) := \sum_{k \in \mathbb{Z}} e^{2i\pi t(k/2^n)} \left(x\left(\frac{k+1}{2^n} \right) - x\left(\frac{k}{2^n} \right) \right).$$

We have

$$\lim_{n \to \infty} Z_n(t) = \int_{\mathbb{R}} e^{2i\pi\nu t} \, dx(\nu)$$

(limit in $L_{\mathbb{C}}^2(P)$) because

$$Z_n(t) = \int_{\mathbb{R}} f_n(t, \nu) \, dx(\nu),$$

where

$$f_n(t, \nu) = \sum_{k \in \mathbb{Z}} e^{2i\pi t(k/2^n)} 1_{(k/2^n,(k+1)/2^n]}(\nu),$$

and therefore, by isometry,

$$E\left| Z_n(t) - \int_{\mathbb{R}} e^{2i\pi\nu t}\, dx(\nu) \right|^2 = \int_{\mathbb{R}} |e^{2i\pi\nu t} - f_n(t, \nu)|^2 \, \mu(d\nu),$$

a quantity which tends to zero when n tends to infinity (by dominated convergence, using the fact that μ is a bounded measure). On the other hand, by definition of φ,

$$Z_n(t) \xrightarrow{\varphi} f_n(t, \nu).$$

Since, for fixed t, $\lim_{n\to\infty} Z_n(t) = \int_{\mathbb{R}} e^{2i\pi\nu t}\, dx(\nu)$ in $L^2_{\mathbb{C}}(P)$ and $\lim_{n\to\infty} f_n(t, \nu) = e^{2i\pi\nu t}$ in $L^2_{\mathbb{C}}(\mu)$,

$$\int_{\mathbb{R}} e^{2i\pi\nu t}\, dx(\nu) \xrightarrow{\varphi} e^{2i\pi\nu t}.$$

But, by definition of φ,

$$X(t) \xrightarrow{\varphi} e^{2i\pi\nu t}.$$

Therefore $X(t) = \int_{\mathbb{R}} e^{2i\pi\nu t}\, dx(\nu)$.

5. We now prove uniqueness. Suppose that there exists another spectral decomposition $d\tilde{x}(\nu)$. Denote by $\mathcal{G}$ the set of finite linear combinations of complex exponentials. Since by hypothesis

$$\int_{\mathbb{R}} e^{2i\pi\nu t}\, dx(\nu) = \int_{\mathbb{R}} e^{2i\pi\nu t}\, d\tilde{x}(\nu) \quad (= X(t))$$

we have

$$\int_{\mathbb{R}} f(\nu)\, dx(\nu) = \int_{\mathbb{R}} f(\nu)\, d\tilde{x}(\nu)$$

for all $f \in \mathcal{G}$, and therefore, for all $f \in L^2_{\mathbb{C}}(\mu) \cap L^2_{\mathbb{C}}(\tilde{\mu}) \subseteq L^2_{\mathbb{C}}(\frac{1}{2}(\mu+\tilde{\mu}))$ because $\mathcal{G}$ is dense in $L^2_{\mathbb{C}}(\frac{1}{2}(\mu + \tilde{\mu}))$ (Corollary A.3.1). In particular, with $f = 1_{(a,b]}$,

$$x(b) - x(a) = \tilde{x}(b) - \tilde{x}(a). \qquad \square$$

3.4 Fourier Analysis of the Trajectories

More details can be obtained as to the continuity properties (in quadratic mean) of the increments of the spectral decomposition. For instance, it is right-continuous in quadratic mean, and it admits a left-hand limit in quadratic mean at any point $\nu \in \mathbb{R}$. If such limit is denoted by $x(\nu-)$, then, for all $a \in \mathbb{R}$,

$$E[|x(a) - x(a-)|^2] = \mu(\{a\}).$$

Proof The right-continuity follows from the continuity of the (finite) measure μ:

$$\lim_{h\downarrow 0} E[|x(a+h) - x(a)|^2] = \lim_{h\downarrow 0} \mu((a, a+h]) = \mu(\varnothing) = 0.$$

As for the existence of left-hand limits, it is guaranteed by the Cauchy criterion, since for all $a \in \mathbb{R}$,

$$\lim_{h,h'\downarrow 0, h<h'} E[|x(a-h) - x(a-h')|^2] = \lim_{h,h'\downarrow 0, h<h'} \mu((a-h', a-h]) = 0.$$

Finally,

$$E[|x(a) - x(a-)|^2] = \lim_{h\downarrow 0} E[|x(a) - x(a-h)|^2] = \lim_{h\downarrow 0} \mu((a-h, a]) = \mu(\{a\}).$$

□

Theorem 3.4.2 *Let $\{X(t)\}_{t\in\mathbb{R}}$ be WSS stochastic process continuous in quadratic mean. It is real if and only if its spectral decomposition is hermitian symmetric, that is, for all $[a, b] \subset \mathbb{R}$,*

$$x(b) - x(a) = (x(-a_-) - x(-b_-))^*$$

Proof If the stochastic process is real,

$$X(t) = \int_{\mathbb{R}} e^{2i\pi\nu t}\, dx(\nu) = \left(\int_{\mathbb{R}} e^{2i\pi\nu t}\, dx(\nu)\right)^*$$
$$= \int_{\mathbb{R}} e^{-2i\pi\nu t}\, dx^*(\nu) = \int_{\mathbb{R}} e^{2i\pi\nu t}\, dx^*(-\nu)$$

and therefore, by uniqueness of the spectral decomposition, $dx(\nu) = dx^*(-\nu)$. Similarly, if $dx(\nu) = dx^*(-\nu)$,

$$X(t) = \int_{\mathbb{R}} e^{2i\pi\nu t}\, dx(\nu)$$

$$= \int_{\mathbb{R}} e^{2i\pi\nu t}\, dx^*(-\nu) = \left(\int_{\mathbb{R}} e^{2i\pi\nu t}\, dx(\nu)\right)^* = X(t)^*$$

and therefore the process is real. □

Theorem 3.4.3 *Let* $\{X(t)\}_{t\in\mathbb{R}}$ *be a centered* WSS *stochastic process continuous in quadratic mean. Then*

$$H_C(x(\nu); \nu \in \mathbb{R}) = H_C(X(t); t \in \mathbb{R})$$

and both Hilbert subspaces are identical with

$$\left\{\int_{\mathbb{R}} g(\nu)\, dx(\nu);\ g \in L^2_{\mathbb{C}}(\mu)\right\}.$$

Proof 1. For all $\nu \in \mathbb{R}$, $x(\nu) \in H_{\mathbb{C}}(X(t); t \in \mathbb{R})$ (by definition of $x(\nu)$; see the proof of Theorem 3.4.1). Therefore,

$$H_C(x(\nu); \nu \in \mathbb{R}) \subseteq H_{\mathbb{C}}(X(t); t \in \mathbb{R}).$$

On the other hand, for all $t \in \mathbb{R}$, $X(t) = \int_{\mathbb{R}} e^{-2i\pi\nu t}\, dx(\nu) \in H_{\mathbb{C}}(x(\nu); \nu \in \mathbb{R})$. Therefore

$$H_{\mathbb{C}}(X(t); t \in \mathbb{R}) \subseteq H_{\mathbb{C}}(x(\nu); \nu \in \mathbb{R}).$$

2. Clearly $H := \{\int_{\mathbb{R}} g(\nu)\, dx(\nu);\ g \in L^2_{\mathbb{C}}(\mu)\} \subseteq H \subseteq H_C(x(\nu))$. Moreover $H_{\mathbb{C}}(X(t); t \in \mathbb{R}) \subseteq H$ since H contains all the variables $X(t) = \int_{\mathbb{R}} e^{-2i\pi\nu t}\, dx(\nu)$. Therefore

$$H_{\mathbb{C}}(X(t); t \in \mathbb{R}) \subseteq H \subseteq H_C(x(\nu))$$

and the conclusion follows from Part 1 of the proof. □

WSS random fields

Consider a WSS random field $\{X(t)\}_{t\in\mathbb{R}^m}$ with mean $E[X(t)] = m$ and *continuous* covariance function

$$C(\tau) := \mathrm{cov}\,(X(t+\tau), X(t)) := E\left[(X(t+\tau) - m)(X(t) - m)^T\right].$$

Theorem 3.4.4 *There exists a unique finite measure μ on $\mathbb{R}^m$ such that the covariance function of the wide-sense stationary random field $\{X(t)\}_{t \in \mathbb{R}^m}$ admits the representation*

$$C(\tau) = \int_{\mathbb{R}^m} e^{2i\pi \langle \nu, \tau \rangle} \, \mu(d\nu)$$

(where $\langle \cdot, \cdot \rangle$ is the scalar product of $\mathbb{R}^m$), called the power spectral measure of the random field.

The proof of this is the same as in the case of stochastic processes, *mutatis mutandis*.

Theorem 3.4.5 *Let $\{X(t)\}_{t \in \mathbb{R}^m}$ be a centered random field with power spectral measure μ. There exists a random pseudo-measure $\{x(A)\}_{A \in \mathcal{B}(\mathbb{R}^m)}$ whose structural measure is the power spectral measure μ of the random field, and such that for all t, P-a.s.,*

$$X(t) = \int_{\mathbb{R}^m} e^{2i\pi \langle \nu, \tau \rangle} \, x(d\nu).$$

The proof is the same as that of Theorem 3.4.1, except that we do not define a process $\{x(\nu)\}_{\nu \in \mathbb{R}}$ but rather a random field $\{x(A)\}_{A \in \mathcal{B}(\mathbb{R}^m)}$. Looking back at the proof of Theorem 3.4.1, step 3 thereof, $x(A)$ is now defined as the random variable in $H(X)$ that corresponds in the isometry to the function $1_A \in L^2_{\mathbb{C}}(\mu)$. Only step 4 has to be changed slightly, defining this time

$$Z_n(t) = \sum_{k=(k_1,\ldots,k_m) \in \mathbb{Z}^m} e^{2i\pi \langle t, \frac{k}{2^n} \rangle} x(A_k^n)$$

where $A_k^n = \prod_{j=1}^m (\frac{k_j}{2^n}, \frac{k_j+1}{2^n}]$.

A Plancherel–Parseval Formula

The following result is the analog of the Plancherel–Parseval formula of classical Fourier analysis.

Theorem 3.4.6 *Let $f : \mathbb{R} \to \mathbb{C}$ be in $L^1_{\mathbb{C}}(\mathbb{R})$ with Fourier transform $\widehat{f}$. Let $\{X(t)\}_{t \in \mathbb{R}}$ be a centered WSS stochastic process with power spectral measure μ and Cramér–Khintchin spectral decomposition $dx(\nu)$. Then:*

$$\int_{\mathbb{R}} \widehat{f}(\nu)^* dx(\nu) = \int_{\mathbb{R}} f(t)^* X(t) \, dt. \tag{3.26}$$

Proof The function $\widehat{f}$ is bounded and continuous (as the Fourier transform of an integrable function) and μ is a finite measure, we have that $\widehat{f} \in L^2_{\mathbb{C}}(\mu)$, and

$$\sum_n \widehat{f}\left(\frac{k}{2^n}\right) 1_{(\frac{k}{2^n},\frac{k+1}{2^n}]} \to \widehat{f} \text{ in } L^2_{\mathbb{C}}(\mu).$$

Therefore (all limits in the following sequence of equalities are in $L^2_{\mathbb{C}}(P)$):

$$\int_{\mathbb{R}} \widehat{f}(\nu)^* \, dx(\nu) = \lim_{n\to\infty} \sum_{-n2^n}^{n2^n-1} \widehat{f}\left(\frac{k}{2^n}\right)^* \left(x\left(\frac{k+1}{2^n}\right) - x\left(\frac{k}{2^n}\right)\right)$$

$$= \lim_{n\to\infty} \sum_{-n2^n}^{n2^n-1} \left(\int_{\mathbb{R}} f^*(t)e^{+2i\pi(k/2^n)t} \, dt\right) \left(x\left(\frac{k+1}{2^n}\right) - x\left(\frac{k}{2^n}\right)\right)$$

$$= \lim_{n\to\infty} \int_{\mathbb{R}} f^*(t) \sum_{-n2^n}^{n2^n-1} \left[e^{+2i\pi(k/2^n)t}\left(x\left(\frac{k+1}{2^n}\right) - x\left(\frac{k}{2^n}\right)\right)\right] dt$$

$$= \lim_{n\to\infty} \int_{\mathbb{R}} f^*(t)X_n(t) \, dt,$$

where

$$X_n(t) = \sum_{-n2^n}^{n2^n-1} e^{+2i\pi(k/2^n)t} \left(x\left(\frac{k+1}{2^n}\right) - x\left(\frac{k}{2^n}\right)\right) \to X(t) \text{ in } L^2_{\mathbb{C}}(P).$$

The announced result will then follow once we prove that

$$\lim_{n\to\infty} \int_{\mathbb{R}} f^*(t)X_n(t) \, dt = \int_{\mathbb{R}} f^*(t)X(t) \, dt,$$

where the limit is in $L^2_{\mathbb{C}}(P)$. In fact, with $Y_n(t) = X(t) - X_n(t)$,

$$E\left[\left|\int_{\mathbb{R}} f(t)Y_n(t) \, dt\right|^2\right] = \int_{\mathbb{R}}\int_{\mathbb{R}} f(t)f(s)^* E\left[Y_n(t)Y_n(s)^*\right] dt \, ds.$$

But for all $t \in \mathbb{R}$, $\lim_{n\uparrow\infty} Y_n(t) = 0$ (in $L^2_{\mathbb{C}}(P)$) and therefore $\lim_{n\uparrow\infty} E\left[Y_n(t)Y_n(s)^*\right] = 0$. Moreover $E\left[Y_n(t)Y_n(s)^*\right]$ is uniformly bounded in n. Therefore, by dominated convergence,

$$\lim_{n \uparrow \infty} \int_{\mathbb{R}} \int_{\mathbb{R}} f(t) f(s)^* E\left[Y_n(t) Y_n(s)^*\right] dt \, ds = 0.$$

$\square$

Example 3.4.1 (Convolutional filtering) Let $h \in L^1_{\mathbb{C}}(\mathbb{R})$ and let $\widehat{h}$ be its Fourier transform. Then

$$\int_{\mathbb{R}} h(t - s) X(s) \, ds = \int_{\mathbb{R}} \widehat{h}(\nu) e^{2i\pi\nu t} \, dx(\nu) \tag{3.27}$$

Proof It suffices to apply (3.26) to the function $s \mapsto h^*(t - s)$, whose Fourier transform is $\widehat{h}(\nu)^* e^{-2i\pi\nu t}$. $\square$

Shannon–Nyquist Theorem for Stochastic Processes

We proceed to give the stochastic version of the Shannon–Nyquist theorem. We start with a definition.

Definition 3.4.1 A WSS stochastic process $\{X(t)\}_{t \in \mathbb{R}}$ is said to be base-band of bandwidth $2B$ ($B > 0$), or base-band (B), if the support of its spectral power measure is contained in the frequency interval $[-B, +B]$.

Theorem 3.4.7 *Consider a centered WSS stochastic processes $\{X(t)\}_{t \in \mathbb{R}}$ that is base-band (B). Then, for any $T > 1/(2B)$,*

$$X(t) = \lim_{N \uparrow \infty} \sum_{n=-N}^{+N} X(nT) \frac{\sin\left(\frac{\pi}{T}(t - nT)\right)}{\frac{\pi}{T}(t - nT)}, \tag{3.28}$$

where the limit is in the quadratic mean.

Proof We have

$$X(t) = \int_{[-B, +B]} e^{2i\pi\nu t} \, dx(\nu).$$

Now,

$$e^{2i\pi\nu t} = \lim_{N \uparrow \infty} \sum_{n=-N}^{+N} e^{2i\pi\nu nT} \frac{\sin\left(\frac{\pi}{T}(t - nT)\right)}{\frac{\pi}{T}(t - nT)},$$

where the limit is uniform in $[-B, +B]$ and bounded (Theorem 2.1.4). Therefore the above limit is also in $L^2_{\mathbb{C}}(\mu)$ because μ is a finite measure. Consequently

$$X(t) = \lim_{N \uparrow \infty} \int_{[-B,+B]} \left\{ \sum_{n=-N}^{+N} e^{2i\pi\nu nT} \frac{\sin\left(\frac{\pi}{T}(t-nT)\right)}{\frac{\pi}{T}(t-nT)} \right\} dx(\nu),$$

where the limit is in $L^2_{\mathbb{C}}(P)$. The result then follows by expanding the integral with respect to the sum.   □

A version of the Shannon–Nyquist theorem for random fields is available, with a similar proof (Exercise 3.5.20).

3.4.2 Linear Operations

A function $g : \mathbb{R} \to \mathbb{C}$ in $L^2_{\mathbb{C}}(\mu)$ defines a linear operation on the centered WSS stochastic process $\{X(t)\}_{t \in \mathbb{R}}$ (called the *input*) by associating with it the centered stochastic process (called the *output*)

$$Y(t) = \int_{\mathbb{R}} e^{2i\pi\nu t} g(\nu) \, dx(\nu). \tag{3.29}$$

On the other hand, the calculation of the covariance function

$$C_Y(\tau) = E[Y(t)Y(t+\tau)^*]$$

of the output gives (isometry formula for the Doob–Wiener integral),

$$C_Y(\tau) = \int_{\mathbb{R}} e^{2i\pi\nu\tau} |g(\nu)|^2 \, \mu_X(d\nu),$$

where μ_X is the power spectral measure of the input. The power spectral measure of the output process is therefore

$$\mu_Y(d\nu) = |g(\nu)|^2 \, \mu_X(d\nu). \tag{3.30}$$

This is similar to the formula obtained when $\{Y(t)\}_{t \in \mathbb{R}}$ is the output of a stable convolutional filter with impulse response h and transmittance $\hat{h}$: $\mu_Y(d\nu) = |\hat{h}(\nu)|^2 \mu_X(d\nu)$. We therefore say that g is the *transmittance* of the "filter" (3.29). Note however that this filter is not necessarily of the convolutional type, since g may well not be the Fourier transform of an integrable function (for instance it may be unbounded, as the next example shows).

Example 3.4.2 (Differentiation) Let $\{X(t)\}_{t \in \mathbb{R}}$ be a WSS stochastic processes with spectral measure μ_X such that

$$\int_{\mathbb{R}} |\nu|^2 \, \mu_X(d\nu) < \infty. \tag{3.31}$$

Then

$$\lim_{h \to 0} \frac{X(t+h) - X(t)}{h} = \int_{\mathbb{R}} (2i\pi\nu) e^{2i\pi\nu t} dx(\nu),$$

where the limit is in the quadratic mean. The linear operation corresponding to the transmittance $g(\nu) = 2i\pi\nu$ is therefore the *differentiation in quadratic mean*.

Proof Let $h \in \mathbb{R}$. From the equality

$$\frac{X(t+h) - X(t)}{h} - \int_{\mathbb{R}} (2i\pi\nu) e^{2i\pi\nu t} dx(\nu) = \int_{\mathbb{R}} e^{2i\pi\nu t} \left(\frac{e^{2i\pi\nu h} - 1}{h} - 2i\pi\nu \right) dx(\nu)$$

we have, by isometry,

$$\lim_{h \to 0} E\left[\left| \frac{X(t+h) - X(t)}{h} - \int_{\mathbb{R}} (2i\pi\nu) e^{2i\pi\nu t} dx(\nu) \right|^2 \right]$$

$$= \lim_{h \to 0} \int_{\mathbb{R}} \left| \frac{e^{2i\pi\nu h} - 1}{h} - 2i\pi\nu \right|^2 \mu_X(d\nu).$$

In view of hypothesis (3.31) and since $\left| \frac{e^{2i\pi\nu h}-1}{h} - 2i\pi\nu \right|^2 \leq 4\pi^2\nu^2$, the latter limit is 0, by dominated convergence. $\square$

"A line spectrum corresponds to a combination of sinusoids." More precisely:

Theorem 3.4.8 *Let $\{X(t)\}_{t \in \mathbb{R}}$ be a centered WSS stochastic processes with spectral measure*

$$\mu_X(d\nu) = \sum_{k \in \mathbb{Z}} P_k \, \varepsilon_{\nu_k}(d\nu),$$

where ε_{ν_k} is the Dirac measure at $\nu_k \in \mathbb{R}$, $P_k \in \mathbb{R}_+$ and $\sum_{k \in \mathbb{Z}} P_k < \infty$. Then

$$X(t) = \sum_{k \in \mathbb{Z}} U_k e^{2i\pi\nu_k t}$$

where $\{U_k\}_{k \in \mathbb{Z}}$ is a sequence of centered uncorrelated square-integrable complex variables, and $E[|U_k|^2] = P_k$.

Proof The function

$$g(\nu) = \sum_{k\in\mathbb{Z}} 1_{\{\nu_k\}}(\nu)$$

is in $L^2_{\mathbb{C}}(\mu_X)$, as well as the function $1 - g$. Also $\int_{\mathbb{R}} |1 - g(\nu)|^2 \mu_X(d\nu) = 0$, and in particular $\int_{\mathbb{R}} (1 - g(\nu))e^{2i\pi\nu t}\, dx(\nu) = 0$. Therefore

$$X(t) = \int_{\mathbb{R}} g(\nu)e^{2i\pi\nu t}\, dx(\nu)$$

$$= \sum_{k\in\mathbb{Z}} e^{2i\pi\nu_k t}(x(\nu_k) - x(\nu_k-)).$$

We conclude by defining $U_k = x(\nu_k) - x(\nu_k-)$.   □

Linear Transformations of Gaussian Processes

We call *linear transformation* of the WSS stochastic process $\{X(t)\}_{t\in\mathbb{R}}$ a transformation of it into the second-order process (not WSS in general)

$$Y(t) = \int_{\mathbb{R}} g(\nu, t)\, dx(\nu), \tag{3.32}$$

where

$$\int_{\mathbb{R}} |g(t,\nu)|^2 \mu_X(d\nu) < \infty \quad \text{for all } t \in \mathbb{R}.$$

Theorem 3.4.9 *Every linear transformation of a Gaussian* WSS *stochastic process yields a Gaussian stochastic process.*

Proof Let $\{X(t)\}_{t\in\mathbb{R}}$ be centered, Gaussian, WSS, with Cramér–Khinchin decomposition $dx(\nu)$. For each $\nu \in \mathbb{R}$, the random variable $x(\nu)$ is in $H_{\mathbb{R}}(X)$, by construction. Now, if $\{X(t)\}_{t\in\mathbb{R}}$ is a Gaussian process, $H_{\mathbb{R}}(X)$ is a Gaussian subspace. But (Theorem 3.4.3) $H_{\mathbb{R}}(X) = H_{\mathbb{R}}(x)$. Therefore the process (3.32) is in $H_{\mathbb{C}}(X)$, hence Gaussian.   □

Example 3.4.3 (Convolutional filtering of a WSS Gaussian process) In particular, if $\{X(t)\}_{t\in\mathbb{R}}$ is a Gaussian WSS process with Cramér–Khinchin decomposition $dx(\nu)$, and if $g \in L^2_{\mathbb{C}}(\mu_X)$, the process

$$Y(t) = \int_{\mathbb{R}} e^{2i\pi\nu t} g(\nu)\, dx(\nu)$$

is Gaussian. A particular case is when $g = \widehat{h}$, the Fourier transform of a filter with integrable impulse response h; the signal $\{Y(t)\}_{t\in\mathbb{R}}$ is the one obtained by convolutional filtering of $\{X(t)\}_{t\in\mathbb{R}}$ with this filter.

3.4.3 Multivariate WSS Stochastic Processes

The Power Spectral Matrix

Let $\{X(t)\}_{t\in\mathbb{R}}$ be a stochastic process with values in $E := \mathbb{C}^L$, where L is an integer greater than or equal to 2: $X(t) = (X_1(t), \ldots, X_L(t))$. This process is assumed to be of the second order, that is:

$$E[||X(t)||^2] < \infty \quad \text{for all } t \in \mathbb{R},$$

and centered. Furthermore it will be assumed that it is wide-sense stationary, in the sense that the mean vector of $X(t)$ and the cross-covariance matrix of the vectors $X(t + \tau)$ and $X(t)$ do not depend upon t. The matrix-valued function C defined by

$$C(\tau) = \text{cov } (X(t + \tau), X(t)) \tag{3.33}$$

is called the (matrix) *covariance function* of the stochastic process. Its general entry is

$$C_{ij}(\tau) = \text{cov}(X_i(t), X_j(t + \tau)).$$

Therefore, each of the processes $\{X_i(t)\}_{t\in\mathbb{R}}$ is a WSS stochastic process, but, furthermore, they are *stationarily correlated* or "jointly WSS". The vector-valued stochastic process $\{X(t)\}_{t\in\mathbb{R}}$ is then called a *multivariate* WSS *stochastic process*.

Example 3.4.4 (Signal plus noise) The following model frequently appears in signal processing:

$$Y(t) = S(t) + B(t),$$

where $\{S(t)\}_{t\in\mathbb{R}}$ and $\{B(t)\}_{t\in\mathbb{R}}$ are two *uncorrelated* centered WSS stochastic process with respective covariance functions C_S and C_B. Then, $\{(Y(t), S(t))^T\}_{t\in\mathbb{R}}$ is a bivariate WSS stochastic process. In fact, owing to the assumption of non-correlation:

$$C(\tau) = \begin{pmatrix} C_S(\tau) + C_B(\tau) & C_S(\tau) \\ C_S(\tau) & C_S(\tau) \end{pmatrix}.$$

Theorem 3.4.10 *Let $\{X(t)\}_{t\in\mathbb{R}}$ be a L-dimensional multivariate WSS stochastic process. For all r, s ($1 \leq r, s \leq L$) there exists a finite complex measure μ_{rs} such that*

$$C_{rs}(\tau) = \int_{\mathbb{R}} e^{2i\pi\nu\tau} \mu_{rs}(d\nu). \tag{3.34}$$

Proof (The case $r = 1$, $s = 2$). Let us consider the stochastic processes

$$Y(t) = X_1(t) + X_2(t), \qquad Z(t) = iX_1(t) + X_2(t).$$

These are WSS stochastic processes with respective covariance functions

$$C_Y(\tau) = C_1(\tau) + C_2(\tau) + C_{12}(\tau) + C_{21}(\tau),$$
$$C_Z(\tau) = -C_1(\tau) + C_2(\tau) + iC_{12}(\tau) - iC_{21}(\tau).$$

From these two equalities we deduce

$$C_{12}(\tau) = \frac{1}{2} \{[C_Y(\tau) - C_1(\tau) - C_2(\tau)] - i[C_Z(\tau) - C_1(\tau) + C_2(\tau)]\},$$

from which the result follows with

$$\mu_{12} = \frac{1}{2} \{[\mu_Y - \mu_1 - \mu_2] - i[\mu_Z - \mu_1 + \mu_2]\}.$$

$\square$

The matrix

$$M := \{\mu_{ij}\}_{1 \le i, j \le k}$$

(whose entries are finite complex measures) is the *interspectral power measure matrix* of the multivariate WSS stochastic process $\{X(t)\}_{t \in \mathbb{R}}$. It is clear that for all $z = (z_1, \ldots, z_k) \in \mathbb{C}^k$, $U(t) = z^T X(t)$ defines a WSS stochastic process with spectral measure $\mu_U = z M z^{\dagger}$ († means transpose conjugate).

The link between the interspectral measure μ_{12} and the Cramér–Khintchine decompositions $dx_1(\nu)$ and $dx_2(\nu)$ is the following:

$$E[x_1(\nu_2) - x_1(\nu_1))(x_2(\nu_4) - x_2(\nu_3))^*] = \mu_{12}((\nu_1, \nu_2] \cup (\nu_3, \nu_4]).$$

This is a particular case of the following: for all functions $g_i : \mathbb{R} \to \mathbb{C}$, $g_i \in L^2_{\mathbb{C}}(\mu_i)$ $(i = 1, 2)$

$$E\left[\left(\int_{\mathbb{R}} g_1(\nu) dx_1(\nu)\right)\left(\int_{\mathbb{R}} g_2(\nu) dx_2(\nu)\right)^*\right] = \int_{\mathbb{R}} g_1(\nu) g_2(\nu)^* \mu_{12}(d\nu). \tag{3.35}$$

Indeed, equality (3.35) is true for $g_1(\nu) = e^{2i\pi t_1 \nu}$, $g_2(\nu) = e^{2i\pi t_2 \nu}$, since it then reduces to

$$E[X_1(t)X_2(t)^*] = \int_\pi e^{2i\pi(t_1-t_2)\nu} \mu_{12}(d\nu).$$

This is therefore verified for $g_1 \in \mathcal{E}$, $g_2 \in \mathcal{E}$, where $\mathcal{E}$ is the set of finite linear combinations of functions of the type $\nu \to e^{2i\pi t \nu}$, $t \in \mathbb{R}$. But $\mathcal{E}$ is dense in $L_\mathbb{C}^2(\mu_i)$ ($i = 1, 2$), and therefore the equality (3.35) is true for all $g_i \in L_\mathbb{C}^2(\mu_i)$ ($i = 1, 2$).

Theorem 3.4.11 *The interspectral measure μ_{12} is absolutely continuous with respect to each of the spectral measures μ_1 and μ_2.*

Proof This means that $\mu_{12}(A) = 0$ whenever $\mu_1(A) = 0$ or $\mu_2(A) = 0$. Indeed,

$$\mu_{12}(A) = E\left[\left(\int_A dZ_1\right)\left(\int_A dZ_2\right)^*\right]$$

and $\mu_1(A) = 0$ implies $\int_A dZ_1 = 0$ since

$$E\left[\left|\int_A dZ_1\right|^2\right] = \mu_1(A).$$

$\square$

Therefore, each of the spectral measures μ_{ij} is absolutely continuous with respect to the trace of the power spectral measure matrix

$$\text{Tr } M = \sum_{j=1}^{k} \mu_j.$$

By the Radon–Nikodym theorem there exists a function $g_{ij} : \mathbb{R} \to \mathbb{C}$ such that

$$\mu_{ij}(A) = \int_A g_{ij}(\nu) \text{ Tr } M(d\nu).$$

The matrix

$$g(\nu) = \{g_{ij}(\nu)\}_{1\leq i, j\leq k}$$

is called the *canonical spectral density* matrix of $\{X(t)\}_{t\in\mathbb{R}}$. One should insist that it is not required that the stochastic processes $\{X_i(t)\}_{t\in\mathbb{R}}$, $1 \leq i \leq k$, admit power spectral densities.

The correlation matrix $C(\tau)$ has, with the above notations, the representation

$$C(\tau) = \int_{\mathbb{R}} e^{2i\pi\nu\tau} g(\nu) \operatorname{Tr} M(d\nu).$$

If each of the WSS stochastic processes $\{X_i(t)\}_{t\in\mathbb{R}}$ admits a spectral density, $\{X(t)\}_{t\in\mathbb{R}}$ admits an interspectral density matrix

$$f(\nu) = \{f_{ij}(\nu)\}_{1\leq i,j\leq k},$$

that is:

$$C_{ij}(\tau) = \operatorname{cov}\left(X_i(t+\tau), X_j(t)\right) = \int_{\mathbb{R}} e^{2i\pi\nu\tau} f_{ij}(\nu)\, d\nu.$$

Example 3.4.5 (Interferences) Let $\{X(t)\}_{t\in\mathbb{R}}$ be a centered WSS stochastic process with power spectral measure μ_X. Let $h_1, h_2 : \mathbb{R} \to \mathbb{C}$ be integrable functions with respective Fourier transforms $\widehat{h}_1$ and $\widehat{h}_2$. Define for $i = 1, 2$,

$$Y_i(t) = \int_{\mathbb{R}} h_i(t-s)X(s)\, ds$$

The WSS stochastic processes $\{Y_1(t)\}_{t\in\mathbb{R}}$ and $\{Y_2(t)\}_{t\in\mathbb{R}}$ are stationarily correlated. In fact (assuming that they are centered, without loss of generality),

$$\begin{aligned}
E[Y_1(t+\tau)Y_2(t)^*] &= E\left[\left(\int_{\mathbb{R}} h_1(t+\tau-s)X(s)\, ds\right)\left(\int_{\mathbb{R}} h_2(t-s)X(s)\, ds\right)^*\right] \\
&= \int_{\mathbb{R}}\int_{\mathbb{R}} h_1(t+\tau-u)h_2^*(t-v)C_X(u-v)\, du\, dv \\
&= \int_{\mathbb{R}}\int_{\mathbb{R}} h_1(\tau-u)h_2^*(-v)C_X(u-v)\, du\, dv,
\end{aligned}$$

and this quantity depends only upon τ. Replacing $C_X(u-v)$ by its expression in terms of the spectral measure μ_X, one obtains

$$C_{Y_1Y_2}(\tau) = \int_{\mathbb{R}} e^{2i\pi\nu\tau} T_1(\nu)T_2^*(\nu)\, \mu_X(d\nu).$$

The power spectral matrix of the bivariate process $\{Y_1(t), Y_2(t)\}_{t\in\mathbb{R}}$ is therefore

$$\mu_Y(d\nu) = \begin{pmatrix} |T_1(\nu)|^2 & T_1(\nu)T_2^*(\nu) \\ T_1^*(\nu)T_2(\nu) & |T_2(\nu)|^2 \end{pmatrix} \mu_X(d\nu).$$

Band-Pass Stochastic Processes

We appply the above results to *band-pass* WSS stochastic processes.

Let $\{X(t)\}_{t \in \mathbb{R}}$ be a centered WSS stochastic process with power spectral measure μ_X and Cramér–Khinchin decomposition $dx(\nu)$. This process is assumed real, and therefore we have the symmetries

$$\mu_X(-d\nu) = \mu_X(d\nu), \qquad dx(-\nu) = dx(\nu)^*.$$

Definition 3.4.2 The above WSS stochastic process is called *base-band* (B), where $0 < B < +\infty$, if the support of μ_X is contained in the frequency band $[-B, +B]$. It is called *band-pass* (ν_0, B), where $\nu_0 > B > 0$, if the support of μ_X is included in the frequency band $[-\nu_0 - B, -\nu_0 + B] \cup [\nu_0 - B, \nu_0 + B]$.

Similarly to Example 1.1.8 concerning band-pass deterninistic functions, a band-pass stochastic process admits the following *quadrature decomposition*

$$X(t) = M(t) \cos 2\pi\nu_0 t - N(t) \sin 2\pi\nu_0 t, \qquad (3.36)$$

where $\{M(t)\}_{t \in \mathbb{R}}$ and $\{N(t)\}_{t \in \mathbb{R}}$, the *quadrature components*, are real base-band (B) WSS stochastic process. To prove this, let $G(\nu) := -i \operatorname{sign}(\nu)$ $(= 0$ if $\nu = 0)$. The function G be the so-called *Hilbert filter* transmittance. The *quadrature process* associated with $\{X(t)\}_{t \in \mathbb{R}}$ is defined by

$$Y(t) = \int_{\mathbb{R}} G(\nu) e^{2i\pi\nu t} dx(\nu).$$

The right-hand side of the preceding equality is well defined since $\int_{\mathbb{R}} |G(\nu)|^2 \mu_X(d\nu) = \mu_X(\mathbb{R}) < \infty$. Moreover, this stochastic process is real, since its spectral decomposition is hermitian symmetric. The *analytic process* associated with $\{X(t)\}_{t \in \mathbb{R}}$ is, by definition, the stochastic process

$$Z(t) = X(t) + i Y(t) = \int_{\mathbb{R}} (1 + i G(\nu)) e^{2i\pi\nu t} dx(\nu) = 2 \int_{(0,\infty)} e^{2i\pi\nu t} dx(\nu).$$

Taking into account that $|G(\nu)|^2 = 1$, the preceding expressions and the Wiener isometry formulas lead to the following properties (Exercise 3.5.21):

$$\mu_Y(d\nu) = \mu_X(d\nu), \quad C_Y(\tau) = C_X(\tau), \quad C_{XY}(\tau) = -C_{YX}(\tau),$$

$$\mu_Z(d\nu) = 4\, 1_{\mathbb{R}_+}(\nu)\, \mu_X(d\nu), \qquad C_Z(\tau) = 2\{C_X(\tau) + i C_{YX}(\tau)\},$$

and

$$E[Z(t + \tau)Z(t)] = 0. \tag{$*$}$$

Defining the *complex envelope* of $\{X(t)\}_{t \in \mathbb{R}}$ by

$$U(t) = Z(t)e^{-2i\pi\nu_0 t}, \tag{$**$}$$

it follows from this definition that

$$C_U(\tau) = e^{-2i\pi\nu_0\tau}C_Z(\tau), \qquad \mu_U(d\nu) = \mu_Z(d\nu + \nu_0), \tag{$\dagger$}$$

whereas $(*)$ and $(**)$ give

$$E[U(t + \tau)U(t)] = 0. \tag{$\dagger\dagger$}$$

The quadrature components $\{M(t)\}_{t \in \mathbb{R}}$ and $\{N(t)\}_{t \in \mathbb{R}}$ of $\{X(t)\}_{t \in \mathbb{R}}$ are the *real* WSS stochastic processes defined by

$$U(t) = M(t) + iN(t).$$

Since

$$X(t) = \mathrm{Re}\{Z(t)\} = \mathrm{Re}\{U(t)e^{2i\pi\nu_0 t}\},$$

we have the decomposition (3.36). Taking $(\dagger\dagger)$ into account we obtain:

$$C_M(\tau) = C_N(\tau) = \frac{1}{4}\left\{C_U(\tau) + C_U(\tau)^*\right\},$$

and

$$C_{MN}(\tau) = C_{NM}(\tau) = \frac{1}{4i}\left\{C_U(\tau) - C_U(\tau)^*\right\}, \tag{$\diamond$}$$

and the corresponding relations for the spectra

$$\mu_M(d\nu) = \mu_N(d\nu) = \{\mu_X(d\nu - \nu_0) + \mu_X(d\nu + \nu_0)\}\, 1_{[-B,+B]}(\nu).$$

From $(\diamond)$ and the observation that $C_U(0) = C_U(0)^*$ (since $C_U(0) = E[|U(0)|^2]$ is real), we deduce $C_{MN}(0) = 0$, that is to say,

$$E[M(t)N(t)] = 0. \tag{3.37}$$

If, furthermore, the original process has a power spectral measure that is symmetric about ν_0 in the band $[\nu_0 - B, \nu_0 + B]$, the same holds for the spectrum of the analytic process, and, by (†), the complex envelope has a spectral measure symmetric about 0, which implies $C_U(\tau) = C_U(\tau)^*$, and thus, by (◊),

$$E[M(t)N(t + \tau)] = 0. \tag{3.38}$$

In summary :

Theorem 3.4.12 *Let* $\{X(t)\}_{t \in \mathbb{R}}$ *be a centered real* WSS *stochastic process, band-pass* (ν_0, B). *The values of its quadrature components at a given time are uncorrelated. Moreover, if the original stochastic process has a power spectral measure symmetric about* ν_0 *in the band* $[\nu_0 - B, \nu_0 + B]$, *the quadrature component processes are uncorrelated.*

More can be said when the original process is Gaussian. In this case, the quadrature component processes are jointly Gaussian (being obtained from the original Gaussian process by linear operations). In particular, for all $t \in \mathbb{R}$, $M(t)$ and $N(t)$ are jointly Gaussian and uncorrelated, and thus independent.

If moreover the original process has a spectrum symmetric about ν_0 in the band $[\nu_0 - B, \nu_0 + B]$, then, by (3.38), $M(t_1)$ and $N(t_2)$ are, for all $t_1, t_2 \in \mathbb{R}$, two uncorrelated conjointly Gaussian variables, and therefore independent. In other words, the quadrature component processes are two independent centered Gaussian WSS stochastic processes.

3.5 Exercises

Exercise 3.5.1 (HPP is Markov) Prove that the counting process $\{N(t)\}_{t \in \mathbb{R}_+}$ of a homogeneous Poisson process with intensity λ is a continuous-time homogeneous Markov chain, and compute its transition semi-group.

Exercise 3.5.2 (The uniform HMC) Prove that the stochastic process of Example 3.1.7 is a continuous-time Markov chain.

Exercise 3.5.3 (Wide-sense, but not strict, stationarity) Give a simple example of a discrete-time (resp., continuous-time) stochastic process that is wide-sense stationary, but not strictly stationary.

Exercise 3.5.4 (Stationarization of a cyclostationary stochastic process) Let $\{Y(t)\}_{t \in \mathbb{R}}$ be the stochastic process taking its values in $\{-1, +1\}$ defined by

$$Y(t) = Z \times (-1)^n \text{ on } (nT, (n+1)T], \, n \in \mathbb{Z},$$

where T is a positive real number and Z is a random variable equidistributed on $\{-1, +1\}$.

(1) Show that $\{Y(t)\}_{t \in \mathbb{R}}$ is not a stationary (strictly or in the wide sense) stochastic process.

(2) Let now U be a random variable uniformly distributed on $[0, T]$ and independent of Z. Define for all $t \in \mathbb{R}$,

$$X(t) = Y(t - U).$$

Show that $\{X(t)\}_{t \in \mathbb{R}}$ is a wide-sense stationary stochastic process and compute its covarianve function.

Exercise 3.5.5 (Second-order characteristics of a 2-state HMC) Let $\{X_n\}_{n \geq 0}$ be a HMC with state space $E = \{-1, +1\}$ and transition matrix

$$\mathbf{P} = \begin{pmatrix} 1 - \alpha & \alpha \\ \beta & 1 - \beta \end{pmatrix},$$

where $\alpha, \beta \in (0, 1)$.

(1) Compute its stationary distribution π.
(2) Assuming that X_0 is distributed according to π (and therefore the chain is strictly stationary), compute the second order characteristics of this chain.

Exercise 3.5.6 (Correlation coefficient) Let X and Y be two real square integrable random variables with means m_X and m_Y, variances $\sigma_X^2 > 0$ and $\sigma_Y^2 > 0$. Let ρ be their correlation coefficient. Show that the quantity $F(a, b) := E\left[(Y - \widehat{Y})^2\right] = E\left[(Y - (aX + b))^2\right]$ is minimized with respect to a, b when

$$\widehat{Y} = m_Y + \frac{\sigma_{XY}}{\sigma_X^2}(X - m_X),$$

where $\sigma_{XY} := \mathrm{cov}\,(X, Y)$, and that in this case

$$E\left[(\widehat{Y} - Y)^2\right] = \sigma_Y^2 \left(1 - \rho^2\right).$$

Exercise 3.5.7 (Continuity in the quadratic mean) Let $\{X(t)\}_{t \in \mathbb{R}}$ be a centered real second-order stochastic process of covariance function Γ. Prove that the three following conditions are equivalent:

(i) $\{X(t)\}_{t \in \mathbb{R}}$ is continuous in the quadratic mean.
(ii) Γ is continuous (as function of its two arguments).
(iii) Γ is continuous (as function of its two arguments) on the diagonal of $\mathbb{R} \times \mathbb{R}$ (the points of the form (t, t)).

Exercise 3.5.8 (The positive quadrant) This continues Example 2.4.4. Compute the quantity $E\left[X 1_{\{Y > 0\}}\right]$.

Exercise 3.5.9 (Ornstein–Uhlenbeck) Consider the stochastic process defined for all $t \in \mathbb{R}_+$ by

$$X(t) = e^{-\alpha t} W(e^{2\alpha t}),$$

where $\{W(t)\}_{t\in\mathbb{R}_+}$ is a standard Wiener process and α is a positive real number. Show that it is, distributionwise, an Ornstein–Uhlenbeck process.

Exercise 3.5.10 (Mixed moments of Gaussian processes)

1. Let $X = (X_1, \ldots, X_n)^T$ be a centered n-dimensional Gaussian vector with the covariance matrix $\Gamma = \{\sigma_{ij}\}$. Show that

$$E[X_1 X_2 X_3 X_4] = \sigma_{12}\sigma_{34} + \sigma_{13}\sigma_{24} + \sigma_{14}\sigma_{23}$$

   Compute $E\left[|W(t) - W(s)|^4\right]$ for a standard Wiener process $\{W(t)\}_{t\in\mathbb{R}_+}$.
2. Let $\{X(t)\}_{t\in\mathbb{R}}$ be a centered wide-sense stationary Gaussian process with covariance function C_X. Compute the probability that $X(t_1) > X(t_2)$ where $t_1, t_2 \in \mathbb{R}$ are fixed times.
3. Give the mean and the covariance function of the process $\{X(t)^2\}_{t\in\mathbb{R}}$.

Exercise 3.5.11 (Simple transformations of Wiener processes) Let $\{W(t)\}_{t\in\mathbb{R}_+}$ be a standard Wiener process. What can you say about the process $\{X(t)\}_{t\in[0,1]}$ where

A. $X(t) = t\, W\left(\frac{1}{t}\right)$ with $X(0) \equiv 0$? (You will admit continuity of the process at time 0.)
B. $X(t) = W(1) - W(1 - t)$?

Exercise 3.5.12 (Random pseudo-measures)

1. For the random pseudo-measure $\{Z(A)\}_{A\in\mathcal{B}_b(\mathbb{R}^d)}$ of Sect. 3.2.3, show that if A and B are disjoint bounded Borel sets, then $Z(A \cup B) = Z(A) + Z(B)$, almost surely.
2. Generalize (3.9) to the Brownian random pseudo-measure of Example 3.2.3.

Exercise 3.5.13 (Power spectra of a coded signal) Let $\{I_n\}_{n\in\mathbb{Z}}$ be an IID sequences of binary symbols (0 or 1) with $P(I_n = 1) = P(I_n = 0) = \frac{1}{2}$. We define $a_n \in \{-1, +1\}$ by $a_{n+1} = a_n$ if $I_{n+1} = 1$, $a_{n+1} = -a_n$ if $I_n = 0$. Compute the power spectral measure of the WSS stochastic process $\{X(t)\}_{t\in\mathbb{R}}$ defined by

$$X(t) = Z(t - U)$$

where U is a random variable uniformly distributed on $[0, 1]$, independent of the process

$$Z(t) = a_k \text{ on } [k, k+1).$$

Exercise 3.5.14 (A simple ergodic theorem) Let $\{X(t)\}_{t\in\mathbb{R}}$ be a real-valued WSS stochastic process with mean m and covariance function C_X. In order that

$$\lim_{T\uparrow\infty} \frac{1}{T} \int_0^T X(s)\, ds = m_X.$$

holds in the quadratic mean, it is necessary and sufficient that

$$\lim_{T \uparrow \infty} \frac{1}{T} \int_0^T \left(1 - \frac{u}{T}\right) C_X(u) \, du = 0.$$

Show that this condition is satisfied in particular when the covariance function is integrable.

Exercise 3.5.15 (Real WSS stochastic processes) Show that the power spectral measure of a real WSS stochastic process is symmetric.

Exercise 3.5.16 (Product of independent WSS stochastic processes) Let $\{X(t)\}_{t \in \mathbb{R}}$ and $\{Y(t)\}_{t \in \mathbb{R}}$ be a two centered WSS stochastic processes of respective covariance functions C_X and C_Y.

1. Assume the two stochastic processes to be independent. Show that the stochastic process $\{Z(t)\}_{t \in \mathbb{R}}$ defined by $Z(t) = X(t)Y(t)$ is WSS. Give its mean and its covariance function.
2. Same hypothesis as in the previous question, and now $\{X(t)\}_{t \in \mathbb{R}}$ is the harmonic stochastic process of Example 3.3.1. Suppose that $\{Y(t)\}_{t \in \mathbb{R}}$ admits a spectral density f_Y. Give the power spectral density f_Z of $\{Z(t)\}_{t \in \mathbb{R}}$.

Exercise 3.5.17 (Finitesimal derivative of the Wiener process, take 1) Let $\{W(t)\}_{t \in \mathbb{R}_+}$ be a Wiener process, and let $a > 0$. Define

$$B_h(t) := \frac{W(t + a) - W(t)}{a}$$

Show that $\{B_h(t)\}_{t \in \mathbb{R}_+}$ is a WSS process. Compute its mean, its covariance function and its power spectral density.

Exercise 3.5.18 (Finitesimal derivative of the Wiener process, take 2) Let $\{W(t)\}_{t \in \mathbb{R}}$ be a standard Wiener process and for $h > 0$, define $B_h(t)$ as in Exercise 3.5.17. Show that for all $f \in L^2_{\mathbb{C}}(\mathbb{R}_+) \cap L^1_{\mathbb{C}}(\mathbb{R}_+)$,

$$\lim_{h \downarrow 0} \int_{\mathbb{R}_+} f(t) B_h(t) \, dt = \int_{\mathbb{R}_+} f(t) \, dW(t)$$

in the quadratic mean.

Exercise 3.5.19 (Squared white noise approximation) Let $\{X(t)\}_{t \in \mathbb{R}}$ be a wide-sense stationary *centered* Gaussian process with covariance function C_X and with the power spectral density

$$f_X(\nu) = \frac{N_0}{2} 1_{[-B, +B]}(\nu) \, ,$$

where $N_0 > 0$ and $B > 0$.

1. Let $Y(t) = X(t)^2$. Show that $\{Y(t)\}_{t \in \mathbb{R}}$ is a wide-sense stationary process.
2. Compute its power spectral density f_Y.

Exercise 3.5.20 (Shannon–Nyquist for random fields) State and prove an extension of the Shannon–Nyquist theorem (Theorem 3.4.7) to random fields.

Exercise 3.5.21 (Band-pass WSS stochastic processes) Give the detailed proof of the relations relative to spectral measures and covariance functions concerning band-pass stochastic processes following Definition 3.4.2.

Chapter 4
Fourier Analysis of Time Series

Discrete-time wide-sense stationary stochastic processes, also called time series, arise from discrete-time measurements (sampling) of random functions. A particularly mathematically tractable class of such processes consists of the so-called moving averages and auto-regressive (and more generally, ARMA) time series. This chapter begins with the general theory of WSS discrete-time stochastic processes (which essentially reproduces that of WSS continuous-time stochastic processes) and then gives the representation theory of ARMA processes, together with their prediction theory. The last section is concerned with the realization problem: what models fit a given finite segment of autocorrelation function of a time series? The corresponding theory is the basis of parametric spectral analysis.

4.1 Power Spectral Measure

4.1.1 Herglotz's Theorem

Examples of Time Series

For the sake of self-containedness, the definitions concerning WSS stochastic processes will be recalled in the discrete-time framework, with a slight change in the notation. A discrete-time stochastic process, or *time series*, is a family of complex random variables, $\{X_n\}_{n \in \mathbf{T}}$, indexed by $\mathbf{T} = \mathbb{N}$ or $\mathbb{Z}$. (Here, we choose $\mathbf{T} = \mathbb{Z}$, thus guaranteeing the generality of the results.) It is called a *second-order* stochastic process if $E[|X_n|^2] < \infty$ for all $n \in \mathbb{Z}$. The *mean* function $m : \mathbb{Z} \to \mathbb{C}$ and the *covariance* function $\Gamma : \mathbb{Z} \times \mathbb{Z} \to \mathbb{C}$ are then well-defined by $m(n) = E[X_n]$ and $\Gamma(p, n) = \text{cov}(X_p, X_n)$. If

$$E[X_n] \equiv m$$

and

© Springer International Publishing Switzerland 2014
P. Brémaud, *Fourier Analysis and Stochastic Processes*, Universitext,
DOI 10.1007/978-3-319-09590-5_4

$$\text{cov}\,(X_p, X_n) = R(p - n),$$

for some $m \in \mathbb{C}$ and some function $R : \mathbb{Z} \to \mathbb{C}$, the time series is called *wide-sense stationary*. The complex number m is its *mean*, and the function R is also called the *covariance function*.

Definition 4.1.1 A discrete-time *white noise* with variance σ^2 is, by definition, a centered WSS time series $\{\varepsilon_n\}_{n \in \mathbb{Z}}$ with covariance function

$$R(k) = \sigma^2 1_{\{0\}}(k).$$

A discrete-time *Gaussian white noise* is a discrete-time white noise for which the common distribution of the white noise sequence is Gaussian.

Example 4.1.1 Moving average, 1. Let $\{\varepsilon_n\}_{n \in \mathbb{Z}}$ be a discrete-time white noise with variance σ^2, and let $\{c_n\}_{n \in \mathbb{Z}}$ be a sequence of $\ell_{\mathbb{C}}^2(\mathbb{Z})$, that is, a sequence of complex numbers such that

$$\sum_{n \in \mathbb{Z}} |c_n|^2 < \infty.$$

Define

$$X_n = \sum_{k \in \mathbb{Z}} c_k \varepsilon_{n-k}.$$

The sum in the right-hand side converges in the quadratic mean, and therefore the process $\{X_n\}_{n \in \mathbb{Z}}$ is well-defined. It is centered with covariance function

$$R(k) = \sigma^2 \sum_{j \in \mathbb{Z}} c_{j-k} c_j^*.$$

Proof The above assertions follow from the result of Exercise 1.4.21 by writing

$$X_n = \sum_{i \in \mathbb{Z}} c_i \varepsilon_{n-i} \quad \text{and} \quad X_{n+k} = \sum_{j \in \mathbb{Z}} c_j \varepsilon_{n+k-j}.$$

In fact, identifying the random variables A_i and B_j thereof with $c_i \varepsilon_{n-i}$ and $c_j \varepsilon_{n+k-j}$ respectively, and observing that

$$E\left[A_i B_j^*\right] = c_i c_j^* E\left[\varepsilon_{n-i} \varepsilon_{n+k-j}^*\right]$$
$$= \sigma^2 c_i c_j^* 1_{\{n-i=n+k-j\}} = \sigma^2 c_i c_j^* 1_{\{j=i+k\}},$$

we have

$$R(k) = \sigma^2 \sum_{i \in \mathbb{Z}} \sum_{j \in \mathbb{Z}} c_i c_j^* 1_{\{j=i+k\}} = \sigma^2 \sum_{j \in \mathbb{Z}} c_{j-k} c_j^*. \qquad \square$$

Example 4.1.2 Gaussian moving average. If the white noise sequence of the previous example is Gaussian, the moving average process is a Gaussian process. In fact, each variable X_n belongs to the Gaussian Hilbert space generated by the Gaussian white noise sequence.

In Sect. 4.3.1 the class of regular (to be defined then) WSS time series will be identified with the class of moving averages described in Example 4.1.1.

Example 4.1.3 Homogeneous Markov chains, 5. Let $\{X_n\}_{n \in \mathbb{Z}}$ be an irreducible stationary discrete-time HMC with finite state space $E = \{1, 2, \ldots, r\}$, transition matrix $\mathbf{P}$ and (unique) stationary distribution π. Then, for any given function $f : E \to \mathbb{R}$, the stochastic process

$$Y_n := f(X_n)$$

is stationary. Its mean is

$$m_Y = \pi^T f$$

where $f^T = (f(1), f(2), \ldots, f(r))$. Also, as simple calculations reveal,

$$E\left[Y_{k+n} Y_n\right] = f^T D_\pi \mathbf{P}^k f ,$$

where D_π is the diagonal matrix whose diagonal is π. Note that one may express the mean as $m_Y = f^T \pi = f^T D_\pi \mathbf{1}$ where $\mathbf{1}$ is a column vector with all entries equal to 1. In particular

$$m_Y^2 = f^T D_\pi \mathbf{1} \mathbf{1}^T D_\pi f = f^T D_\pi \Pi f$$

where $\Pi := \pi \mathbf{1}^T$ is a square matrix with all lines identical to the stationary distribution vector π. Therefore, the covariance function of $\{Y_n\}_{n \in \mathbb{Z}}$ is

$$R_Y(k) = f^T D_\pi \left(\mathbf{P}^k - \Pi\right) f.$$

Among the basic results of the theory of matrices concerning eigenvalues and eigenvectors, we quote the following, relative to a square matrix A of dimension r. We assume for simplicity, that the eigenvalues are *distinct*.

Let $\lambda_1, \ldots, \lambda_r$ be these eigenvalues and let $u_1, \ldots, u_r$ and $v_1, \ldots, v_r$ be the associated sequences of left- and right-eigenvectors, respectively. Then, $u_1, \ldots, u_r$ form an independent collection of vectors, and so do $v_1, \ldots, v_r$. Also, $u_i^T v_j = 0$ if $i \neq j$. Since eigenvectors are determined up to multiplication by an arbitrary nonnull scalar, one can choose them in such a way that $u_i^T v_i = 1$ for all i, $1 \leq i \leq r$. We then have the spectral decomposition

$$A^n = \sum_{i=1}^{r} \lambda_i^n v_i u_i^T . \qquad (*)$$

Transition matrices are special cases of non-negative matrices (matrices with non-negative coefficients), the theory of which, due to Perron and Frobenius, is well-developed. We now summarize its central result.

Let A be a *non-negative* $r \times r$ matrix. Suppose that it is *primitive*, that is, there exists an integer $m > 0$ such that all the entries of A^m are positive (such as in the case of the transition matrix of an irreducible *aperiodic* Markov chain). There exists a real eigenvalue λ_1 with algebraic as well as geometric multiplicity one such that $\lambda_1 > 0$, and $\lambda_1 > |\lambda_j|$ for any other eigenvalue λ_j. Moreover, the left-eigenvector u_1 and the right-eigenvector v_1 associated with λ_1 can be chosen positive and such that $u_1^T v_1 = 1$.

Let $\lambda_2, \lambda_3, \ldots, \lambda_r$ be the eigenvalues of A other than λ_1 ordered in such a way that

$$\lambda_1 > |\lambda_2| \geq \cdots \geq |\lambda_r|$$

and if $|\lambda_2| = |\lambda_j|$ for some $j \geq 3$, then $m_2 \geq m_j$, where m_j is the algebraic multiplicity of λ_j. If in addition, $A = \mathbf{P}$ is a stochastic matrix, then $\lambda_1 = 1$. The right eigenvector associated with the eigenvalue 1 is in this case $v = \mathbf{1}$, the vector with all entries equal to 1, since $\mathbf{P1} = \mathbf{1}$. If $A = \mathbf{P}$ is an irreducible transition matrix with period $d > 1$, then there are exactly d distinct eigenvalues of modulus 1, namely the d-th roots of unity, and all other eigenvalues have modulus strictly less than 1. If moreover it is assumed that $\mathbf{P}$ is diagonalizable (the geometric and algebraic multiplicities of any given eigenvalue are equal), then a representation such as ($*$) exists:

$$\mathbf{P}^n = \sum_{i=1}^{r} \lambda_i^n v_i u_i^T . \tag{4.1}$$

Example 4.1.4 Two-state Markov chain. Consider the transition matrix on $E = \{1, 2\}$ with transition matrix

$$\mathbf{P} = \begin{pmatrix} 1 - \alpha & \alpha \\ \beta & 1 - \beta \end{pmatrix},$$

where $\alpha, \beta \in (0, 1)$. Its characteristic polynomial $(1 - \alpha - \lambda)(1 - \beta - \lambda) - \alpha\beta$ admits the roots $\lambda_1 = 1$ and

$$\lambda_2 = 1 - \alpha - \beta.$$

The stationary distribution

$$\pi^T = \left(\frac{\beta}{\alpha + \beta}, \frac{\alpha}{\alpha + \beta} \right)$$

is the left eigenvector corresponding to the eigenvalue 1. In this example, the representation (4.1) takes the form

$$\mathbf{P}^n = \frac{1}{\alpha + \beta} \begin{pmatrix} \beta & \alpha \\ \beta & \alpha \end{pmatrix} + \frac{(1 - \alpha - \beta)^n}{\alpha + \beta} \begin{pmatrix} \alpha & -\alpha \\ -\beta & -\beta. \end{pmatrix}$$

Example 4.1.5 A periodic chain. Let this time

$$\mathbf{P} = \begin{pmatrix} 0 & 1 \\ 1 & 0 \end{pmatrix}.$$

This chain has period 2, and the two eigenvalues of modulus 1 are $+1$ and -1. We find that

$$\mathbf{P}^n = \frac{1}{2} \begin{pmatrix} 1 & 1 \\ 1 & 1 \end{pmatrix} + \frac{1}{2}(-1)^n \begin{pmatrix} 1 & -1 \\ -1 & 1 \end{pmatrix}$$

We now introduce the notion of power spectral measure, which is virtually the same as in the continuous-time case.

Absolutely Continuous Spectrum

Let $\{X_n\}_{n \in \mathbb{Z}}$ be a WSS time series with covariance function R satisfying the absolute summability condition

$$\sum_{n \in \mathbb{Z}} |R(n)| < \infty.$$

In particular, the Fourier sum

$$f(\omega) = \frac{1}{2\pi} \sum_{n \in \mathbb{Z}} R(n)e^{-i\omega n}$$

is a 2π-periodic bounded and continuous function. Integrating the right-hand side of the last display term by term (this is allowed because the covariance function is absolutely summable), we obtain the inversion formula,

$$R(n) = \int_{-\pi}^{+\pi} f(\omega)e^{+i\omega n}d\omega.$$

We call the function $f : [-\pi, +\pi] \to \mathbb{R}$ the *power spectral density* (PSD) of the time series. Note that

$$\int_{-\pi}^{+\pi} f(\omega)d\omega = R(0) < \infty$$

It turns out that the PSD is a non-negative function, as we shall soon see from the general result below (Theorem 4.1.1).

Example 4.1.6 Power spectral measure of the discrete-time white noise.

For the discrete-time white noise, we immediately obtain that the spectral power density is a constant:

$$f(\omega) \equiv \frac{\sigma^2}{2\pi}.$$

Example 4.1.7 Moving average, 2. Consider the moving average process of Example 4.1.1. Its PSD is

$$f(\omega) = \frac{\sigma^2}{2\pi} \left| \sum_{k \in \mathbb{Z}} c_j e^{-ik\omega} \right|^2.$$

Proof The Fourier sum $S(\omega) := \sum_{k \in \mathbb{Z}} c_k e^{-ik\omega}$ converges in $L_\mathbb{C}^2([-\pi, +\pi])$. The Fourier sum associated to the sequence $\{c_{j-k}\}_{j \in \mathbb{Z}}$ is $S(\omega)e^{+ik\omega}$. Therefore, from the expression of the covariance function given in Example 4.1.1 and the Plancherel–Parseval identity,

$$R(k) = \sigma^2 \sum_{j \in \mathbb{Z}} c_{j-k} c_j^* = \frac{\sigma^2}{2\pi} \int\limits_{-\pi}^{+\pi} \left| \sum_{j \in \mathbb{Z}} c_j e^{-ij\omega} \right|^2 e^{ik\omega} \, d\omega.$$

$\square$

Example 4.1.8 Homogeneous Markov chains, 6. Recall from Example 4.1.3 the expression of the covariance of the time series $\{Y_n\}_{n \in \mathbb{Z}}$ where $Y_n = f(X_n)$ and $\{X_n\}_{n \in \mathbb{Z}}$ is a stationary irreducible HMC with finite state space $E = \{1, 2, \ldots, r\}$, transition matrix $\mathbf{P}$ and stationary distribution π:

$$R_Y(k) = f^T D_\pi (\mathbf{P}^k - \Pi) f.$$

In order to simplify the notation, we shall suppose that $\mathbf{P}$ is diagonalizable. We then have (Example 4.1.3)

$$(\mathbf{P}^k - \Pi) = \sum_{j=2}^{r} v_j u_j^T \lambda_j^k$$

where u_j is the (up to a multiplicative constant) left-eigenvector and v_j is the (up to a multiplicative constant) right-eigenvector corresponding to the eigenvalue λ_j, and where the multiplicative constants are chosen in such a way that $u_j^T v_j = 1$. Therefore

$$R_Y(k) = f^T D_\pi \left(\sum_{j=2}^{r} v_j u_j^T \lambda_j^k \right) f. \tag{*}$$

We know that all the eigenvalues are of modulus lesser than or equal to 1. If we suppose, as we now do, that the chain is aperiodic, then, besides the eigenvalue $\lambda_1 =$

1, all eigenvalues have a modulus strictly less than 1. In particular, the covariance function is absolutely summable and there exists a power spectral density which is easily computed using the fact that, for $|\lambda| < 1$,

$$\sum_{k \in \mathbb{Z}} \lambda^k e^{-ik\omega} = \frac{1}{|e^{-i\omega} - \lambda|^2},$$

from which it follows, by (4.1), that

$$f_Y(\omega) = \frac{1}{2\pi} \sum_{j=2}^{r} (f^T D_\pi v_j u_j^T f) \frac{1}{|e^{-i\omega} - \lambda_j|^2}.$$

Line Spectrum

As in continuous-time, the case occurs when the covariance function is not summable, and when the power spectral density is a pseudo-density, of the symbolic form (using the Dirac pseudo-function δ)

$$f(\omega) = \sum_{\ell \in \mathbb{Z}} \alpha_\ell \delta(\omega - \omega_\ell),$$

where for all $\ell \in \mathbb{Z}$: $\omega_\ell \in (-\pi, +\pi]$, $\alpha_\ell \in \mathbb{R}_+$, and

$$\sum_{\ell \in \mathbb{Z}} \alpha_\ell < \infty.$$

Example 4.1.9 Harmonic series. Let $\{U_\ell\}_{\ell \geq 1}$ and $\{\Phi_\ell\}_{1 \leq \ell \leq N}$ be as in Example 3.3.1. Define for all $n \in \mathbb{Z}$

$$X_n = \sum_{\ell \geq 1} U_j \cos(2\pi \omega_\ell n + \Phi_\ell),$$

where the ω_ℓ's are arbitrary real numbers in the interval $(-\pi, +\pi]$ and

$$\sum_{\ell \geq 1} E\left[|U_\ell|^2\right] < \infty.$$

Such a process is called a *harmonic* series. It is a WSS time series, centered, with covariance function

$$R(k) = \sum_{\ell \geq 1} E[|U_\ell|^2] \cos(2\pi \omega_\ell k).$$

The proof is similar to that of Example 3.3.1.

The General Case

In discrete time, the analog of Bochner's theorem is *Herglotz's theorem*:

Theorem 4.1.1 *Let $\{X_n\}_{n\in\mathbb{Z}}$ be a* WSS *time series. There exists a unique* finite *measure μ on $((-\pi, +\pi], \mathcal{B}((-\pi, +\pi]))$ such that the covariance function is given by the formula*

$$R(k) = \int_{(-\pi,+\pi]} e^{ik\omega}\mu(d\omega). \tag{4.2}$$

The finite measure μ is called the *power spectral measure* of $\{X_n\}_{t\in\mathbb{Z}}$. If for some non-negative measurable function $f : (-\pi, +\pi] \to \mathbb{R}$,

$$\mu(d\omega) = f(\omega)d\omega,$$

f is called the *power spectral density* (PSD) of $\{X_n\}_{n\in\mathbb{Z}}$. Note that it is integrable over $[-\pi, +\pi]$:

$$\int_{-\pi}^{+\pi} f(\omega)d\omega = R(0) < \infty.$$

Proof First observe that for all complex numbers $z_1, \ldots, z_N$,

$$\sum_{k=1}^{N}\sum_{k=1}^{N} R(k - \ell)z_k z_\ell^* = E\left[\left|\sum_{k=1}^{N} z_k X_k\right|^2\right] \geq 0.$$

In particular the function $f_N : [-\pi, +\pi] \to \mathbb{R}$ defined by

$$f_N(\omega) = \frac{1}{2N}\sum_{k=1}^{N}\sum_{k=1}^{N} R(k - \ell)e^{-ik\omega}e^{+i\ell\omega}$$

is non-negative. There are $N - |m|$ pairs (k, ℓ) for which $k - \ell = m$, and therefore

$$f_N(\omega) = \frac{1}{2}\sum_{|m|<N}\left(1 - \frac{|m|}{N}\right)R(m)e^{-im\omega}.$$

From then on, one may imitate the proof of Theorem 2.2.8 (with $f_N(\omega)$ playing the role of $g(x, A)$) and prove the existence of a finite measure μ such that

$$\int_{[-\pi,+\pi]} e^{in\omega}\mu(d\omega) = \lim_{N\uparrow\infty}\int_{[-\pi,+\pi]} e^{in\omega}\mu_N(d\omega) = R(n).$$

The mass of μ on $-\pi$ can now be transfered to $+\pi$ to obtain a finite measure on $(-\pi, +\pi]$, still denoted by μ, and such that (4.2) holds true.

The uniqueness of the power spectral measure comes from the fact that every bounded continuous function $\omega \to g(\omega)$ can be approximated uniformly on $(-\pi, +\pi]$ by finite linear combinations of exponential functions $\omega \to e^{in\omega}, n \in \mathbb{Z}$ (Stone–Weierstrass theorem, Example 1.3.2). Therefore if there are two such finite measures μ_1 and μ_2, then for any bounded continuous function $g : (-\pi, +\pi] \to \mathbb{C}$,

$$\int_{(-\pi,+\pi]} g(\omega)\mu_1(d\omega) = \int_{(-\pi,+\pi]} g(\omega)\mu_2(d\omega).$$

For any interval $(a, b] \in (-\pi, +\pi]$, there exists a sequence of continuous functions $\{g_M\}_{M\geq 1}$ uniformly bounded by 1 such that for all $\omega \in (-\pi, +\pi], \lim_{M\uparrow\infty} g_M(\omega) = 1_{(a,b]}(\omega)$ (for instance, take for g_M a continuous function equal to 1 on $[a + \frac{1}{M}, b]$, to 0 outside $[a, b + \frac{1}{M}]$ and otherwise linear). In particular, for $\ell = 1, 2$

$$\lim_{M\uparrow\infty} \int_{(-\pi,+\pi]} g_M(\omega)\mu_\ell(d\omega) = \int_{(a,b]} \mu_\ell(d\omega),$$

and therefore $\mu_1((a, b]) = \mu_2((a, b])$, and this is enough to guarantee that the two measures are identical (Theorem A.2.1, taking for π-system S the collection of intervals $(a, b] \in (-\pi, +\pi]$). □

The existence part of Herglotz's theorem can also be deduced from Bochner's theorem using the simple device of Exercise 4.5.4.

The following is an example mixing an absolutely continuous spectrum with a line spectrum.

Example 4.1.10 Homogeneous Markov chains, 7. This is a continuation of Example 4.1.8. Suppose that the chain has a period equal to $d > 1$. In this case, there are $d - 1$ eigenvalues besides $\lambda_1 = 1$ with a modulus equal to 1, and these are precisely the d-th roots of unity besides $\lambda_1 = 1$:

$$\lambda_\ell = e^{+i\omega_\ell}, \quad (2 \leq \ell \leq d)$$

where

$$\omega_\ell = (\ell - 1)\frac{2\pi}{d}.$$

Observing that

$$e^{+ik\omega_\ell} = \int_{(-\pi,+\pi]} e^{+ik\omega}\varepsilon_{\omega_\ell}(d\omega),$$

we find for the complete spectral measure, in the case where the eigenvalues are distinct, or more generally, the transition matrix is diagonalizable,

$$\mu_Y(d\omega) = \sum_{\ell=2}^{d}(f^T D_\pi \, v_\ell u_\ell^T \, f) \, \varepsilon_{\omega_\ell}(d\omega)$$

$$+ \frac{1}{2\pi} \sum_{j=d+1}^{r} (f^T D_\pi \, v_j u_j^T \, f) \frac{1}{|e^{-i\omega} - \lambda_j|^2} d\omega \, .$$

Filtering

The theory of filtering for WSS time series is similar to that for WSS continuous-time stochastic processes. It is based on the following preliminary computations:

Lemma 4.1.1 *Let* $\{X_n\}_{n\in\mathbb{Z}}$ *be a* WSS *time series with mean m, covariance function R and power spectral measure* μ, *and let* $\{a_n\}_{n\in\mathbb{Z}}$ *and* $\{b_n\}_{n\in\mathbb{Z}}$ *be two sequences in* $\ell^1_{\mathbb{C}}(\mathbb{Z})$ *with respective formal z-transforms A(z) and B(z). The sum* $\sum_{n\in\mathbb{Z}} a_n X_n$ *is almost-surely absolutely convergent, and*

$$\sum_{n\in\mathbb{Z}} a_n X_n = \left(\sum_{n\in\mathbb{Z}} a_n\right) \times m \, .$$

Also

$$\mathrm{cov}\left(\sum_{n\in\mathbb{Z}} a_n X_n \, , \, \sum_{n\in\mathbb{Z}} b_n X_n\right) = \sum_{n\in\mathbb{Z}}\sum_{k\in\mathbb{Z}} a_n b_k^* R(n-k) \, ,$$

and

$$\mathrm{cov}\left(\sum_{n\in\mathbb{Z}} a_n X_n \, , \, \sum_{n\in\mathbb{Z}} b_n X_n\right) = \int_{(-\pi,+\pi]} A(e^{i\omega})B(e^{i\omega})^* \mu(d\omega).$$

Proof Schwarz's inequality applied to the square-integrable random variables $X_n - m$ and $X_{n+k} - m$ gives

$$\left|E\left[(X_n - m)(X_{n+k} - m)^*\right]\right| \le E\left[|X_n - m|^2\right]^{\frac{1}{2}} E\left[|X_{n+k} - m|^2\right]^{\frac{1}{2}} \, ,$$

and applied to the random variables $X_n - m$ and 1, it gives

$$\left|E\left[(X_n - m)\right]\right| \le E\left[|X_n - m|^2\right]^{\frac{1}{2}} E\left[1^2\right]^{\frac{1}{2}} \, .$$

In summary, we have the inequalities

$$|R(k)| \leq R(0), \quad E[|X_n - m|] \leq R(0)^{\frac{1}{2}}. \tag{4.3}$$

We have, by Tonelli's theorem and (4.3):

$$E\left[\sum_{n\in\mathbb{Z}} |a_n||X_n|\right] = \sum_{n\in\mathbb{Z}} |a_n| E[|X_n|] \leq R(0)^{\frac{1}{2}} \sum_{n\in\mathbb{Z}} |a_n| < \infty,$$

and therefore, P-a.s.,

$$\sum_{n\in\mathbb{Z}} |a_n||X_n| < \infty.$$

Thus the sum $\sum_{n\in\mathbb{Z}} a_n X_n$ is P-a.s. well-defined, being P-a.s. absolutely convergent. Morever, since $E\left[\sum_{n\in\mathbb{Z}} |a_n||X_n|\right] < \infty$, we may apply Fubini's theorem to obtain

$$E\left[\sum_{n\in\mathbb{Z}} a_n X_n\right] = \sum_{n\in\mathbb{Z}} a_n E[X_n].$$

We now suppose without loss of generality that the series is centered. Again by Tonelli's theorem and (4.3):

$$E\left[\left(\sum_{n\in\mathbb{Z}} |a_n||X_n|\right)\left(\sum_{p\in\mathbb{Z}} |b_p||X_p|\right)\right] = \sum_{n\in\mathbb{Z}}\sum_{p\in\mathbb{Z}} |a_n||b_p| E[|X_n X_p|]$$

$$\leq R(0)\left(\sum_{n\in\mathbb{Z}} |a_n|\right)\left(\sum_{p\in\mathbb{Z}} |b_p|\right) < \infty.$$

Therefore the product

$$\left(\sum_{n\in\mathbb{Z}} a_n X_n\right)\left(\sum_{n\in\mathbb{Z}} b_n X_n\right)^*$$

is in $L^1_{\mathbb{C}}(P)$ and in particular (with $a_n = b_n$) the sums $\sum_{n\in\mathbb{Z}} a_n X_n$ and $\sum_{n\in\mathbb{Z}} b_n X_n$ are in $L^2_{\mathbb{C}}(P)$. Also, as we just saw that $\sum_{n\in\mathbb{Z}}\sum_{p\in\mathbb{Z}} |a_n||b_p| E[|X_n X_p|] < \infty$, we may apply Fubini's theorem to obtain

$$E\left[\left(\sum_{n\in\mathbb{Z}} a_n X_n\right)\left(\sum_{n\in\mathbb{Z}} b_n X_n\right)^*\right] = \sum_{n\in\mathbb{Z}}\sum_{p\in\mathbb{Z}} a_n b_p E[X_n X_p]$$

$$= \sum_{n\in\mathbb{Z}}\sum_{p\in\mathbb{Z}} a_n b_p R(n-p).$$

Write now

$$\sum_{n\in\mathbb{Z}}\sum_{p\in\mathbb{Z}} a_n b_p^* R(n-p) = \sum_{n\in\mathbb{Z}}\sum_{p\in\mathbb{Z}} a_n b_p^* \int_{(-\pi,+\pi]} e^{i(n-p)\omega}\mu(d\omega)$$

$$= \int_{(-\pi,+\pi]} \left(\sum_{n\in\mathbb{Z}} a_n e^{in\omega}\right)\left(\sum_{p\in\mathbb{Z}} b_p e^{ip\omega}\right)^* \mu(d\omega)$$

$$= \int_{(-\pi,+\pi]} A(e^{i\omega})B(e^{i\omega})^* \mu(d\omega) ,$$

where the interchange of the order of integration and summation (Fubini) is justified by the fact that, by Tonelli's theorem,

$$\int_{(-\pi,+\pi]} \sum_{n\in\mathbb{Z}}\sum_{p\in\mathbb{Z}} |a_n||b_p||e^{i(n-p)\omega}|\mu(d\omega) = \int_{(-\pi,+\pi]} \sum_{n\in\mathbb{Z}}\sum_{p\in\mathbb{Z}} |a_n||b_p|\mu(d\omega)$$

$$= \left(\sum_{n\in\mathbb{Z}} |a_n|\right)\left(\sum_{p\in\mathbb{Z}} |b_p|\right)\mu((-\pi,+\pi]) < \infty.$$

$\square$

Remember that a *stable convolutional filter* is characterized by its *impulse response* $\{h_n\}_{n\in\mathbb{Z}}$, an absolutely summable sequence of complex numbers, its z-transform $H(z)$, called the formal *transfer function*, and its Fourier sum $H(e^{-i\omega})$, called the *transmittance*. To the *input* $\{x_n\}_{n\in\mathbb{Z}} \in \ell_\mathbb{C}^1(\mathbb{Z})$ it associates the *output* $\{y_n\}_{n\in\mathbb{Z}} \in \ell_\mathbb{C}^1(\mathbb{Z})$ defined as the convolution

$$y_n = \sum_{k\in\mathbb{Z}} h_k x_{n-k}.$$

$\square$

Theorem 4.1.2 *When the* WSS *time series* $\{X_n\}_{n\in\mathbb{Z}}$ *is the input of a convolutional filter with impulse response* $\{h_n\}_{n\in\mathbb{Z}}$ *in* $\ell_\mathbb{C}^1(\mathbb{Z})$, *the output* $\{Y_n\}_{n\in\mathbb{Z}}$ *defined by formula*

$$Y_n = \sum_{k\in\mathbb{Z}} h_{n-k} X_k$$

is a well-defined WSS *time series with mean* $m_Y = m\sum_{n\in\mathbb{Z}} h_n$ *and power spectral measure*

$$\mu_Y(\omega) = |H(e^{-i\omega})|^2 \mu_X(d\omega).$$

(This formula is called the *fondamental filtering formula* for WSS time series.)

Proof This follows directly from the formulas given in Lemma 4.1.1, with $a_j = h_{n+k-j}$ and $b_j = h_{n-j}$, in which case

$$A(z) = \sum_{j \in \mathbb{Z}} h_{n+k-j} z^j = z^{n+k} \sum_{j \in \mathbb{Z}} h_{n+k-j} z^{-(n+k-j)}$$

$$= z^{n+k} \sum_{\ell \in \mathbb{Z}} h_\ell z^{-\ell} = z^{n+k} H(z^{-1}),$$

and $B(z) = z^n H(z^{-1})$. $\hfill\square$

4.1.2 The Periodogram

Let $\{X_n\}_{n \in \mathbb{Z}}$ be a *centered and real* WSS time series with covariance function R. Given a sample $X_1, \ldots, X_n$ of this time series, it seems natural to estimate $R(m)$ for $|m| \leq n - 1$ by the empirical quantity

$$\hat{R}_n(m) = \frac{1}{n} \sum_{j=1}^{n-|m|} X_j X_{j+m} . \tag{4.4}$$

Taking expectations, we obtain

$$E[\hat{R}_n(m)] = \left(1 - \frac{|m|}{n}\right) R(m). \tag{4.5}$$

If

$$\sum_{m \in \mathbb{Z}} |R(m)| < \infty, \tag{4.6}$$

the time series $\{X_n\}_{n \in \mathbb{Z}}$ has a power spectral density f given by

$$f(\omega) = \frac{1}{2\pi} \sum_{m \in \mathbb{Z}} R(m) e^{-im\omega} = \frac{1}{2\pi} \left(R(0) + \sum_{m \geq 1} R(m) \left(e^{-in\omega} + e^{-in\omega}\right) \right),$$

where we have taken into account the symmetry of R (the time series is real-valued). Replacing R by $\hat{R}_n$ in this latter expression and truncating the sum, we obtain the empirical estimator

$$\hat{f}_n(\omega) = \frac{1}{2\pi} \left(\hat{R}_n(0) + \sum_{m=1}^{n-1} \hat{R}_n(m)(e^{-im\omega} + e^{im\omega}) \right). \tag{4.7}$$

Using (4.4), this estimator can be put into the compact form

$$\hat{f}_n(\omega) = \frac{1}{2\pi n} \left| \sum_{m=1}^{n} X_m e^{-im\omega} \right|^2. \tag{4.8}$$

The function $\hat{f}_n : [-\pi, +\pi] \to \mathbb{R}$ is, by definition, the *periodogram* of $\{X_n\}_{n\in\mathbb{Z}}$.

The estimate $\hat{f}_n(\omega)$ of the power spectral density at pulsation ω is a random number. From (4.7) and (4.5) we deduce

$$2\pi E[\hat{f}_n(\omega)] = R(0) + \sum_{m=1}^{n-1} \left(1 - \frac{|m|}{n}\right) R(m)(e^{-im\omega} + e^{im\omega}).$$

Under the integrability assumption (4.6), by dominated convergence,

$$\lim_{n\uparrow\infty} E[\hat{f}_n(\omega)] = f(\omega). \tag{4.9}$$

The estimator $\hat{f}_n(\omega)$ is therefore, in the standard terminology of Statistics, *asymptotically unbiased*. Unfortunately, it is *not consistent*, in the sense that, in general,

$$\lim_{n\uparrow\infty} E\left[|\hat{f}_n(\omega) - f(\omega)|^2\right] \neq 0$$

as the following simple example shows.

Example 4.1.11 Let $\{\varepsilon_n\}_{n\in\mathbb{Z}}$ be a Gaussian white noise of variance σ^2. The spectral density of such process is constant: $f(\omega) = \sigma^2/2\pi$, and explicit calculations (exercise) show that

$$E\left|\hat{f}_n(\omega) - \frac{\sigma^2}{2\pi}\right|^2 = \sigma^4\left(1 - \frac{1}{n} + \frac{2}{n^2}\left|\frac{1 - e^{-i\omega}}{1 - e^{-in\omega}}\right|^2\right),$$

a quantity that does not tend to zero as n goes to infinity.

If we only seek to estimate integrals of the type

$$I(\phi) = \int_{-\pi}^{+\pi} \phi(\omega) f(\omega)\, d\omega \tag{4.10}$$

(for example, $\int_{\omega_1}^{\omega_2} f(\omega)\, d\omega$, the energy lying between the pulsations ω_1 and ω_2), then the estimator

$$\hat{I}_n(\phi) = \int_{-\pi}^{+\pi} \phi(\omega) \hat{f}_n(\omega)\, d\omega \tag{4.11}$$

has, at first view, a more interesting behaviour. Indeed, when ϕ is continuous on $[-\pi, +\pi]$ and if $\{X_n\}_{n \in \mathbb{Z}}$ is ergodic,

$$\lim_{n \uparrow \infty} \hat{I}_n(\phi) = I(\phi) \text{ almost surely.} \tag{4.12}$$

In other words, the measure $\hat{f}_n(\omega) \, d\omega$ converges weakly to the measure $f(\omega) \, d\omega$. To prove this result, it is sufficient to verify that for all $m \in \mathbb{Z}$

$$\lim_{n \uparrow \infty} \int_{-\pi}^{+\pi} e^{im\omega} \hat{f}_n(\omega) \, d\omega = \int_{-\pi}^{+\pi} e^{im\omega} f(\omega) \, d\omega,$$

that is to say,

$$\lim_{n \uparrow \infty} \hat{R}_n(m) = R(m),$$

which is satisfied under the assumption of ergodicity.

However, the behaviour of this estimator is very chaotic in the sense explained in the following discussion.[1] If the WSS time series $\{X_n\}_{n \in \mathbb{Z}}$ is Gaussian, and under the additional condition

$$\sum_{n \in \mathbb{Z}} |m| \, |R(m)| < \infty, \tag{4.13}$$

it can be shown that for every even bounded Borel function $\phi : [-\pi, +\pi] \to \mathbb{R}^k$ we have

$$\lim E[\hat{I}_n(\phi)] = I(\phi)$$

and convergence in distribution takes place as follows

$$\sqrt{n} \, (\hat{I}_n(\phi) - I(\phi)) \to \mathcal{N} \left(0, \, 4\pi \int_{-\pi}^{+\pi} \phi^\dagger \phi f^2 \, d\omega \right). \tag{4.14}$$

If ϕ is continuous, one can show that

$$\lim_{n \uparrow \infty} \hat{I}_n(\phi) = I(\phi), \text{ a.s.} \tag{4.15}$$

In particular, if $\Delta_1, \ldots, \Delta_k$ are disjoint intervals of $[-\pi, +\pi]$ and if ϕ is a continuous approximation of $(1_{\Delta_1}, \ldots, 1_{\Delta_k})^T$, we see that the random variables

---

[1] Details and proofs can be found in [Dacunha-Castelle and Duflo], Vol. II, Chap. 1; or [Rosenblatt], Chap. V.

$$\int\limits_{\Delta_1} \hat{f}_n(\omega)\, d\omega, \ldots, \int\limits_{\Delta_k} \hat{f}_n(\omega)\, d\omega$$

are asymptotically independent. This corresponds to a very chaotic behaviour of the periodogram.

To the two defects mentioned above (inconsistency and chaotic behaviour) one must add a third, which is the *lack of resolution*: if there are two spectral lines, or peaks in the PSD, that are very close one from the other, the periodogram has a tendency to identify them all as one peak. More generally, the periodogram method has a tendency to smoothe the power spectral density. This is why one prefers, in certain applications, the parametric methods called *high resolution methods*, the mathematical principles of which will be presented in Sect. 4.4.

4.1.3 The Cramér–Khinchin Decomposition

The following result corresponds to the Cramér–Khinchin decomposition in continuous time.

Theorem 4.1.3 *Let* $\{X_n\}_{n\in\mathbb{Z}}$ *be a centered complex* WSS *time series with power spectral measure* μ. *Then there exists a unique process* $\{Z(\omega)\}_{\omega\in(-\pi,+\pi]}$ *with centered square-integrable orthogonal increments and structural measure* μ, *such that*

$$X_n = \int\limits_{(-\pi,+\pi]} e^{in\omega}\, dZ(\omega), \qquad (4.16)$$

where the above integral is a Doob integral.

We then say that $Z := \{Z(\omega)\}_{\omega\in(-\pi,+\pi]}$ (or for short, $dZ(\omega)$) is the *Cramér–Khinchin spectral decomposition* of $\{X_n\}_{n\in\mathbb{Z}}$.

By "uniqueness", the following is meant: If there exists another spectral decomposition $\tilde{Z}$, then for all $a, b \in \mathbb{R}$, $a \le b$, $Z(b) - Z(a) = \tilde{Z}(b) - \tilde{Z}(a)$.

Proof The proof is the same, *mutatis mutandis*, as that of Theorem 3.4.1. However, we shall repeat the arguments for the sake of self-containedness.

1. Denote by $\mathcal{H}(X)$ the vector subspace of $L^2_{\mathbb{C}}(P)$ formed by the finite complex linear combinations of the type $\sum_{k=1}^{K} \lambda_k X_k$. Define $L^2_{\mathbb{C}}((-\pi, +\pi], \mu)$ to be the collection of all measurable functions $f : (-\pi, +\pi] \to \mathbb{C}$ such that

$$\int\limits_{(-\pi,+\pi]} |f(\omega)|^2\, \mu(d\omega) < \infty.$$

Let φ be the mapping of $\mathcal{H}(X)$ into $L^2_{\mathbb{C}}((-\pi, +\pi], \mu)$ defined by

$$\varphi : \sum_{k=1}^{K} \lambda_k X_k \to \sum_{k=1}^{K} \lambda_k e^{ik\omega}.$$

Using Herglotz's theorem, we verify that it is a linear isometry from $H(X)$ into $L_{\mathbb{C}}^2((-\pi, +\pi], \mu)$:

$$E\left[\left|\sum_{k=1}^{K} \lambda_k X_k\right|^2\right] = \sum_{k=1}^{K}\sum_{\ell=1}^{K} \lambda_k \lambda_\ell^* E\left[X_k X_\ell^*\right] = \sum_{k=1}^{K}\sum_{\ell=1}^{K} \lambda_k \lambda_\ell^* R(k-\ell)$$

$$= \sum_{k=1}^{K}\sum_{\ell=1}^{K} \lambda_k \lambda_\ell^* \int_{(-\pi, +\pi]} e^{i(k-\ell)\omega} \mu(d\omega)$$

$$= \int_{(-\pi, +\pi]} \left(\sum_{k=1}^{K}\sum_{\ell=1}^{K} \lambda_k \lambda_\ell^* e^{i(k-\ell)\omega}\right) \mu(d\omega)$$

$$= \int_{(-\pi, +\pi]} \left|\sum_{k=1}^{K} \lambda_k e^{ik\omega}\right|^2 \mu(d\omega).$$

2. This isometric linear mapping can be uniquely extended to an isometric linear mapping (that we shall continue to call φ) from $H(X)$, the closure of $\mathcal{H}(X)$, into $L_{\mathbb{C}}^2((-\pi, +\pi], \mu)$ (Theorem 1.3.7). Since the combinations $\sum_{k=1}^{K} \lambda_k e^{ik\omega}$ are dense in $L_{\mathbb{C}}^2((-\pi, +\pi], \mu)$ when μ is a finite measure, φ is *onto*. Therefore, it is a linear isometric bijection between $H(X)$ and $L_{\mathbb{C}}^2((-\pi, +\pi], \mu)$.

3. For $\omega_0 \in (-\pi, +\pi]$, define $Z(\omega_0)$ to be the random variable in $H(X)$ corresponding in this isometry to the function $1_{(-\pi, \omega_0]}$ (which belongs to $L_{\mathbb{C}}^2((-\pi, +\pi], \mu)$ because μ is a finite measure). First, we observe that

$$E[Z(\omega_2) - Z(\omega_1)] = 0$$

since $H(X)$ is the closure in $L_{\mathbb{C}}^2(P)$ of a collection of centered random variables. Also, by isometry,

$$E[(Z(\omega_2) - Z(\omega_1))(Z(\omega_4) - Z(\omega_3))^*] = \int_{(-\pi, +\pi]} 1_{(\omega_1, \omega_2]}(\omega) 1_{(\omega_3, \omega_4]}(\omega) \mu(d\omega)$$

$$= \mu((\omega_1, \omega_2] \cap (\omega_3, \omega_4]).$$

Therefore $\{Z(\omega)\}_{\omega \in (-\pi, +\pi]}$ is a process with second-order increments having μ for structural measure. We can therefore define the Doob integral $\int_{(-\pi, +\pi]} f(\omega)\, dZ(\omega)$ for all $f \in L_{\mathbb{C}}^2((-\pi, +\pi], \mu)$.

4. Let now

$$Z_{\ell,n} := \sum_{k=-2^n}^{2^n-1} e^{i\ell(k/2^n)}\left(Z\left(\frac{k+1}{2^n}\pi\right) - Z\left(\frac{k}{2^n}\pi\right)\right).$$

We have

$$\lim_{n\to\infty} Z_{\ell,n} = \int_{(-\pi,+\pi]} e^{i\ell\omega}\,dZ(\omega)$$

(limit in $L^2_{\mathbb{C}}(P)$). Indeed,

$$Z_{\ell,n} = \int_{(-\pi,+\pi]} f_{\ell,n}(\omega)\,dZ(\omega),$$

where

$$f_{\ell,n}(\omega) = \sum_{k=-2^n}^{2^n-1} e^{i\ell(k/2^n)} 1_{A_{n,k}}(\omega),$$

where $A_{n,k} := (k\pi/2^n, (k+1)\pi/2^n]$, and therefore, by isometry,

$$E\left|Z_{\ell,n} - \int_{(-\pi,+\pi]} e^{i\ell\omega}\,dZ(\omega)\right|^2 = \int_{(-\pi,+\pi]} |e^{i\ell\omega} - f_{\ell,n}(\omega)|^2 \mu(d\omega),$$

a quantity which tends to zero when n tends to infinity (by dominated convergence, using the fact that μ is a bounded measure). Also, by definition of φ,

$$Z_{\ell,n} \xrightarrow{\varphi} f_{\ell,n}(\omega).$$

Since, for fixed ℓ, $\lim_{n\to\infty} Z_{\ell,n} = \int_{(-\pi,+\pi]} e^{i\ell\omega}\,dZ(\omega)$ in $L^2_{\mathbb{C}}(P)$ and $\lim_{n\to\infty} f_{\ell,n}(\omega) = e^{i\ell\omega}$ in $L^2_{\mathbb{C}}((-\pi,+\pi],\mu)$, we have by isometry:

$$\int_{(-\pi,+\pi]} e^{i\ell\omega}\,dZ(\omega) \xrightarrow{\varphi} e^{i\ell\omega}.$$

But, by definition of φ,

$$X_\ell \xrightarrow{\varphi} e^{i\ell\omega}.$$

Therefore we necessarily have (4.16).

5. We now prove uniqueness. Suppose that there exists another spectral decomposition $d\tilde{Z}(\omega)$. Denote by $\mathcal{G}$ the set of finite linear combinations of complex exponentials. Since by hypothesis

$$\int\limits_{(-\pi,+\pi]} e^{i\omega\ell}\,dZ(\omega) = \int\limits_{(-\pi,+\pi]} e^{i\ell\omega}\,d\tilde{Z}(\omega) \quad (= X_\ell)$$

we have

$$\int\limits_{(-\pi,+\pi]} f(\omega)\,dZ(\omega) = \int\limits_{(-\pi,+\pi]} f(\omega)\,d\tilde{Z}(\omega)$$

for all $f \in \mathcal{G}$, and therefore, for all $f \in L^2_{\mathbb{C}}((-\pi,+\pi],\mu) \cap L^2_{\mathbb{C}}((-\pi,+\pi],\tilde{\mu}) \subseteq L^2_{\mathbb{C}}((-\pi,+\pi],\frac{1}{2}(\mu+\tilde{\mu}))$ because $\mathcal{G}$ is dense in $L^2_{\mathbb{C}}((-\pi,+\pi],\frac{1}{2}(\mu+\tilde{\mu}))$ (Theorem A.33). In particular, with $f = 1_{(a,b]}$,

$$Z(b) - Z(a) = \tilde{Z}(b) - \tilde{Z}(a).$$

$\square$

Linear Operations

Let $\{X_n\}_{n\in\mathbb{Z}}$ be a centered WSS time series with spectral measure μ_X and Cramér–Khinchin spectral decomposition Z_X. For every function $\tilde{f}(\omega)$ in $L^2_{\mathbb{C}}(\mu_X)$, we can define a sequence

$$Y_n = \int\limits_{(-\pi,+\pi]} \tilde{f}(\omega)e^{in\omega}\,dZ_X(\omega), \qquad (4.17)$$

where the integral in the right hand side is a Doob integral. From the properties of this type of integral, we obtain that the time series $\{Y_n\}_{n\in\mathbb{Z}}$ is *centered* and that

$$E[Y_{n+m}Y_n^*] = \int\limits_{\pi} \tilde{f}(\omega)e^{i(n+m)\omega}\tilde{f}(\omega)^* e^{-in\omega}\,\mu_X(d\omega).$$

This time series is therefore WSS with spectral measure

$$\mu_Y(d\omega) = |\tilde{f}(\omega)|^2\,\mu_X(d\omega),$$

and its Cramér–Khinchin spectral decomposition is, by (4.17),

$$dZ_Y(\omega) = \tilde{f}(\omega)\,dZ_X(\omega).$$

We then say: $\{Y_n\}_{n\in\mathbb{Z}}$ is obtained by homogeneous linear filtering of $\{X_n\}_{n\in\mathbb{Z}}$ by a filter with transmittance $\tilde{f}$.

Stable convolutional filtering is a particular case since the transmittance, that is the Fourier sum associated with the absolutely summable impulse response, is uniformly bounded and therefore is square-integrable with respect to the (finite) measure μ_X. Next example features a very simple linear operation that is not representable by a stable convolutional filter.

Example 4.1.12 The pure delay, 2. Take $\tilde{f}(\omega) = e^{im\omega}$. Then

$$Y_n = \int_{(-\pi,+\pi]} e^{im\omega} e^{in\omega} \, dZ_X(\omega) = \int_{(-\pi,+\pi]} e^{i(n+m)\omega} \, dZ_X(\omega) = X_{n+m}.$$

The following theorem was stated and proven in continuous time. The same proof avails for the discrete-time case, and we therefore omit it.

Theorem 4.1.4 *Every linear operation on a Gaussian* WSS *time series yields a Gaussian* WSS *time series.*

Multivariate WSS Time Series

Let $\{X_n\}_{n\in\mathbb{Z}}$ be a time series with values in $E := \mathbb{C}^L$, where L is an integer greater than or equal to 2: $X_n = (X_{n,1}, \ldots, X_{n,L})$. It is assumed to be of the second order, that is:

$$E[||X_n||^2] < \infty \quad \text{for all } n \in \mathbb{Z},$$

and centered. Furthermore it will be assumed that it is wide-sense stationary, in the sense that the mean vector of X_n and the cross-covariance matrix of the vectors X_{n+k} and X_n do not depend upon n. The matrix-valued function R defined by

$$R(k) = \text{cov } (X_{n+k}, X_n) := E\left[X_n^T X_{n+k}\right] \tag{4.18}$$

is called the (matrix) *covariance function* of the time series. Its general entry is

$$R_{ij}(k) = \text{cov}(X_{n,i}, X_{n+k,j}).$$

Therefore, each of the series $\{X_{n,i}\}_{n\in\mathbb{Z}}$ is a WSS time series, but, furthermore, they are *stationarily correlated* or "jointly WSS". The vector-valued time series $\{X_n\}_{n\in\mathbb{Z}}$ is then called a *multivariate* WSS *time series.*

We have the following result:

Theorem 4.1.5 *If* $\{X_n\}_{n\in\mathbb{Z}}$ *is a L-dimensional multivariate* WSS *time series, then for all* r, s *$(1 \le i, j \le L)$ there exists a finite complex measure* μ_{ij} *on* $(-\pi, +\pi])$ *such that*

$$R_{ij}(k) = \int_{(-\pi,+\pi]} e^{i\omega k} \mu_{ij}(d\omega).$$

The proof is similar to that of Theorem 3.4.10.
The matrix

$$M := \{\mu_{ij}\}_{1 \leq i, j \leq L}$$

(whose entries are finite complex measures) is the *interspectral power measure matrix* of the multivariate WSS time series $\{X_n\}_{n \in \mathbb{Z}}$. It is clear that for all $z = (z_1, \ldots, z_L) \in \mathbb{C}^L$, $U_n = z^T X_n$ defines a WSS time series with spectral measure $\mu_U = z\, M\, z^\dagger$.

The link between the interspectral measure μ_{12} and the Cramér–Khintchine decompositions Z_1 and Z_2 is the following:

$$E[dZ_1(\omega)\, dZ_2(\omega')^*] = \mu_{12}(d\omega)\, 1_{\{\omega = \omega'\}},$$

a symbolic notation for

$$E[Z_1(\omega_2) - Z_1(\omega_1))(Z_2(\omega_4) - Z_2(\omega_3))^*] = \mu_{12}((\omega_1, \omega_2] \cap (\omega_3, \omega_4]).$$

The proof is the same as that of the similar result in continuous-time of Sect. 3.4.3.

4.2 ARMA Processes

4.2.1 Prediction Theory

Let

$$P(z) = \sum_{k=0}^{p} a_k z^k, \qquad Q(z) = \sum_{\ell=0}^{q} a_\ell z^\ell$$

be two polynomials with complex coefficients, and of respective degrees p and q.

Definition 4.2.1 A centered WSS time series $\{X_n\}_{n \in \mathbb{Z}}$ satisfying the recurrence

$$\sum_{k=0}^{p} a_k X_{n-k} = \sum_{\ell=0}^{q} b_\ell \varepsilon_{n-\ell}, \tag{4.19}$$

where $\{\varepsilon_n\}_{n \in \mathbb{Z}}$ is a discrete-time white noise with variance $\sigma^2 > 0$, is called an *autoregressive moving average* (ARMA) process.

If $P \equiv 1$, that is, if

$$X_n = \sum_{\ell=0}^{q} b_\ell \varepsilon_{n-\ell},$$

it is called a *moving average* (MA) process. If $Q \equiv 1$, that is, if

$$\sum_{k=0}^{p} a_k X_{n-k} = \varepsilon_n,$$

it is called an *autoregressive* (AR) process.

We now introduce two conditions that will occasionally be taken as hypotheses.

(HP)   *P has no roots inside the closed unit disk* centered at the origin.

(HQ)   *Q has no roots inside the closed unit disk* centered at the origin.

Theorem 4.2.1 *Under condition (HP), the above ARMA process admits the power spectral density,*

$$f_X(\omega) = \frac{\sigma^2}{2\pi} \left| \frac{Q(e^{-i\omega})}{P(e^{-i\omega})} \right|^2. \tag{4.20}$$

Proof Let μ_X be the power spectral measure of $\{X_n\}_{n\in\mathbb{Z}}$. Applying the fundamental filtering formula to both sides of (4.19), we obtain

$$|P(e^{-i\omega})|^2 \mu_X(d\omega) = \frac{\sigma^2}{2\pi} |Q(e^{-i\omega})|^2 d\omega,$$

where σ^2 is the power of the white noise. In particular, if the polynomial P has no roots on the unit circle centered at 0, $|P(e^{-i\omega})|^2 > 0$ for all ω, and therefore

$$\mu_X(d\omega) = \frac{\sigma^2}{2\pi} \left| \frac{Q(e^{-i\omega})}{P(e^{-i\omega})} \right|^2 d\omega.$$

$\square$

It is traditional to write (4.19) in the *symbolic* form

$$P(z)X_n = Q(z)\varepsilon_n \tag{4.21}$$

This means that the left-hand side is the process $\{X_n\}_{n\in\mathbb{Z}}$ passed through the filter with the transfer function P, with a similar interpretation for the right-hand side. One can also write (4.19) in the form involving the Cramér–Khintchin decompositions of the process and of the white noise:

$$\int_{(-\pi,+\pi]} P(e^{-i\omega})e^{in\omega} dZ_X(\omega) = \int_{(-\pi,+\pi]} Q(e^{-i\omega})e^{in\omega} dZ_\varepsilon(\omega). \tag{4.22}$$

We now consider the following problem: Given a number $\sigma^2 > 0$ and two polynomials with complex coefficients

$$P(z) = 1 + \sum_{k=1}^{p} a_k z^k, \qquad Q(z) = 1 + \sum_{k=1}^{q} b_k z^k,$$

where P has no root on the unit circle centered at 0, we define on $(-\pi, +\pi]$ the function

$$f(\omega) = \frac{\sigma^2}{2\pi} \left| \frac{Q(e^{-i\omega})}{P(e^{-i\omega})} \right|^2.$$

Note that f is bounded and therefore integrable on $(-\pi, +\pi]$. We ask the question: Does there exist an ARMA process admitting the function f above for power spectral density? The answer is yes, and the proof is by construction, as follows.

Proof With the goal of exhibiting such a process, introduce an arbitrary white noise $\{\varepsilon_n\}_{n\in\mathbb{Z}}$ with variance σ^2, and denote by Z_ε its Cramér–Khintchin spectral decomposition. Define

$$X_n = \int_{(-\pi,+\pi]} e^{in\omega} \frac{Q(e^{-i\omega})}{P(e^{-i\omega})} \, dZ_\varepsilon(\omega).$$

The right-hand side of the above equality is well-defined since $g(\omega) = Q(e^{-i\omega})/P(e^{-i\omega})$ is square-integrable with respect to Lebesgue measure on $(-\pi, +\pi]$, and therefore $\int_{(-\pi,+\pi]} |g(\omega)|^2 \mu_\varepsilon(d\omega) = \frac{\sigma^2}{2\pi} \int_{-\pi}^{+\pi} |g(\omega)|^2 \, d\omega < \infty$. Moreover, $\{X_n\}_{n\in\mathbb{Z}}$ is a WSS process with spectral decomposition

$$dZ_X(\omega) = \frac{Q(e^{-i\omega})}{P(e^{-i\omega})} \, dZ_\varepsilon(\omega).$$

Passing $\{X_n\}_{n\in\mathbb{Z}}$ through a filter of transmittance $P(e^{-i\omega})$ leads to a WSS time series $\{U_n\}_{n\in\mathbb{Z}}$ with spectral decomposition

$$dZ_U(\omega) = P(e^{-i\omega}) \, dZ_X(\omega)$$
$$= Q(e^{-i\omega}) \, dZ_\varepsilon(\omega).$$

On the other hand, passing $\{\varepsilon_n\}_{n\in\mathbb{Z}}$ through a filter of transmittance $Q(e^{-i\omega})$ leads to a WSS time series $\{V_n\}_{n\in\mathbb{Z}}$ with spectral decomposition

$$dZ_V(\omega) = Q(e^{-i\omega}) \, dZ_\varepsilon(\omega).$$

Thus we have $dZ_V(\omega) = dZ_U(\omega)$, which implies $\{V_n\}_{n\in\mathbb{Z}} \equiv \{U_n\}_{n\in\mathbb{Z}}$, that is,

$$P(z)X_n = Q(z)\varepsilon_n. \qquad \square$$

The Prediction Problem

From now on, we assume that conditions (HP) and (HQ) are both satisfied. (As we shall see in Sect.4.2.2, this is not much of a restriction.) For standardization, and without loss of generality, we may assume that

$$P(0) = Q(0) = 1. \tag{4.23}$$

Recall that $H(X; n)$ is the Hilbert subspace of $L^2_{\mathbb{C}}(P)$ generated by $X_n, X_{n-1}, X_{n-2} \ldots$ (By definition, $H(X; n)$ is the set of all finite linear combinations of random variables among $X_n, X_{n-1}, X_{n-2} \ldots$ plus the random variables that are limits in quadratic mean of such finite linear combinations.) We are interested in the problem of *linear prediction*, that of finding, for $m \in \mathbb{N}_+$, and all $n \in \mathbb{Z}$ the projection of X_n on $H(X; n - m)$

$$\widehat{X}_{n;n-m} := P_{n-m} X_n := P\left(X_n \mid H(X; n - m)\right),$$

and it is called the *m-step (linear) predictor* of X_n. Recall that, by definition, $\widehat{X}_{n;n-m}$ realizes the minimum of the quadratic criterion

$$E\left[|X_n - Z|^2\right]$$

among all elements $Z \in H(X; n - m)$.

In view of solving the prediction problem, we first show how to obtain the ARMA process from the white noise, and the white noise from the ARMA process.

The Analysis (or Whitening) Filter

We start with the *analysis problem*, that of obtaining the white noise $\{\varepsilon_n\}_{n \in \mathbb{Z}}$ from the process $\{X_n\}_{n \in \mathbb{Z}}$. Let r_2 be the smallest root modulus of Q strictly larger than 1 ($r_2 = \infty$ if there is no root outside the closed unit disk centered at 0). Let $D(0; r_2) := \{z \in \mathbb{C}; |z| < r_2\}$. The function

$$W(z) = \frac{P(z)}{Q(z)}$$

is analytic inside $D(0; r_2)$. Inside this open disk it admits a power series expansion

$$\frac{P(z)}{Q(z)} = 1 + \sum_{n \geq 1} w_n z^n$$

[noting that $w_0 = 1$ in view of (4.23)] that is absolutely convergent. Since $D(0; r_2)$ contains the unit circle centered at 0, we have

$$\sum_{n \geq 0} |w_n| < \infty.$$

Defining $w_n = 0$ for $n < 0$, we can interpret $\{w_n\}_{n \in \mathbb{Z}}$ as the impulse response of a *causal* stable convolutional filter.

Similarly, the filter with transfer function $\frac{1}{Q}$ is a causal stable filter. Applying this filter to both sides of $P(z)X_n = Q(z)\varepsilon_n$, we therefore obtain

$$\frac{P(z)}{Q(z)} X_n = \varepsilon_n,$$

that is to say

$$\varepsilon_n = X_n + \sum_{k \geq 1} w_k X_{n-k}. \tag{4.24}$$

The filter with transfer function

$$W(z) = \sum_{n \geq 0} w_n z^n = 1 + \sum_{n \geq 1} w_n z^n$$

is called the *analysis filter* (or *whitening filter*).

The Synthesis (or Generating) Filter

We now consider the *synthesis problem*, that of obtaining the process $\{X_n\}_{n \in \mathbb{Z}}$ from the white noise $\{\varepsilon_n\}_{n \in \mathbb{Z}}$ just constructed. Then, under the condition stated above for the polynomial P, letting r_1 be the smallest root modulus of P strictly larger than 1 ($r_1 = \infty$ if there is no root outside the closed unit disk centered at 0), the function

$$H(z) = \frac{Q(z)}{P(z)}$$

admits an absolutely convergent power series expansion inside $D(0; r_1)$:

$$H(z) = 1 + \sum_{n \geq 1} h_n z^n$$

[noting that $h_0 = 1$ in view of (4.23)] inside $D(0; r_1)$, and in particular, since $D(0; r_1)$ contains the unit circle centered at 0,

$$\sum_{n \geq 0} |h_n| < \infty$$

Defining $h_n = 0$ for $n < 0$, we interpret $\{h_n\}_{n \in \mathbb{Z}}$ as the impulse response of a stable and *causal* convolutional filter. Similarly, the filter with transfer function $\frac{1}{P(z)}$ is a stable filter. Applying this filter to both sides of $P(z)X_n = Q(z)\varepsilon_n$, we therefore obtain

$$X_n = \frac{Q(z)}{P(z)}\varepsilon_n,$$

that is to say

$$X_n = \varepsilon_n + \sum_{k \geq 1} h_k \varepsilon_{n-k}. \tag{4.25}$$

The filter with transfer function

$$H(z) = \sum_{n \geq 0} h_n z^n$$

is called the *synthesis filter* (or *generating filter*).

Solution of the Prediction Problem

Definition 4.2.2 A sequence $\{\varepsilon_n\}_{n \in \mathbb{Z}}$ is called an *innovations sequence* for the centered WSS time series $\{X_n\}_{n \in \mathbb{Z}}$ if
   (i) it is a white noise and
   (ii) $H(X; n) = H(\varepsilon; n)$ for all $n \in \mathbb{Z}$.

Theorem 4.2.2 *We assume that both conditions (HP) and (HQ) are satisfied. Then, for all $n \in \mathbb{Z}$,*

$$H(\varepsilon; n) = H(X; n). \tag{4.26}$$

In particular, since $\{\varepsilon_n\}_{n \in \mathbb{Z}}$ is also a white noise, it is an innovations sequence for the time series.

Proof The proof immediately follows from (4.24) and (4.25) and the following lemma (Exercise 4.5.10).   □

Lemma 4.2.1 *Let $\{Y_n\}_{n \in \mathbb{Z}}$ be the output of a stable causal filter when the input is a WSS discrete-time stochastic process $\{X_n\}_{n \in \mathbb{Z}}$. Then, for all $n \in \mathbb{Z}$, $Y_n \in H(X; n)$.*

In the situation of interest (where (HP) and (HQ) are satisfied), we therefore have for all $n \in \mathbb{Z}$, $\varepsilon_n \in H(X; n)$ and $X_n \in H(\varepsilon; n)$, which implies (4.26)   □
We now solve the prediction problem, starting with a simple example.

Example 4.2.1 Prediction of AR processes. We first treat the prediction problem for the simpler AR processes

$$X_n + \sum_{k=1}^{p} a_k X_{n-k} = \varepsilon_n,$$

where $P(z) = 1 + \sum_{k=1}^{p} a_k z^k$ has all its roots outside the closed unit disc centered at 0 and $\{\varepsilon_n\}_{n\in\mathbb{Z}}$ is a white noise of variance σ^2. Then

$$\widehat{X}_{n,n-1} = -\sum_{k=1}^{p} a_k X_{n-k}. \tag{4.27}$$

Proof We note that (α) being a linear combination of $X_{n-1}, \ldots, X_{n-p}$, the random variable $\widehat{X}_{n,n-1}$ is in $H(X; n-1)$, and
   (β) $X_n - \widehat{X}_{n,n-1}$ is orthogonal to $H(X; n-1)$. In fact,

$$X_n - \widehat{X}_{n,n-1} = \varepsilon_n \perp H(\varepsilon; n-1)$$

and therefore by (4.24),

$$X_n - \widehat{X}_{n,n-1} \perp H(X; n-1).$$

By the projection principle of Hilbert spaces (see Sect. 1.3.2), properties (α) and (β) characterise $X_{n,n-1}$ as the projection of X_n onto $H(X; n-1)$, the one-step predictor of X_n.   □

We now treat the "general" case. We seek to calculate the one-step prediction $P(X_n \mid H(X; n-1)) = \widehat{X}_{n,n-1}$ of an ARMA process in the canonical form

$$P(z)X_n = Q(z)\varepsilon_n,$$

where, as before, both P and Q have no roots inside the closed unit disc, $P(0) = 1$, $Q(0) = 1$, and $\{\varepsilon_n\}_{n\in\mathbb{Z}}$ is a white noise with variance σ^2. By (4.25) and (4.24)

$$X_n = \varepsilon_n - \sum_{k\geq 1} w_k X_{n-k}$$

whence we deduce as above for the AR model that

$$\widehat{X}_{n,n-1} = -\sum_{k\geq 1} w_k X_{n-k} \tag{4.28}$$

is the projection of X_n onto $H(X; n-1)$.
   The preceding results show that

$$\varepsilon_n = X_n - \widehat{X}_{n,n-1}.$$

The noise is therefore the one-step *prediction error* process, and this is why it is called the *innovations* sequence, or innovations process: It is "what you get minus what you expected". The *prediction gain* is the quantity

$$G = \frac{E|X_n|^2}{E|\varepsilon_n|^2}. \tag{4.29}$$

Now, $E[|X_n|^2] = \frac{\sigma^2}{2\pi} \int_{-\pi}^{+\pi} f_X(\omega)d\omega$ and $E|\varepsilon_n|^2 = \sigma^2$ and therefore

$$G = \frac{1}{2\pi} \int\limits_{-\pi}^{+\pi} \left| \frac{Q(e^{-i\omega})}{P(e^{-i\omega})} \right|^2 d\omega. \tag{4.30}$$

Another method for computing the innovations gain uses the representation $X_n = \varepsilon_n + \sum_{k \geq 0} w_k \varepsilon_{n-k}$, from which it follows, since $\{\varepsilon_n\}_{n \in \mathbb{Z}}$ is a white noise with variance σ^2, that

$$E[|X_n|^2] = \sigma^2 \left(1 + \sum_{n \geq 0} |w_n|^2\right),$$

and therefore

$$G = 1 + \sum_{n \geq 1} |w_n|^2. \tag{4.31}$$

4.2.2 Regular ARMA Processes

The existence of the ARMA process when P has no roots on the unit circle centered at 0 was proven in Sect. 4.2.1. For the prediction problem, we have considered ARMA processes with a power spectral density for which both polynomials P and Q have no roots in the closed unit disk centered at the origin. The purpose of the current subsection is to show that this is not too stringent a restriction.

Theorem 4.2.3 *Let* $f : (-\pi, +\pi] \to \mathbb{R}_+$ *be a function of the form*

$$f(\omega) = \left| \frac{Q'(e^{-i\omega})}{P'(e^{-i\omega})} \right|^2,$$

where P' *and* Q' *are polynomials with complex coefficients such that* P' *has no root on the unit circle centered at 0, and in particular,*

$$\int\limits_{-\pi}^{+\pi} f_X(\omega)d\omega < \infty$$

(this inequality is necessary if we want f_X to be a power spectral density). Then, one can always find a non-negative number σ^2 and polynomials with complex coefficients P and Q, where P is without roots inside the closed *unit disk centered at the origin and Q is without roots inside the* open *unit disk centered at the origin, such that*

$$\left|\frac{Q'(e^{-i\omega})}{P'(e^{-i\omega})}\right|^2 = \frac{\sigma^2}{2\pi}\left|\frac{Q(e^{-i\omega})}{P(e^{-i\omega})}\right|^2.$$

Moreover, this representation is unique if we impose that $P(0) = Q(0) = 1$.

Proof The proof is contained in *Féjer's lemma* (Theorem 1.2.3). $\qquad\qquad\square$

Definition 4.2.3 If P and Q are such that

1. $P(0) = Q(0) = 1$;
2. P has no roots in the closed unit disk $\{z \; ; \; |z| \le 1\}$;
3. Q has not roots in the open unit disk $\{z \; ; \; |z| < 1\}$;

and if $\{\varepsilon_n\}_{n\in\mathbb{Z}}$ is an innovations sequence for $\{X_n\}_{n\in\mathbb{Z}}$, the representation

$$P(z)X_n = Q(z)\varepsilon_n$$

is called *canonical*.

Theorem 4.2.4 *Let $\{X_n\}_{n\in\mathbb{Z}}$ be a WSS time series with spectral density of the form*

$$f_X(\omega) = \frac{\sigma^2}{2\pi}\left|\frac{Q(e^{-i\omega})}{P(e^{-i\omega})}\right|^2$$

where P is a polynomial without roots inside the closed *unit disk and Q is a polynomial without roots inside the* open *unit disk, and $P(0) = Q(0) = 1$. Then $\{X_n\}_{n\in\mathbb{Z}}$ admits a canonical representation.*

Proof Let

$$\psi(\omega) = \frac{P(e^{-i\omega})}{Q(e^{-i\omega})} 1_{\{Q(e^{-i\omega})>0\}}.$$

This function defines a linear operation

$$Y_n = \int\limits_{(-\pi,+\pi]} e^{in\omega}\, dZ_Y(\omega) \to \int\limits_{(-\pi,+\pi]} e^{in\omega}\psi(\omega)\, dZ_Y(\omega)$$

whose domain is the collection of WSS time series $\{Y_n\}_{n\in\mathbb{Z}}$ with spectral measure μ_Y such that

$$\int\limits_{(-\pi,+\pi]} |\psi(\omega)|^2 \, \mu_Y(d\omega) < \infty.$$

This domain contains, in particular, the process $\{X_n\}_{n\in\mathbb{Z}}$ under consideration, since

$$\int\limits_{(-\pi,+\pi]} |\psi(\omega)|^2 \frac{\sigma^2}{2\pi} \left|\frac{Q(e^{-i\omega})}{P(e^{-i\omega})}\right|^2 d\omega = \frac{\sigma^2}{2\pi}.$$

The WSS time series $\{\varepsilon_n\}_{n\in\mathbb{Z}}$ obtained from $\{X_n\}_{n\in\mathbb{Z}}$ in this way admits the Cramér–Khinchin spectral decomposition $\psi(\omega)\,dZ_X(\omega)$ and (therefore) the power spectral density

$$f_\varepsilon(\omega) = |\psi(\omega)|^2 f_X(\omega) = \frac{\sigma^2}{2\pi}.$$

It is therefore a white noise with variance σ^2. The process $\{A_n\}_{n\in\mathbb{Z}} = \{Q(z)\varepsilon_n\}_{n\in\mathbb{Z}}$ admits the spectral decomposition

$$Q(e^{-i\omega})\,dZ_\varepsilon(\omega) = Q(e^{i\omega})\psi(\omega)\,dZ_X(\omega)$$
$$= P(e^{-i\omega})\,1_{\{Q(e^{-i\omega})>0\}}\,dZ_X(\omega).$$

Since $\mu_X(\{\omega \mid Q(e^{-i\omega}) = 0\}) = 0$,

$$1_{\{Q(e^{-i\omega})>0\}}\,dZ_X(\omega) = dZ_X(\omega),$$

and therefore $\{A_n\}_{n\in\mathbb{Z}}$ admits the spectral decomposition $P(e^{-i\omega})\,dZ_X(\omega)$. The process $\{B_n\}_{n\in\mathbb{Z}} = \{P(z)X_n\}_{n\in\mathbb{Z}}$ admits the same decomposition, and therefore $\{B_n\}_{n\in\mathbb{Z}} \equiv \{A_n\}_{n\in\mathbb{Z}}$, that is, $P(z)X_n \equiv Q(z)\varepsilon_n$.

It now remains to show that $\{\varepsilon_n\}_{n\in\mathbb{Z}}$ is an innovations sequence for $\{X_n\}_{n\in\mathbb{Z}}$. For this, we first show that for all $n \in \mathbb{Z}$, $\varepsilon_n \perp H(X; n-1)$, that is

$$E[X_m \varepsilon_n^*] = 0 \quad \text{for all } m < n.$$

By Doob–Wiener's isometry this latter quantity is

$$E\left[\left(\int\limits_{(-\pi,+\pi]} e^{im\omega}\,dZ_X(\omega)\right)\left(\int\limits_{(-\pi,+\pi]} e^{in\omega}\psi(\omega)\,dZ_X(\omega)\right)^*\right] = \int\limits_{(-\pi,+\pi]} e^{i(m-n)\omega}\psi(\omega)\,\mu_X(d\omega),$$

that is,

$$\frac{\sigma^2}{2\pi} \int\limits_{-\pi}^{+\pi} e^{i(m-n)\omega} \frac{Q(e^{-i\omega})}{P(e^{-i\omega})} \, d\omega,$$

or again, up to a multiplicative constant,

$$\oint_C z^{n-m+1} \frac{Q(z)}{P(z)} \, dz,$$

where the integral is a line integral along the unit circle C centred at the origin taken anticlockwise. When $m < n$, $z^{n-m+1}Q(z)/P(z)$ is analytic on an open disc of radius strictly greater than 1 (the root of P with smallest modulus has a modulus strictly greater than 1), and therefore, by Cauchy's theorem, this latter integral is null. □

The next example shows that, in practice, all we need in order to find the canonical representation is the *Féjer identity*

$$(z - \beta)\left(z - \frac{1}{\beta^*}\right) = -\frac{1}{\beta^*} z|z - \beta|^2, \tag{4.32}$$

true for all $\beta \in \mathbb{C}$, $\beta \neq 0$, and for all $z \in \mathbb{C}$ such that $|z| = 1$. The verification is left as Exercise 4.5.12.

Example 4.2.2 Using Féjer's identity. We want to find the representation in canonical ARMA form of the power spectral density

$$f(\omega) = \frac{5 - 4\cos\omega}{10 - 6\cos\omega}.$$

(We verify that $f(\omega) \geq 0$ and that $\int_{-\pi}^{+\pi} f(\omega) \, d\omega < \infty$, and therefore f is a power spectral density.)

By setting $z = e^{-i\omega}$ we have

$$f(\omega) = \frac{5 - 2(z + z^{-1})}{10 - 3(z + z^{-1})} = \frac{2(z - \frac{1}{2})(z - 2)}{3(z - \frac{1}{3})(z - 3)}.$$

By Féjer's identity, for all z such that $|z| = 1$ we have

$$(z - 2)\left(z - \frac{1}{2}\right) = -\frac{1}{2} z|z - 2|^2,$$

and

$$(z - 3)\left(z - \frac{1}{3}\right) = -\frac{1}{3} z|z - 3|^2,$$

and therefore, for $z = e^{-i\omega}$

$$f(\omega) = \frac{|z-2|^2}{|z-3|^2} = \frac{4}{9}\left|\frac{1-\frac{1}{2}z}{1-\frac{1}{3}z}\right|^2 .$$

The canonical representation of $\{X_n\}$ is therefore

$$\sigma^2 = \frac{8}{9}\pi, \qquad P(z) = 1 - \frac{1}{3}z, \qquad Q(z) = 1 - \frac{1}{2}z,$$

that is to say,

$$X_n - \frac{1}{3}X_{n-1} = \varepsilon_n - \frac{1}{2}\varepsilon_{n-1},$$

where $\mathrm{Var}\,(\varepsilon_n) = \frac{8}{9}\pi$.

Example 4.2.3 Non-canonical representation and anticipative noise. It is sometimes believed, or at least inadvertently stated, that there exists no WSS time series $\{X_n\}_{n\in\mathbb{Z}}$ satifying an ARMA relation for which P would have roots inside the unit disc. This assertion is false (we gave a construction of such a process in Sect. 4.2.1), but it is interesting to study the argument that is sometimes invoked to contradict the existence of $\{X_n\}_{n\in\mathbb{Z}}$ under non canonical conditions. It runs as follows. Consider the recurrence

$$X_n - 2X_{n-1} = \varepsilon_n, \tag{4.33}$$

(in particular $P(z) = 1 - 2z$ has a root inside the open unit disk, and therefore the conditions are not canonical) which can be explicitly solved from the initial condition X_0:

$$X_n = 2^n X_0 + 2^{n-1}\varepsilon_1 + 2^{n-2}\varepsilon_2 + \cdots + \varepsilon_n. \tag{4.34}$$

To 'prove' that $\{X_n\}_{n\in\mathbb{Z}}$ is not stationary we calculate the variance σ_n^2 of X_n starting from (4.34), which, if X_0 and $(\varepsilon_1, \ldots, \varepsilon_n)$ were uncorrelated, gives

$$\sigma_n^2 = 4^n \sigma_0^2 + 2(4^{n-1} + \cdots + 1)\sigma^2$$
$$= 4^n \sigma_0^2 + \frac{4}{3}(4^n - 1)\sigma^2,$$

and thus

$$\lim_{n\uparrow\infty} \sigma_n^2 = \infty,$$

which contradicts stationarity. In view of the above discussion, one could believe that an AR process $\{X_n\}_{n\in\mathbb{Z}}$ with representation $P(z)X_n = \varepsilon_n$, where $\{\varepsilon_n\}_{n\in\mathbb{Z}}$ is a white noise, can not be WSS if P has roots inside the open unit circle.

The invalid point in the argument is the assumption of the non-correlation of X_0 and $(\varepsilon_1, \ldots, \varepsilon_n)$. As a matter of fact, there exist WSS time series $\{X_n\}_{n\in\mathbb{Z}}$ such that

(4.33) holds (again, see the construction of Sect. 4.2.1), but in that case, the white noise $\{\varepsilon_n\}_{n\in\mathbb{Z}}$ that was constructed is *not* the innovations sequence of $\{X_n\}_{n\in\mathbb{Z}}$, and X_0 and $(\varepsilon_1, \ldots, \varepsilon_n)$ are correlated. In the representation (4.33) $\{X_n\}_{n\in\mathbb{Z}}$ *anticipates* $\{\varepsilon_n\}_{n\in\mathbb{Z}}$. The essential notion here is the *causality* of the filter $1/P$ rather than the stability of P. In (4.34), when a value of X_0 is observed, the values $\varepsilon_1, \varepsilon_2, \ldots$ no longer correspond to an uncorrelated sequence, and in fact the values are "arranged" in such a way that $\{X_n\}_{n\in\mathbb{Z}}$ has a stationary behaviour.

Multivariate ARMA Processes

We consider now multivariate time series $\{X_n\}_{n\in\mathbb{Z}}$, where $X_n = (X_{n,1}, \ldots, X_{n,m})$ is a m-dimensional square-integrable random vector, satisfying the difference equation

$$X_n + a_1 X_{n-1} + \cdots + a_p X_{n-p} = \varepsilon_n + b_1 \varepsilon_{n-1} + \cdots + b_q \varepsilon_{n-} \tag{4.35}$$

where p and q are positive integers, $a_1, \ldots, a_p, b_1, \ldots, b_q$ are $m \times m$ matrices, and $\{\varepsilon_n\}_{n\in\mathbb{Z}}$ is an IID sequence of m-dimensional square-integrable random vectors, centered and with covariance matrix Σ (an *m-dimensional white noise*). This is formally denoted by

$$P(z)X_n = Q(z)\varepsilon_n \tag{4.36}$$

where

$$P(z) = I + \sum_{k=1}^{p} a_k z^k \text{ and } Q(z) = I + \sum_{k=1}^{q} b_k z^k$$

are polynomials in z with matrix coefficients. The time series $\{X_n\}_{n\in\mathbb{Z}}$ is called a ARMA(p, q) m-dimensional time series.

Theorem 4.2.5 *Suppose that*

$$\det P(z) \neq 0 \text{ for all } z \in \mathbb{C} \text{ such that } |z| \leq 1. \tag{4.37}$$

Then (4.35) has exactly one stationary distribution

$$X_n = \varepsilon_n + \sum_{k\geq 1} h_k \varepsilon_{n-k} \tag{4.38}$$

where the $m \times m$ matrices h_k are uniquely determined by

$$H(z) := \sum_{k\geq 0} h_k z^k = P^{-1}(z)Q(z) \quad (|z| \leq 1). \tag{4.39}$$

Proof Condition (4.37) guarantees that the inverse matrix $P^{-1}(z)$ is well defined on $\{z \in \mathbb{C} ; |z| \leq 1\}$. In fact, since $\det P(z)$ is a polynomial in z, (4.37) holds true for all $z \in \mathbb{C}$ such that $|z| \le (1 + \varepsilon)$ for some $\varepsilon > 0$. Each of the elements of $P^{-1}(z)$ is a rational function with no singularities in $\{|z| \le (1 + \varepsilon)\}$, and therefore $P^{-1}(z)$ admits a power series expansion

$$P^{-1}(z) = \sum_{k \ge 0} \alpha_k z^k .$$

In particular $\alpha_k (1 + \frac{1}{2}\varepsilon)^k$ tends to 0 componentwise as $k \uparrow \infty$ and consequently there exists a finite positive constant C such that the components of α_k are bounded by $C(1 + \frac{1}{2}\varepsilon)^k$, which in turn implies the absolute summability of the componebnts of the matrix α_k. If $\{X_n\}_{n \in \mathbb{Z}}$ is a WSS stationary sequence verifying (4.36), we can apply to both sides of this equality $P(z)^{-1}$ to obtain $X_n = P^{-1}(z)Q(z)$ from which representation (4.38) folows immediately.

Conversely, if (4.38) holds with the matrices h_k's defined by (4.39), then

$$P(z)X_n = P(z)H(z)\varepsilon_n = P(z)P^{-1}(z)Q(z)\varepsilon_n = Q(z)\varepsilon_n ,$$

showing that $P^{-1}(z)Q(z)\varepsilon_n$ is a stationary solution of (4.36).

The first part of the proof shows that it is in fact the *unique* stationary distribution of (4.36).   □

Similarly to the unidimensional case, the matrix filter $H(z)$ is called the generatiiong filter of the series. The following result, introducing the whitening matrix filter $W(z)$, is proven in the same way as the previous theorem.

Theorem 4.2.6 *Suppose that*

$$\det Q(z) \ne 0 \text{ for all } z \in \mathbb{C} \text{ such that } |z| \le 1 \qquad (4.40)$$

and if $\{X_n\}_{n \in \mathbb{Z}}$ is a stationary solution of (4.36), then

$$\varepsilon_n = \sum_{k \ge 0} w_k X_{n-k} , \qquad (4.41)$$

where the matrices w_k are uniquely determined by

$$W(z) := \sum_{k \ge 0} w_k z^k = Q^{-1}P(z) .$$

4.2.3 Long-Range Dependence

As in the case of continuous-time stochastic processes, we observe that the "classical" models of WSS time series, such as the ARMA process, have a covariance function which is geometrically bounded, that is:

$$R_X(|m|) \le C\alpha^{|m|},$$

where $C > 0$ and $0 < \alpha < 1$. One encounters WSS time series for which exponential decay is replaced by polynomial decay. More precisely,

$$R_X(|m|) \sim C|m|^{2d-1} \quad \text{as } m \to \infty,$$

where $C > 0$ and $d < \frac{1}{2}$. One then makes the distinction between *intermediate memory* and *long memory* according to whether $d < 0$, in which case

$$\sum_{k \in \mathbb{Z}} |R(k)| < \infty,$$

or $0 \le d < \frac{1}{2}$, in which case

$$\sum_{k \in \mathbb{Z}} |R(k)| = \infty.$$

A *long-range dependent* time series is one whose covariance function is not summable. In particular a long-memory time series is long-range dependent. The interest in long-range dependence processes arose in hydrology and economics and has been revived by experimental studies on traffic flows in communications networks, which also revealed long-range dependence phenomena.

Example 4.2.4 Increments of self-similar processes. Let $\{Y(t)\}_{t \in \mathbb{R}_+}$ be a self-similar stochastic process with stationary increments and self-similarity parameter $H > 0$ (in particular $Y(0) = 0$). Define the discrete-time incremental process $\{X_n\}_{n \in \mathbb{N}}$ by

$$X_n = Y(n) - Y(n-1)$$

for $n \ge 1$, and $X_0 = Y(0) = 0$. Its covariance function is, for $k \ge 0$,

$$
\begin{aligned}
R(k) &= \operatorname{cov}(X_1, X_k) \\
&= \frac{1}{2} E\left[\left(\sum_{j=1}^{k+1} X_j \right)^2 + \left(\sum_{j=2}^{k} X_j \right)^2 - \left(\sum_{j=1}^{k} X_j \right)^2 + \left(\sum_{j=2}^{k+1} X_j \right)^2 \right] \\
&= \frac{1}{2} E\left[(Y(k+1) - Y(0))^2 \right] + \frac{1}{2} E\left[(Y(k-1) - Y(0))^2 \right] \\
&\qquad - E\left[(Y(k) - Y(0))^2 \right].
\end{aligned}
$$

Therefore, by self-similarity, for $k \ge 0$,

$$R(k) = \frac{1}{2}\sigma^2 \left[(k+1)^{2H} - 2k^{2H} + (k-1)^{2H} \right] \tag{4.42}$$

and $R(-k) = R(k)$. Defining $g(x) = (1+x)^{2H} - 2 + (1-x)^{2H}$, we can write for $k \geq 0$ $R(k) = \frac{1}{2}\sigma^2 k^{2H} g(k^{-1})$.

If $0 < H < 1$, $H \neq \frac{1}{2}$, the first non-null term of the Taylor expansion of g at 0 is $2H(2H-1)x^2$ and therefore

$$\lim_{k \uparrow \infty} R(k)/\sigma^2 2H(2H-1)k^{2H-2} = 1.$$

For $\frac{1}{2} < H < 1$, this implies the non-summability of the covariance function:

$$\sum_{k \in \mathbb{Z}} R(k) = \infty,$$

and the incremental process therefore has long memory.

For $H = \frac{1}{2}$, the incremental process is a white noise.

For $0 < H < \frac{1}{2}$, the covariance function is summable, and

$$\sum_{k \in \mathbb{Z}} R(k) = 0.$$

This is not an interesting case from the point of view of modelization.[2]

The case $H = 1$ is also not interesting since the covariance function is then a constant. As for the case $H > 1$, it must be excluded since $g(k^{-1})$ diverges to infinity and therefore R given by (4.42) cannot be a covariance function.

We shall quote a result[3] giving the spectral density of the incremental process when $0 < H < \frac{1}{2}$

$$f(\omega) = 2K(1 - \cos \omega) \sum_{j \in \mathbb{Z}} |2\pi j + \omega|^{-2H-1}$$

where

$$K = \frac{\sigma^2}{2\pi} \Gamma(2H+1) \sin(\pi H)$$

and therefore, around the frequency $\omega = 0$,

$$f(\omega) = K|\omega|^{1-2H} + O\left(|\omega|^{2 \wedge (3-2H)}\right).$$

This behaviour of the spectral density around the null pulsation ($f(\omega) = K|\omega|^\beta$ for some $\beta \in (0, 1)$) is typical of long-range dependence.

---

[2] See the discussion in [Beran], p. 50–54.

[3] Ya. G. Sinai (1976), "Self-similar probability distributions", *Theory Probab. Appl.*, pp. 21, 64–80.

Fractionally Integrated White Noise

In order to introduce the fractionally integrated white noise, we begin with some preliminary remarks. Let $\{\varepsilon_n\}_{n\in\mathbb{Z}}$ be a white noise with variance σ^2, and consider the auto-regressive time series $\{X_n\}_{n\in\mathbb{Z}}$

$$(1 - \alpha z)X_n = \varepsilon_n$$

with $0 < \alpha < 1$, whose covariance decays exponentially. However, as α increases to 1 "the tail of the covariance thickens" and one might want to check the limiting case $\alpha = 1$ from the point of view of long dependence.

A stochastic process $\{X_n\}_{n\in\mathbb{Z}}$ such that

$$X_n - X_{n-1} = \varepsilon_n$$

is called an *integrated white noise*. For instance, taking for white noise sequence

$$\varepsilon_n := W(n) - W(n - 1)$$

where $\{W(t)\}_{t\in\mathbb{R}}$ is a Wiener process of variance σ^2, the corresponding integrated white noise is $X_n := W(n)$ which is not a WSS. This is also in accord with the general theory of ARMA sequences (not given here) which tells that a time series $\{X_n\}_{n\in\mathbb{Z}}$ satisfying

$$P(z)X_n = Q(z)\varepsilon_n$$

for some polynomials P and Q can not be WSS if the roots of P on the unit circle are not also roots of Q.

The search for long dependence with relatively simple models will therefore follow a different line of thought. We now introduce the *fractionally integrated white noise*.

Definition 4.2.4 A *fractional integrated white noise* $\{X_n\}_{n\in\mathbb{Z}}$ is a WSS discrete time process such that

$$(1 - z)^d X_n = \varepsilon_n, \tag{4.43}$$

where $\{\varepsilon_n\}_{n\in\mathbb{Z}}$ is a white noise with variance σ^2 and

$$d \neq 0, \quad -\frac{1}{2} < d < \frac{1}{2}. \tag{4.44}$$

The meaning of (4.43) is that if the time series $\{X_n\}_{n\in\mathbb{Z}}$ is the input of a filter with transfer function $(1 - z)^d$, the output is then the white noise $\{\varepsilon_n\}_{n\in\mathbb{Z}}$. Of course, if such a WSS process exists we expect that

$$X_n = (1 - z)^{-d}\varepsilon_n,$$

in the sense that the time series is the output of a filter with transmittance $(1-z)^{-d}$ when the input is the white noise $\{\varepsilon_n\}_{n\in\mathbb{Z}}$. We shall examine these claims more precisely.

We must first study the transfer functions $(1-z)^d$ and $(1-z)^{-d}$. To begin with, for all d satisfying (4.44) the function $z \to (1-z)^d$ is holomorphic in the domain $\{z : |z| < 1\}$ and therefore admits a power series expansion

$$(1-z)^d = \sum_{n\geq 0} \beta_n z^n$$

in a neighbourhood of the origin. Here, by the expression (1.6) of the coefficient of the power series expansion,

$$\beta_n = \frac{1}{2i\pi} \oint_C \frac{(1-z)^d}{z^{n+1}}\, dz,$$

where C is a closed path without multiple points that lies within the interior of the ring of convergence, for example the unit circle, taken in the anti-clockwise sense. By the method of residues, we obtain

$$\beta_n = \prod_{0<k\leq n} \frac{k-1-d}{k} = \frac{\Gamma(n-d)}{\Gamma(n+1)\Gamma(-d)}, \qquad (4.45)$$

where Γ is the gamma function defined by

$$\Gamma(x) = \int_0^{+\infty} t^{x-1} e^{-t}\, dt \text{ if } x > 0,$$

$$= +\infty \text{ if } x = 0,$$

$$= x^{-1}\Gamma(x+1) \text{ if } x < 0$$

From the generalized Stirling's equivalence

$$\Gamma(x) \sim \sqrt{2\pi}\, e^{-x+1}(x-1)^{x-1/2} \quad \text{as } x \to \infty$$

we obtain

$$\beta_n \sim \frac{n^{-d-1}}{\Gamma(-d)} \quad \text{as } n \to \infty. \qquad (4.46)$$

Replacing d by $-d$ in the above we find that

$$(1-z)^{-d} = \sum_{n\geq 0} \gamma_n z^n \qquad (4.47)$$

in a neighbourhood of the origin, where

$$\gamma_n = \frac{\Gamma(n+d)}{\Gamma(n+1)\Gamma(+d)} = \prod_{0 < k \le n} \frac{k-1+d}{k} \tag{4.48}$$

and, in particular,

$$\gamma_n \sim \frac{n^{+d-1}}{\Gamma(d)} \quad \text{as } n \to \infty. \tag{4.49}$$

We must distinguish between the case where d is positive and the case where d is negative.

CASE WHERE $0 < d < +\frac{1}{2}$. In view of (4.46) and (4.49) we then have

$$\sum_{n \ge 0} |\beta_n| < \infty$$

(and, a fortiori, $\sum_{n \ge 0} |\beta_n|^2 < \infty$) and

$$\sum_{n \ge 0} |\gamma_n|^2 < \infty$$

(but $\sum_{n \ge 0} |\gamma_n| = \infty$).

CASE WHERE $-\frac{1}{2} < d < 0$. In this case

$$\sum_{n \ge 0} |\beta_n|^2 < \infty$$

(but $\sum_{n \ge 0} |\beta_n| = \infty$) and

$$\sum_{n \ge 0} |\gamma_n| < \infty$$

(and, a fortiori, $\sum_{n \ge 0} |\gamma_n|^2 < \infty$).

In both cases, that is, for all $d \ne 0$, $-\frac{1}{2} < d < \frac{1}{2}$, the transmittance $(1 - e^{-i\omega})^d$ is in $L_{\mathbb{C}}^2([-\pi, +\pi], d\omega)$. Also

$$(1 - e^{-i\omega})^d = \lim_{n \uparrow \infty} \left\{ \sum_{k=1}^{n} \beta_k e^{ik\omega} \right\} \tag{4.50}$$

and

$$(1 - e^{-i\omega})^{-d} = \lim_{n\uparrow\infty}\left\{\sum_{k=1}^{n}\gamma_k e^{ik\omega}\right\}, \tag{4.51}$$

where for all $d \neq 0$, $-\frac{1}{2} < d < \frac{1}{2}$, both limits are $L^2_{\mathbb{C}}(\mathbb{R})$. In the first case ($0 < d < \frac{1}{2}$) the limit in (4.50) is also pointwise and uniform in ω, in the second case ($-\frac{1}{2} < d < 0$) the same is true of the limit in (4.51).

These preliminary results having been established, we can now turn to the statement and proof of the existence and uniqueness of a WSS solution $\{X_n\}_{n\in\mathbb{Z}}$ of Eq. (4.43).

Theorem 4.2.7 *If $d \neq 0$, $-\frac{1}{2} < d < \frac{1}{2}$, there exists a centered WSS time series $\{X_n\}_{n\in\mathbb{Z}}$ satisfying*

$$(1 - z)^d X_n = \varepsilon_n, \tag{4.52}$$

where $\{\varepsilon_n\}_{n\in\mathbb{Z}}$ is an arbitrary white noise of variance σ^2. The meaning of the last equality is

$$\sum_{k\geq 0}\beta_k X_{n-k} = \varepsilon_n, \tag{4.53}$$

where the coefficients β_k are given by (4.45) and where the series converges almost surely absolutely if $0 < d < +\frac{1}{2}$, and in the quadratic mean if $d \neq 0$, $-\frac{1}{2} < d < 0$. Moreover,

$$X_n = (1 - z)^{-d}\varepsilon_n,$$

that is to say,

$$X_n = \sum_{k\geq 0}\gamma_k \varepsilon_{n-k}, \tag{4.54}$$

where the coefficients γ_k are given by (4.48) and where the series converges almost surely absolutely if $-\frac{1}{2} < d < 0$, and in the quadratic mean if $d \neq 0$, $-0 < d < \frac{1}{2}$.

Proof Since $\{\varepsilon_n\}_{n\in\mathbb{Z}}$ is a white noise and $\sum_{n\geq 0}|\gamma_n|^2 < \infty$ the series in (4.54) converges in $L^2_{\mathbb{C}}(P)$. Moreover, for all $N \geq 0$

$$\sum_{k=0}^{N}\gamma_k \varepsilon_{n-k} = \int_{(-\pi,+\pi]}\left\{\sum_{k=0}^{N}\gamma_k e^{-ik\omega}\right\}e^{in\omega}\,dZ_\varepsilon(\omega), \tag{4.55}$$

where Z_ε is the Cramér–Khinchin decomposition of the white noise $\{\varepsilon_n\}_{n\in\mathbb{Z}}$. In particular

$$E[|dZ_\varepsilon(\omega)|^2] = \frac{\sigma^2}{2\pi}\,d\omega.$$

By (4.51) taken in the L^2-sense, and from the isometry formula for Doob–Wiener integrals by letting $N \uparrow \infty$ in (4.55),

$$X_n = \sum_{k \geq 0} \gamma_k \varepsilon_{n-k} = \int_{(-\pi, +\pi]} (1 - e^{-i\omega})^{-d} e^{in\omega} \, dZ_\varepsilon(\omega). \qquad (4.56)$$

Note that in the case where $\{\gamma_n\}_{n \in \mathbb{Z}}$ is in $\ell_{\mathbb{C}}^1(\mathbb{Z})$, $\sum_{k \geq 0} \gamma_k \varepsilon_{n-k}$ is an almost surely absolutely convergent sum. The $L_{\mathbb{C}}^2(P)$ limit and the P–a.s. limit are P–a.s. equal when they both exist. Having defined $\{X_n\}_{n \in \mathbb{Z}}$ as above we must now show that (4.52) is satisfied, at least in the L^2 sense. For this define

$$Y_n^N = \sum_{k=0}^{N} \beta_k X_{n-k}.$$

In view of (4.56),

$$Y_n^N = \int_{(-\pi, +\pi]} \left\{ \sum_{k=0}^{N} \beta_k e^{-ik\omega} \right\} (1 - e^{-i\omega})^{-d} e^{in\omega} \, dZ_\varepsilon(\omega).$$

Now, in both of the cases $0 < d < +\frac{1}{2}$ and $-\frac{1}{2} < d < 0$

$$\lim_{N \uparrow \infty} \left\{ \sum_{k=0}^{N} \beta_k e^{-ik\omega} \right\} (1 - e^{-i\omega})^{-d} = 1$$

in $L_{\mathbb{C}}^2([-\pi, +\pi], d\omega)$, and therefore, by the isometry formula for Wiener integrals, Y_n^N converges to

$$\int_{(-\pi, +\pi]} e^{in\omega} \, dZ_\varepsilon(\omega) = \varepsilon_n \quad \text{in } L_{\mathbb{C}}^2(P),$$

that is, $\sum_{k \geq 0} \beta_k X_{n-k} = \varepsilon_n$. Note also that when $\sum_{k \geq 0} |\beta_k| < \infty$ the convergence of the above series is absolute P–a.s., as we argued a few lines above. $\qquad \square$

Theorem 4.2.8 *The fractional integrated white noise of Theorem 4.2.7 admits a power spectral density*

$$f_X(\omega) = \frac{\sigma^2}{2\pi} \left| 2 \sin\left(\frac{\omega}{2}\right) \right|^{-2d}. \qquad (4.57)$$

Its covariance function R_X is given by the formula

$$\frac{R_X(m)}{R_X(0)} = \frac{\Gamma(m+d)\Gamma(1-d)}{\Gamma(m-d+1)\Gamma(d)}, \tag{4.58}$$

where the variance $R_X(0)$ of X_n is

$$R_X(0) = \sigma^2 \frac{\Gamma(1-2d)}{\Gamma^2(1-d)}. \tag{4.59}$$

By Stirling's equivalence formula,

$$\frac{R_X(|m|)}{R_X(0)} \sim \frac{\Gamma(1-d)}{\Gamma(d)} |m|^{2d-1} \text{ as } |m| \to \infty, \tag{4.60}$$

and therefore the fractionally integrated white noise is long-range dependent or, more precisely, according to the alternative classification mentioned above, if $-\frac{1}{2} < d < 0$ the fractionally integrated white noise has long memory, and if $0 < d < -\frac{1}{2}$ it has an intermediate memory. Note also that, from (4.57),

$$f_X(\omega) \sim \frac{\sigma^2}{2\pi} \frac{1}{|\omega|^{2d}} \text{ as } |\omega| \to 0. \tag{4.61}$$

Proof From (4.56) and Wiener's isometry formula we obtain the power spectral density of $\{X_n\}$

$$f_X(\omega) = \frac{\sigma^2}{2\pi} |1 - e^{-i\omega}|^{-2d},$$

that is to say, precisely (4.57). The value of the covariance function at m is the mth Fourier coefficient of the power spectral density, and formulas (4.58) and (4.59) from the computation of these coefficients, with the help of the identity[4]

$$\int_0^\pi \cos(mx) \sin^{\nu-1}(x)\,dx = \frac{\pi \cos(\pi m/2)\Gamma(\nu+1)2^{1-\nu}}{\nu\Gamma((\nu+m+1)/2)\Gamma((\nu-m+1)/2)}.$$

$\square$

Theorem 4.2.9 *If we interpret the Eq. (4.52) as*

$$\sum_{k\geq0} \beta_k X_{n-k} = \varepsilon_n \tag{4.62}$$

$((1-z)^d = \sum_{k\geq0} \beta_k z^k)$, *where the series is required to be convergent in $L^2_{\mathbb{C}}(P)$, there is an unique WSS solution of (4.52) if we constrain this solution to have a power spectral density.*

---

[4] Gradshtein and Ryzhik (1965), p. 372.

Proof Since

$$\sum_{k=0}^{N} \beta_k X_{n-k} = \int_{(-\pi,+\pi]} \left(\sum_{k=0}^{N} \beta_k e^{-ik\omega} \right) e^{in\omega} \, dZ_X(\omega) \qquad (4.63)$$

converges in $L^2_{\mathbb{C}}(P)$ *by assumption*, and $\sum_{k=0}^{N} \beta_k e^{-ik\omega}$ converges in $L^2_{\mathbb{C}}(\mu_X)$ to some function denoted $g(\omega)$ that belongs to $L^2_{\mathbb{C}}(\mu_X)$. Passing to the limit $N \uparrow \infty$ in (4.63) gives

$$\sum_{k\geq 0} \beta_k X_{n-k} = \int_{(-\pi,+\pi]} g(\omega) e^{in\omega} \, dZ_X(\omega).$$

Equating the spectral measures of both sides of (4.53) we have

$$|g(\omega)|^2 f_X(\omega) \, d\omega = \frac{\sigma^2}{2\pi} \, d\omega.$$

Since $f_X(\omega) \, d\omega \ll d\omega$,

$$|g(\omega)|^2 = |1 - e^{-i\omega}|^{2d}$$

(the limit of $\sum_{k=0}^{N} \beta_k e^{-ik\omega}$ in $L^2_{\mathbb{C}}((-\pi,+\pi], d\omega)$), and

$$f_X(\omega) = \frac{\sigma^2}{2\pi} |1 - e^{-i\omega}|^{-2d}.$$

From the representation

$$\varepsilon_n = \int_{(-\pi,+\pi]} (1 - e^{-i\omega})^d e^{in\omega} \, dZ_X(\omega)$$

we obtain

$$\sum_{k=0}^{N} \gamma_k \varepsilon_{n-k} = \int_{(-\pi,+\pi]} \left(\sum_{k=0}^{N} \gamma_k e^{-ik\omega} \right) (1 - e^{-i\omega})^d e^{in\omega} \, dZ_X(\omega). \qquad (4.64)$$

Observe that

$$\left(\sum_{k=0}^{N} \gamma_k e^{-ik\omega} \right) (1 - e^{-i\omega})^d$$

converges to unity in $L^2_{\mathbb{C}}(\mu_X)$ since $\sum_{k=0}^{N} \gamma_k e^{-ik\omega}$ converges to $(1 - e^{-i\omega})^{-d}$ in $L^2_{\mathbb{C}}((-\pi,+\pi], d\omega)$. Therefore passing to the limit in (4.64) yields, in $L^2_{\mathbb{C}}(P)$,

$$X_n = \sum_{k \geq 0} \gamma_k \varepsilon_{n-k},$$

and this uniquely determines $\{X_n\}_{n \in \mathbb{Z}}$.   □

ARIMA Processes

The class of fractionally integrated white noises has only two parameters (σ^2 and d) which can be tuned to obtain a desired aspect of the covariance function, and for this reason, their modelling value is limited. A more versatile class of time series, the ARIMA processes was introduced to act on the small lag behaviour of the covariance function.

Definition 4.2.5 Let $d \in (-\frac{1}{2}, +\frac{1}{2})$, and let P and Q be two polynomials of orders p and q respectively, without roots in the closed unit disk centered at the origin and such that $P(0) = Q(0) = 1$. A WSS time series $\{X_n\}_{n \in \mathbb{Z}}$ such that

$$P(z)(1 - z)^d X_n = Q(z)\varepsilon_n , \qquad (4.65)$$

where $\{\varepsilon_n\}_{n \in \mathbb{Z}}$ is a white noise with variance σ^2 is called an ARIMA(p, d, q) time series.

(The terminology is not quite standard due to the assumptions on the roots of P and Q.) Clearly, this means that $\{(1 - z)^d X_n\}_{n \in \mathbb{Z}}$ is an ARMA(p, q) time series. If Q has no root inside the closed unit disk centered at 0, defining $Y_n := \frac{P(z)}{Q(z)} X_n$ we have that

$$(1 - z)^d X_n = \varepsilon_n$$

and

$$P(z)X_n = Q(z)Y_n .$$

Therefore an ARIMA(p, d, q) time series can be viewed as an ARMA(p, q) time series driven by fractional white noise.

4.3 General Prediction Theory

4.3.1 Wold's Decomposition

The theory of prediction for ARMA processes will now be extended to a more general situation. Let $\{X_n\}_{n \in \mathbb{Z}}$ be a centered WSS time series. The Hilbert subspace of $L^2_{\mathbb{C}}(P)$ generated by the random variables $X_n, X_{n-1}, X_{n-2}, \ldots$ and by the whole time series are denoted by $H(X; n)$ and $H(X)$ respectively. They are called respectively the *linear past* at time n and the *linear history* of the time series. The Hilbert subspace of $L_{\mathbb{C}}^2(P)$

$$H(X; -\infty) := \cap_{n \in \mathbb{Z}} H(X; n)$$

is called the *linear past at infinity* of the time series. In general, $\{0\} \subseteq H(X; -\infty) \subseteq H(X)$.

Definition 4.3.1 If $H(X; -\infty) = H(X)$, the WSS time series is called *purely deterministic* or *singular*. If $H(X; -\infty) = \{0\}$, it is called *purely non-deterministic* or *regular*.

We have the following general decomposition of a WSS time series as the sum of two WSS time series, respectively regular and singular.[5]

Theorem 4.3.1 *Let* $\{X_n\}_{n \in \mathbb{Z}}$ *be a centered* WSS *time series. It can be decomposed in a unique manner as*

$$X_n = U_n + V_n$$

where $\{U_n\}_{n \in \mathbb{Z}}$ *and* $\{V_n\}_{n \in \mathbb{Z}}$ *are jointly* WSS *time series such that*
  (a) $H(U) \perp H(V)$,
  (b) $\{U_n\}_{n \in \mathbb{Z}}$ *is purely non-deterministic (regular), and*
  (c) $\{V_n\}_{n \in \mathbb{Z}}$ *is purely deterministic (singular).*

Proof Denote by P_n and $P_{-\infty}$ the orthogonal projection operators on $H(X; n)$ and $H(X; -\infty)$ respectively. Let

$$V_n = P_{-\infty} X_n, \qquad U_n = X_n - P_{-\infty} X_n.$$

We first observe that, the operator $S : H(X) \to H(X)$ defined by $SX_n = X_{n-1}$ is an isometry and that $P_n S = S P_{n+1}$ (Exercise 4.5.8). In particular, letting $n \downarrow -\infty$, $P_{-\infty} S = S P_{-\infty}$. Applying S to the equalities $U_n = X_n - P_{-\infty} X_n$ and $V_n = P_{-\infty} X_n$ yields $SU_n = U_{n-1}$ and $SV_n = V_{n-1}$. As S is isometric, this implies that $\{U_n\}_{n \in \mathbb{Z}}$ and $\{V_n\}_{n \in \mathbb{Z}}$ are jointly WSS.

(a) From the definitions of U_n and V_n, for all n, $U_n \perp H(X; -\infty)$, and therefore, for all m, n, $U_n \perp V_m$ since $V_m \in H(X; -\infty)$ for all m. This proves (a).

Among the above observations, we shall keep record of the following ones for easy reference:

$$H(U; n) \subseteq H(X; n) \tag{α}$$

$$H(U; n) \perp H(X; -\infty) \tag{β}$$

$$H(V; n) \subseteq H(X; -\infty) \tag{γ}$$

(b) Letting $n \downarrow -\infty$ in (α) and (γ) gives $H(U; -\infty) \subseteq H(X; -\infty)$ and $H(U; -\infty) \perp H(X; -\infty)$. This is possible only if $H(U; -\infty) = \{0\}$. This proves (b).

---

[5] H.O.A Wold, *A Study in the Analysis of Time Series*, Almquist and Wiksell, Upsalla (1938).

(c) From (a), $H(U;n) \perp H(V;n)$. Therefore, since $X_n = U_n + V_n$, $H(X;n) \subseteq H(U;n) \oplus H(V;n)$. But $H(U;n)$ and $H(V;n)$ are in $H(X;n)$, and therefore

$$H(X;n) = H(U;n) \oplus H(V;n).$$

In particular
$$H(X;-\infty) \subseteq H(U;n) \oplus H(V;n).$$

This together with (α) gives

$$H(X;-\infty) \subseteq H(V;n).$$

Therefore, in view of (γ),

$$H(X;-\infty) = H(V;n).$$

From this, we deduce that $H(X;-\infty) = H(V;-\infty) = H(V;n)$. This proves (c).

We now prove uniqueness. Let $X_n = U_n' + V_n'$ be another decomposition [satisfying (a), (b) and (c)]. Since $H(V';n) = H(V')$, we have

$$H(X;n) = H(U';n) \oplus H(V';n) = H(U';n) \oplus H(V')$$

and therefore

$$H(X;-\infty) = H(U';-\infty) \oplus H(V') = H(V').$$

Since $V_n' \in H(V') = H(X;-\infty)$ and $U_n' \perp H(V') = H(X;-\infty)$, we have that $P_{-\infty}X_n = P_{-\infty}U_n' + P_{-\infty}V_n' = V_n'$, that is $V_n' = V_n$. This proves uniqueness. $\square$

Example 4.3.1 A white noise is regular whereas a harmonic series is singular (Exercise 4.5.11).

4.3.2 Innovations Sequence

Lemma 4.3.1 *Let $\{X_n\}_{n\in\mathbb{Z}}$ be a WSS time series with a PSD f that is almost everywhere positive with respect to Lebesgue measure. Then there exists a white noise $\{\varepsilon_n\}_{n\in\mathbb{Z}}$ such that*

$$X_n = \sum_{k\in\mathbb{Z}} c_k \varepsilon_{n-k}, \qquad (*)$$

where $\sum_{k\in\mathbb{Z}} |c_k|^2 < \infty$.

Proof Since the spectral density is non-negative and Lebesgue integrable on $[-\pi, +\pi]$, it has the form

$$f(\omega) = \frac{1}{2\pi} |g(\omega)|^2 , \tag{\dagger}$$

where g is Lebesgue square-integrable on $[-\pi, +\pi]$ and can therefore be represented as a Fourier series

$$g(\omega) = \sum_{k \in \mathbb{Z}} c_k e^{-ik\omega}$$

where $\sum_{k \in \mathbb{Z}} |c_k|^2 < \infty$. Define for all $C \in \mathcal{B}([-\pi, +\pi])$

$$\tilde{Z}(C) = \int_C \frac{1}{g(\omega)} Z_X(d\omega)$$

and check that $\tilde{Z}$ is a random pseudo-measure with the Lebesgue measure as structural measure. In particular,

$$\varepsilon_n = \int_{(-\pi, +\pi]} e^{in\omega} \tilde{Z}(d\omega)$$

defines a white moise sequence with unit variance. Moreover,

$$X_n = \int_{(-\pi, +\pi]} e^{in\omega} Z_X(d\omega) = \int_{(-\pi, +\pi]} e^{in\omega} g(\omega) \tilde{Z}(d\omega)$$

that is

$$X_n = \int_{(-\pi, +\pi]} e^{in\omega} \left(\sum_{k \in \mathbb{Z}} c_k e^{-ik\omega} \right) \tilde{Z}(d\omega)$$

$$= \sum_{k \in \mathbb{Z}} c_k \int_{(-\pi, +\pi]} e^{in\omega} e^{-ik\omega} \tilde{Z}(d\omega) = \sum_{k \in \mathbb{Z}} c_k \varepsilon_{n-k} .$$

$\square$

We make two observations. First, the factorization ($\dagger$) is not unique. any other function g' such that $g'(\omega) = H(e^{-i\omega}) g(\omega)$ where $|H(e^{-i\omega})|^2 = 1$ will do (take for instance $H(z)$ to be the transfer function of an all-pass filter). Secondly, the representation ($*$) involves in general a summation over $\mathbb{Z}$. When the prediction problem (that of obtaining the projection of X_{n+m} on $H(X; n)$ for $m > 0$) will be considered, we will see that the latter fact is an hindrance, and we shall seek representations of the kind

$$X_n = \sum_{k \geq 0} c_k \varepsilon_{n-k} . \tag{4.66}$$

The following theorem tells us that this is possible in the purely non-deterministic case. Kolmogorov's criterion of regularity (Theorem 4.3.3) will tell us when a time series is regular and what to do to obtain decomposition (4.66).

Theorem 4.3.2 *Let* $\{X_n\}_{n \in \mathbb{Z}}$ *be a non-trivial (not identically null) centered* WSS *time series. It is purely non-deterministic (regular) if and only there exists a unique (up to a multiplicative constant) innovations sequence (see Definition 4.2.2)* $\{\varepsilon_n\}_{n \in \mathbb{Z}}$ *for it and a sequence* $\{c_n\}_{n \in \mathbb{Z}}$ *in* $\ell^2_{\mathbb{C}}$ *such that the representation (4.66) holds. Moreover one can choose* $c_0 = 1$ *and*

$$\varepsilon_n = X_n - P_{n-1} X_n. \tag{4.67}$$

Proof Suppose the existence of an innovations sequence $\{\varepsilon_n\}_{n \in \mathbb{Z}}$ for $\{X_n\}_{n \in \mathbb{Z}}$. In particular $H(X; n) = H(\varepsilon; n)$, and therefore $H(X; -\infty) \subseteq H(\varepsilon; n)$ for all $n \in \mathbb{Z}$, which in turn gives $H(X; -\infty) \subseteq H(\varepsilon; -\infty)$. This implies, since a white noise is purely non-deterministic, that

$$H(X; -\infty) = \{0\},$$

that is, $\{X_n\}_{n \in \mathbb{Z}}$ is purely non-deterministic.

We now turn to the proof of necessity. We therefore suppose that $\{X_n\}_{n \in \mathbb{Z}}$ is purely non-deterministic. Define ε_n as in (4.67). We have $S\varepsilon_n = \varepsilon_{n-1}$ and therefore $\{\varepsilon_n\}_{n \in \mathbb{Z}}$ is WSS since S is an isometry. By definition $\varepsilon_n \in H(X; n)$ and $\varepsilon_n \perp H(X; n-1)$. Therefore ε_n is orthogonal to ε_k for all $k \leq n-1$, that is: $\{\varepsilon_n\}_{n \in \mathbb{Z}}$ is a white noise. By construction [see (4.67)], $H(X; n) = L(\varepsilon_n) \oplus H(X; n-1)$, where $L(\varepsilon_n) = \{\lambda \varepsilon_n; \lambda \in \mathbb{C}\}$. Iterating this identity gives

$$H(X; n) = L(\varepsilon_n) \oplus L(\varepsilon_{n-1}) \oplus \cdots \oplus L(\varepsilon_{n-j}) \oplus H(X; n-j-1).$$

Therefore there exists complex coefficients $c_0(n), c_1(n), \ldots c_j(n)$ such that

$$X_n = \sum_{k=0}^{j} c_k(n) \varepsilon_{n-k} + P_{n-j-1} X_n. \tag{*}$$

Since the original serises is assumed regular, $\lim_{j \uparrow \infty} P_{n-j-1} X_n = P_{-\infty} X_n = 0$, so that

$$X_n = \sum_{k=0}^{\infty} c_k(n) \varepsilon_{n-k}$$

where of course the convergence is in $L^2_{\mathbb{C}}(P)$. In particular, $\sum_{k=0}^{\infty} |c_k(n)|^2 < \infty$. The coefficients $c_k(n)$ must now be shown to be independent of n. For this apply S to both sides of (*), remembering that $S\varepsilon_n = \varepsilon_{n-1}$ and that $SP_{k+1} = P_k S$, to obtain

$$X_{n-1} = \sum_{k=0}^{j} c_k(n)\varepsilon_{n-1-k} + P_{n-1-j-1}X_n$$

$$= \sum_{k=0}^{j} c_k(n-1)\varepsilon_{n-1-k} + P_{n-1-j-1}X_n .$$

Uniqueness of the decomposition $(*)$ implies that $c_k(n) = c_k(n-1)$ for all n, all j. We have therefore obtained a representation of the type (4.66). Replacing if necessarily ε by $\alpha\varepsilon$, and c_n by αc_n, we still have a representation of the type (4.66) and $\{\varepsilon_n\}_{n\in\mathbb{Z}}$ is still an innovations sequence for the time series. We can in particular choose $c_0 = 1$ and in this case the filter with transfer function

$$H(z) = 1 + \sum_{k=1}^{\infty} c_k z^k$$

is called the (canonical) generating filter of the time series, since the latter can be generated by applying a white noise to it. It is canonical in two ways: $c_0 = 1$ and most importantly is the fact that it is a causal filter.

From (4.66), we have $H(X; n-1) \subseteq H(\varepsilon; n-1)$. This together with $\varepsilon_n \perp H(\varepsilon; n-1)$ implies that $\varepsilon_n \perp H(X; n-1)$. Therefore, from (4.66) (with $c_0 = 1$) and the projection principle (Theorem 1.3.10)

$$P_{n-1}X_n = \sum_{k=1}^{\infty} c_k\varepsilon_{n-k} . \tag{4.68}$$

Subtracting the latter equality from (4.66) gives (4.67), which proves uniqueness. $\square$

Representation (4.66) with $\{\varepsilon_n\}_{n\in\mathbb{Z}}$ being an innovations sequence for the time series also gives a solution to the m-step prediction problem, that of obtaining $\widehat{X}_{n+m,n} := P_{n-m}X_n$, the projection of X_n on the Hilbert space $H(X; n-1)$ generated by the random variables $X_{n-m}, X_{n-m-1}, \ldots$. With an argument similar to that for proving (4.68), we have

$$\widehat{X}_{n+m,n} = \sum_{k=m}^{\infty} c_k\varepsilon_{n-k} . \tag{*}$$

But a solution in the form $(*)$ is not quite satisfactory as they are in terms of the innovations and not in terms of the time series itself. We shall come back to this in the next subsection.

4.3.3 Kolmogorov's Regularity Criterion

Theorem 4.3.3 *A non degenerate* WSS *time series that is regular admits a PSD f, and*

$$\int_{-\pi}^{+\pi} \ln f(\omega)\, d\omega > -\infty.$$

(4.69)

(In particular $f > 0$ almost everywhere with respect to the Lebesgue measure.)
   Conversely if a WSS *time series admits a PSD satisfying (4.69), it is regular.*

Proof We saw that a centered WSS time series $\{X_n\}_{n\in\mathbb{Z}}$ that is regular has a spectral density of the form

$$f(\omega) = \frac{1}{2\pi}\left|\varphi(e^{-i\omega})\right|^2$$

where

$$\varphi(z) = \sum_{k\geq 0} c_k z^k$$

(*)

and

$$\{c_k\}_{k\geq 0} \in \ell_{\mathbb{C}}^2(\mathbb{N}).$$

(**)

The function φ is analytic in the open unit disk centered at the origin, and because of (**), it belongs to the *Hardy class* H^2 of functions that are analytic in the open unit disk centered at the origin and satisfy

$$\sup_{0\leq r<1} \int_{-\pi}^{+\pi} \left|\varphi(re^{-i\omega})\right|^2 d\omega < \infty,$$

since for all $r \in [0, 1)$,

$$\int_{-\pi}^{+\pi} \left|\varphi(re^{-i\omega})\right|^2 d\omega = 2\pi \sum_{k\in\mathbb{N}} |c_k|^2 r^{2k} \leq 2\pi \sum_{k\in\mathbb{N}} |c_k|^2 < \infty.$$

It is known from the theory of functions of a complex variable[6] that if φ is moreover not identically null, then

$$\int_{-\pi}^{+\pi} \ln\left(\left|\varphi(e^{-i\omega})\right|\right) d\omega > -\infty$$

---

[6] See for instance [Helson], Chap. 3.

which is equivalent to (4.69).

Conversely, let the PSD of a time series satisfy condition (4.69). Again from the theory of functions of the complex variable, we have that there exists φ of the form ($*$) that is in the Hardy class H^2 and such that almost-everywhere (with respect to the Lebesgue measure)

$$f(\omega) = \frac{1}{2\pi} \left| \varphi(e^{-i\omega}) \right|^2 ,$$

that is

$$f(\omega) = \frac{1}{2\pi} \left| \sum_{k \in \mathbb{N}} c_k e^{-ik\omega} \right|^2$$

where $\sum_{k \in \mathbb{Z}} |c_k|^2 < \infty$. This implies the existence of a discrete time white noise $\{\varepsilon_n\}_{n \in \mathbb{Z}}$ such that $X_n = \sum_{k \in \mathbb{N}} c_k \varepsilon_{n-k}$, and therefore the time series is regular. $\square$

Since $\ln f(\omega) \leq f(\omega)$ and $\int_{-\pi}^{+\pi} f(\omega)\, d\omega + E\left[|X_0|^2 \right] < \infty$ the integral can only diverge with the value $-\infty$. This happen for instance if the spectral density is null on an interval (not reduced to a point, of course).

An explicit construction of φ in a special case[7] (yet general enough for most situations) is as follows. Define

$$h(z) := \frac{1}{2\pi} \sum_{n \in \mathbb{Z}} R(n) z^n ,$$

where R is the covariance function of the time series. In particular, $f(\omega) = h(e^{-i\omega})$. Suppose that the function $\ln h$ is analytic in the donut $\rho < |z| < \rho^{-1}$ for some positive $\rho < 1$. Let

$$\ln h(z) = \sum_{n \in \mathbb{Z}} b_n z^n$$

be its Laurent expansion inside this donut. Since the donut contains the unit circle centered at the origin, $\{b_n\}_{n \in \mathbb{Z}} \in \ell_{\mathbb{C}}^1(\mathbb{Z}) \subset \ell_{\mathbb{C}}^2(\mathbb{Z})$, and the b_n's are the Fourier coefficients of $\ln h(e^{-i\omega})$:

$$b_n = \frac{1}{2\pi} \int_{-\pi}^{+\pi} \ln h(e^{-i\omega}) e^{in\omega}\, d\omega ,$$

and in particular

$$b_0 = \frac{1}{2\pi} \int_{-\pi}^{+\pi} \ln h(e^{-i\omega})\, d\omega . \qquad (4.70)$$

---

[7] [Whittle].

Since $\omega \to \ln h(e^{-i\omega})$ is real-valued and even, the b_n's are real-valued and $b_n = b_{-n}$ for all $n \in \mathbb{Z}$. From the Laurent expansion of $\ln h(z)$ we may write

$$h(z) = e^{c_0} \exp\left(\sum_{n=1}^{+\infty} b_n z^n\right) \exp\left(\sum_{n=-\infty}^{-1} b_n z^n\right)$$

or, using the symmetry property of the b_n's

$$h(z) = e^{b_0} G(z) G(z^{-1})$$

where

$$G(z) := \exp\left(\sum_{n=1}^{+\infty} b_n z^n\right). \tag{*}$$

But $G(z)$ is analytic in $|z| < \rho^{-1}$ and therefore has a power series expansion

$$G(z) = \sum_{n \geq 0} g_n z^n$$

with $g_0 = G(0) = 1$ and $\{g_n\}_{n \geq 0} \in \ell^1_{\mathbb{C}}(\mathbb{Z}) \subset \ell^2_{\mathbb{C}}(\mathbb{Z})$. From (*) it is also clear that $G(z)^{-1}$ is analytic in $|z| < \rho^{-1}$ and therefore $G(z)$ has no zero in $|z| < \rho^{-1}$ and in particular on $|z| = 1$. Finally

$$f(\omega) = h(e^{-i\omega}) = e^{b_0} |G(e^{-i\omega})|^2.$$

We now make the connection with the objects in Lemma 4.3.1 by letting $g(\omega) = 2\pi e^{b_0} |G(e^{-i\omega})|^2$, so that

$$\varepsilon_n = \int_{(-\pi, +\pi]} e^{in\omega} \frac{1}{G(e^{-i\omega})} Z_X(d\omega),$$

and therefore with $sum_{n \geq 0} \alpha_n z^n := G(z)^{-1} =,$

$$\varepsilon_n = G(z)^{-1} X_n = \sum_{k \geq 0} \alpha_{n-k} X_{n-k}$$

Finally, with ε_n (of variance 1) replaced by $(2\pi e^{b_0})^{\frac{1}{2}} \varepsilon_n$ we obtain a representation of the type

$$\varepsilon_n = W(z) X_n = \sum_{k \geq 0} w_{n-k} X_{n-k},$$

where $w_0 = 1$ and $\{\varepsilon_n\}_{n\in\mathbb{Z}}$ is a white noise of variance $2\pi e^{c_0}$. Writing the last equation as

$$X_n = \varepsilon_n + \sum_{k\geq 1} w_{n-k} X_{n-k}$$

and noting that $\varepsilon_n \perp H(X; n-1)$ and $\sum_{k\geq 1} w_{n-k} X_{n-k} \in H(X; n-1)$, we have, by the projection principle that

$$\widehat{X}_{n,n-1} := P_{n-1} X_n = \sum_{k\geq 1} w_{n-k} X_{n-k}$$

and moreover the one-step prediction error $X_n - \widehat{X}_{n,n-1}$ is equal to ε_n of variance $2\pi e^{c_0}$. Therefore, in view of the expression (4.70) of b_0, the error variance is

$$E\left[|X_n - \widehat{X}_{n,n-1}|^2\right] = 2\pi \exp\left\{\frac{1}{2\pi} \int_{-\pi}^{+\pi} \ln(f(\omega))d\omega\right\}. \tag{4.71}$$

4.4 Realization Theory for Spectral Analysis

4.4.1 The Levinson–Durbin Algorithm

Realization theory is concerned with the following problem: Given a segment of covariance function of a *real* centered WSS time series

$$R(0), R(1), \ldots, R(M),$$

in what different ways can it be continued to a sequence

$$R(0), R(1), \ldots, R(M), R(M+1), \ldots$$

that is the covariance function of a WSS time-series, called a *realization* of the initial segment?

This theory is at the basis of a method of spectral analysis (estimation of the power spectral density) known as *parametric* spectral analysis which is in certain applications an alternative to the "classical" methods of *nonparametric* spectral analysis based on the periodogram.[8]

---

[8] For a general view of parametric and non-parametric spectral analysis, see [Stoica and Moses]. For a review of classical spectral analysis, see [Brockwell and Davis].

Realization theory starts with the study of a one-step prediction algorithm with finite memory, called the *Levinson–Durbin algorithm.*[9]

Finite Memory Prediction

The *one-step prediction of memory size* M is, by definition, the projection of X_n on the subspace generated by the M preceding values $X_{n-1}, \ldots, X_{n-M}$. In other words, we seek to obtain the linear regression of X_n on $X_{n-1}, \ldots, X_{n-M}$. A theoretical solution of this problem is known (Example 1.3.2). Here, we shall derive an algorithm that performs efficiently the inversion of the covariance matrix of the $(X_n, X_{n-1}, \ldots, X_{n-M})$ needed to obtain the regression vector.

Let $(X_0, \ldots, X_M)$ be a vector of square-integrable *real centered* random variables with covariance matrix

$$
R_M = \begin{pmatrix}
R(0) & R(1) & \cdots & R(M-1) & R(M) \\
R(1) & R(0) & \cdots & R(M-2) & R(M-1) \\
\vdots & \vdots & \vdots & \vdots & \vdots \\
R(M-1) & R(M-2) & \cdots & R(0) & R(1) \\
R(M) & R(M-1) & \cdots & R(1) & R(0)
\end{pmatrix},
$$

which will be denoted $R_M = T(R(0), \ldots, R(M))$, where T is for Töplitz. Indeed, a (real) matrix of this form and which is in addition non-negative definite is called a (real) *Töplitz matrix*. This situation occurs when $X_0, \ldots, X_M$ are samples from a centered real WSS time series $\{X_n\}_{n \in \mathbb{Z}}$ with covariance function R.

Let us note at this point that the reversed vector $(X_M, \ldots, X_0)$ has the same second-order properties (that is, the same covariance matrix) as the forwards vector $(X_0, \ldots, X_M)$. This particular structure will be exploited. For the time being, we are interested in the calculation of the regression vector

$$
a_M = (a_1(M), \ldots, a_M(M))
$$

of X_M on $X_0, \ldots, X_{M-1}$. We shall write

$$
X_M^f(M) = \sum_{i=1}^{M} a_i(M) X_{M-i}
$$

for the projection of X_M onto the Hilbert subspace of $L^2_{\mathbb{R}}(P)$ generated by $X_0, \ldots, X_{M-1}$ and

$$
\epsilon_M^f(M) = X_M - X_M^f(M)
$$

[9] J. Durbin, "The fitting of time series nodel", *Rev. Inst. Internat. Statist.*, **28**, 233–244.

for the corresponding prediction error. As we know (Example 1.3.2) every solution
of the system

$$R_{M-1}a(M) = r_M, \tag{4.72}$$

where

$$r_M = (R(1), \ldots, R(M))^T,$$

provides a regression vector $a(M)$ and the quadratic error $P_M = E|\epsilon_M^f(M)|^2$ is then
given by $P_M = E[(X_M - X_M^f(M))X_M]$, that is to say,

$$P_M = R(0) - \sum_{i=1}^{M} a_i(M)R(i) = R(0) - a(M)^T r_M. \tag{4.73}$$

The relations (4.72) and (4.73) constitute the *Yule–Walker equations*. If R_{M-1}
is *invertible*, Cramér's rule applied to this linear system in the $M + 1$ unknowns
$a_1(M), \ldots, a_M(M)$, P_M, gives in particular

$$P_M = \frac{\det R_M}{\det R_{M-1}}. \tag{4.74}$$

The solution of (4.72) requires the inversion (or the pseudo-inversion) of the
matrix R_{M-1}. This operation can be performed economically by the Levinson–
Durbin algorithm, which exploits the Töplitz structure of the covariance matrix, itself
a consequence of the second order reversibility of the vector $X_0, \ldots, X_M$ mentioned
above.

The property of reversibility of a centered WSS time series naturally leads one to
consider both the forward prediction and the backward prediction. Thus, for a WSS
time series $\{X_n\}_{n \in \mathbb{Z}}$ we introduce the *forward* and *backward* one-step prediction
of order N

$$X_n^f(N) = P(X_n|X_{n-1}, \ldots, X_{n-N}), \qquad X_n^b(N) = P(X_n|X_{n+1}, \ldots, X_{n+N})$$

(where "P" stands for projection) as well as the associated errors of prediction

$$\epsilon_n^f(N) = X_n - X_n^f(N), \qquad \epsilon_n^b(N) = X_n - X_n^b(N).$$

In terms of the regression vector

$$a(N) = (a_1(N), \ldots, a_N(N))$$

we have

$$X_n^f(N) = \sum_{i=1}^{N} a_i(N) X_{n-i}, \qquad X_n^b(N) = \sum_{i=1}^{N} a_i(N) X_{n+i} \qquad (4.75)$$

(The second equality follows from the first equality and from the second-order reversibility of the process).

Theorem 4.4.1 *There exists a constant K_N such that*

$$|K_N| \leq 1, \qquad (4.76)$$

and such that the forwards and backwards prediction errors satisfy the recurrence relations

$$\epsilon_n^f(N) = \epsilon_n^f(N-1) - K_N \epsilon_{n-N}^b(N-1), \qquad (4.77)$$

$$\epsilon_{n-N}^b(N) = \epsilon_{n-N}^b(N-1) - K_N \epsilon_n^f(N-1). \qquad (4.78)$$

The constant K_N is called the N-th *reflection coefficient* of order N.

Proof The prediction errors $\epsilon_n^f(N-1)$ and $\epsilon_{n-N}^b(N-1)$ are relative to the projections of X_n and X_{n-N}, respectively, onto the same subspace, that generated by $X_{n-1}, \ldots, X_{n-N+1}$. They are thus both orthogonal to the latter, and, in particular,

$$\beta_n^f(N) := \epsilon_n^f(N-1) - K_N \epsilon_{n-N}^b(N-1) \perp X_{n-k} \quad (1 \leq k \leq N-1)$$

for any constant K_N. If we choose K_N so that $\beta_n^f(N)$ is orthogonal to X_{n-N}, then $\beta_n^f(N) = \epsilon_n^f(N)$ (Indeed, on the one hand $\hat{X}_n = X_n - \beta_n^f(N)$ is a linear combination of $X_{n-1}, \ldots, X_{n-N}$, and on the other hand, $X_n - \hat{X}_n = \beta_n^f(N)$ is orthogonal to $X_{n-1}, \ldots, X_{n-N}$, and therefore, by the projection principle, $\hat{X}_n = P(X_n | X_{n-1}, \ldots, X_{n-N}) = X_n^f(N)$). We have therefore obtained (4.77). Equation (4.78) is obtained in the same way by considering the time-reversed process. The constant K_N is the same, as it depends only upon the geometry in $L_{\mathbb{R}}^2(P)$ of $\{X_n\}_{n \in \mathbb{Z}}$, which is the same as that of $\{X_{-n}\}_{n \in \mathbb{Z}}$. The proof of (4.76) will be given in the course of the discussion to follow that leads to the Levinson–Durbin algorithm. $\square$

If we rewrite the Eqs. (4.77) and (4.78) in the form

$$X_n^f(N) = X_n^f(N-1) + K_N(X_{n-N} - X_{n-N}^b(N-1)),$$

$$X_{n-N}^b(N) = X_{n-N}^b(N-1) + K_N(X_n - X_n^f(N-1))$$

we obtain, by substituting the expressions (4.75), the following equalities which allow us to pass from the regression vector of order $N-1$ to the regression vector of order N: For $1 \leq k \leq N-1$

$$a_k(N) = a_k(N-1) - K_N a_{N-k}(N-1), \qquad (4.79)$$

and for $k = N$,

$$a_N(N) = K_N. \qquad (4.80)$$

From (4.78) we deduce

$$E[\epsilon_n^f(N)X_{n-N}] = E[\epsilon_n^f(N-1)X_{n-N}] - K_N E[\epsilon_{n-N}^b(N-1)X_{n-N}].$$

As $\epsilon_n^f(N) \perp X_{n-N}$ the left hand side of this equality is zero. On the other hand,

$$E[\epsilon_{n-N}^b(N-1)X_{n-N}] = P_{N-1}.$$

Therefore, we obtain

$$P_{N-1}K_N = E[\epsilon_n^f(N-1)X_{n-N}],$$

and, after replacing in the last equality $\epsilon_n^f(N-1)$ by its expression

$$X_n - \sum_{i=1}^{N-1} a_i(N-1)X_{n-i} ,$$

we obtain

$$P_{N-1}K_N = R(N) - \sum_{i=1}^{N-1} a_i(N-1)R(N-i). \qquad (4.81)$$

From (4.77) we deduce

$$E[\epsilon_n^f(N)X_n] = E[\epsilon_n^f(N-1)X_n] - K_N E[\epsilon_{n-N}^b(N-1)X_n],$$

that is to say, since $E[\epsilon_n^f(i)X_n] = P_i$,

$$P_N = P_{N-1} - K_N E[\epsilon_{n-N}^b(N-1)X_n].$$

On replacing $\epsilon_{n-N}^b(N-1)$ in this equality by its expression

$$X_{n-N} - \sum_{i=1}^{N-1} a_i(N-1)X_{n-N+i} ,$$

and taking account of (4.78), we obtain

$$P_N = P_{N-1}(1 - K_N^2). \tag{4.82}$$

It is clear that as the depth of the memory increases, the quadratic error of prediction decreases, that is to say, $P_{N-1} \geq P_N$. From this observation and from equality (4.82) we deduce condition (4.76) on the reflection coefficient.

A zero prediction error at order $N - 1$ ($P_{N-1} = 0$) means that

$$X_N = \sum_{i=1}^{N-1} a_i(N - 1)X_{n-i} = P(X_n | X_{n-1}, \dots, X_{n-N+1}).$$

Thus we have, *a fortiori*,

$$X_n = P(X_n | X_{n-1}, \dots, X_{n-(N+i)})$$

for all $i \geq 0$. In particular, for all $i \geq -1$,

$$P_{N+i} = 0$$

and

$$a(N + i) = (a_1(N + i), \dots, a_{N+i}(N + i))$$
$$= (a_1(N - 1), \dots, a_{N-1}(N - 1), 0, \dots, 0).$$

Equations (4.79), (4.80), (4.81) and (4.82), together with the previous remark, constitute the Levinson–Durbin algorithm, allowing us to calculate recursively the regression vectors $a(i)$ ($1 \leq i \leq M$) from the data $R(0), \dots, R(M)$. It remains to initialize this algorithm with $a(1)$ and P_1. We calculate (the scalar) $a(1)$ by minimising

$$P_1(a) = E[(X_n - aX_{n-1})^2]$$
$$= a^2 R(0) - 2aR(1) + R(0).$$

This gives

$$a(1) = \frac{R(1)}{R(0)}, \qquad P_1 = \frac{R(0)^2 - R(1)^2}{R(0)}.$$

The algorithm is summarized below.
Data: $R(0), \dots, R(M)$.
Initialisation

$$a(1) = \frac{R(1)}{R(0)}, \quad P_1 = \frac{R(0)^2 - R(1)^2}{R(0)}.$$

Loop: $P_{N-1}, a(N - 1)$ are given, obtain $P_N, a(N)$ as follows:
If $P_{n-1} > 0$ do

$$K_N = \left\{ R(N) - \sum_{i=1}^{N-1} a_i(N-1)R(N-i) \right\} \frac{1}{P_{N-1}},$$

$$a_i(N) = a_i(N-1) - K_N a_{N-i}(N-1) \quad (1 \le i \le N-1)$$

$$a_N(N) = K_N,$$

$$P_N = (1 - K_N^2)P_{N-1}.$$

If $P_{N-1} = 0$, do

$$K_N = 0$$

$$a_i(N) = a_i(N-1) \quad (1 \le i \le N-1)$$

$$a_N(N) = 0,$$

$$P_N = 0.$$

Lattice Structure and Stability

Equations (4.77) and (4.78) constitute what is called the *lattice structure* of the pre-

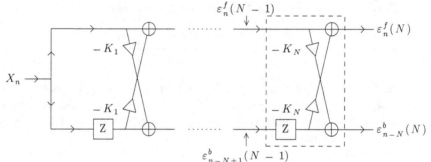

diction.

The *error filter* associated with the forward one-step prediction with memory N, that passes from the input X_n to the output $\epsilon_n^f(N)$, has for transfer function

$$E_N(z) = 1 - \sum_{i=1}^{N} a_i(N)z^i = E_N^f(z),$$

and the error filter associated with the backwards one step prediction with memory N, which passes from the input X_n to the output $\epsilon_n^b(N)$, has for transfer function

$$E_N^b(z) = 1 - \sum_{i=1}^{N} a_i(N)z^{-i} = E_N(z^{-1}).$$

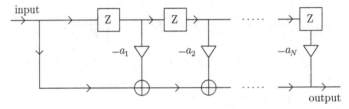

The recurrence equations

$$E_N(z) = E_{N-1}(z) - K_N z^N E_{N-1}(z^{-1}), \tag{4.83}$$

$$z^N E_N(z^{-1}) = z^N E_{N-1}(z^{-1}) - K_N E_{N-1}(z), \tag{4.84}$$

are equivalent to the equalities (4.77) and (4.78).

Lemma 4.4.1 *Let E_N and E_{N-1} be two polynomial transfer functions linked by equalities (4.83) and (4.84), K_N being a real constant. Suppose that the polynomial E_{N-1} is strictly stable (that is, all its roots have modulus strictly greater than unity). Then, if $|K_N| < 1$ the polynomial E_N is strictly stable; and if $|K_N| = 1$, it has all its roots on the unit circle (it is then said to be purely oscillatory).*

Proof Consider the rational filter with transfer function

$$H(z) = z \, \frac{z^{N-1} E_{N-1}(z^{-1})}{E_{N-1}(z)} .$$

As E_{N-1} is a polynomial with real coefficients, if ζ_i is one of its roots, ζ_i^* is also a root, and $1/\zeta_i^*$ is then a root of $z^{N-1} E_{N-1}(z^{-1})$. This shows that H is an all-pass transfer function to within a multiplicative factor z. Now let z_i be a root of E_N. From Eq. (4.83) we have

$$0 = E_N(z_i)$$
$$= E_{N-1}(z_i) - K_N z_i^N E_{N-1}(z_i^{-1}),$$

and, dividing by $K_N E_{N-1}(z_i)$, we obtain

$$|H(z_i)| = \frac{1}{K_N} .$$

The conclusion then follows from the all-pass lemma (Theorem 1.2.2).   □

Theorem 4.4.2 *The prediction filters E_N are either purely oscillatory or strictly stable.*

Proof For $N = 1$ we have $E_1(z) = 1 - K_1 z$, and since $|K_1| \le 1$, the corresponding filter is stable. The result then follows by induction, using Lemma 4.4.1.   □

(The oscillatory case). If $|K_N| = 1$ then $P_N = 0$, which implies that the prediction is error-free, and therefore

$$X_n = \sum_{i=1}^{N} a_i(N) X_{n-i}.$$

This implies that the process $\{X_n\}_{n \in \mathbb{Z}}$ is harmonic with N spectral lines, all being effective (that is of non-null power) when $P_{N-1} > 0$.

The preceding results show that a lattice structure leads to a strictly stable filter when the reflection coefficients K_i all have moduli strictly less than unity. Conversely:

Theorem 4.4.3 *Given a strictly stable polynomial filter, we can associate with it a lattice structure of which all the reflection coefficients have moduli strictly less than unity.*

Proof Let

$$E_M(z) = 1 - \sum_{i=1}^{M} a_i(M) z^i$$

be a polynomial transfer function of which all the roots have moduli strictly greater than unity. We know that it is possible to associate with it an autoregressive WSS time series $\{X_n\}_{n \in \mathbb{Z}}$ of order M satisfying

$$E_M(z) X_n = \epsilon_n, \tag{4.85}$$

where $\{\epsilon_n\}_{n \in \mathbb{Z}}$ is a discrete-time white noise and E_M is the one step forwards prediction error filter with memory depth M. The Levinson–Durbin algorithm permits us to obtain the lattice representation of E_M characterised by its reflection coefficients $K_1, \ldots, K_M$ with moduli strictly less than unity. $\square$

We have therefore proved that every strictly stable polynomial filter admits a lattice representation with reflection coefficients that have moduli strictly less than unity.

To determine these reflection coefficients from the coefficients of E_M we start from the relations

$$(1 - K_N^2) z^N E_{N-1}(z^{-1}) = K_N E_N(z) + z^N E_N(z^{-1}) \quad (2 \leq N \leq M),$$

which is obtained by adding to the second equality of (4.84) the first equality of (4.83) multiplied by K_N. Identification of the coefficients of the polynomials in the latter equality leads to the relations

$$a_i(N-1) = \frac{a_i(N) + K_N a_{N-i}(N)}{1 - K_N^2},$$

whence the algorithm below.

The Conversion Algorithm

Data: A regression vector $a_M = (a_1(M), \ldots, a_M(M))$ such that the polynomial

$$E_M(z) = 1 - \sum_{i=1}^{M} a_i(M)z^i$$

is strictly stable.

Loop: The regression vector

$$a_1(M - k), \ldots, a_{M-k}(M - k) = K_{M-k}$$

being computed, obtain $a(M - k - 1)$ and K_{M-k-1} by

$$a_i(M - k - 1) = \frac{a_i(M - k) + K_{M-k}a_{M-k-i}(M - k)}{1 - K_{M-k}^2}$$

$$(1 \le i \le M - k - 1)$$

$$K_{M-k-1} = a_{M-k-1}(M - k - 1).$$

The Levinson–Durbin algorithm can be considered as a mapping

$$R_M \to (E_M(z), P_M),$$

where R_M is a real Töplitz matrix, E_M is a stable polynomial transfer function, and P_M a non-negative real number. When R_M is strictly positive E_M is strictly stable and P_M is strictly positive. In this case the algorithm is invertible:

The Inverse Levinson–Durbin Algorithm

Data: A polynomial

$$E_M(z) = 1 - \sum_{i=1}^{M} a_i(M)z^i$$

that is strictly stable, and a real number $P_M > 0$.

Loop:
DATA: $K_M, \ldots, K_1, a(M), \ldots, a(1)$.
1. RESTITUTION OF $P_0, P_1, \ldots, P_M$:
DATA: P_M.
LOOP: P_N is available. Do

$$P_{N-1} = P_N/(1 - K_N^2).$$

2. RESTITUTION OF R_M:
DATA: $R(0) = P_0$.
LOOP: $R(N-1)$ is available. Do:

$$R(N) = P_{N-1}K_N + \sum_{i=1}^{N-1} a_i(N-i)R(N-i).$$

Continuation of a Covariance Segment

Let $R_M = T(R(0), \ldots, R(M))$ be a strictly positive Töplitz matrix. Can we find $R(M+1)$ such that $R_{M+1} = T(R(0), \ldots, R(M), R(M+1))$ is also a Töplitz matrix? In this case, the matrix R_{M+1} (or the scalar $R(M+1)$) is said to *continue* the Töplitz matrix R_M.

Let us assume that $R(M+1)$ continues R_M. Application of the Levinson–Durbin algorithm to R_{M+1} leads, in particular, to the relation [See Eq. (4.81)]

$$R(M+1) = P_M K_{M+1} - \sum_{i=1}^{M} a_i(M)R(M+1-i), \qquad (4.86)$$

where $|K_{M+1}| \leq 1$. Thus we have obtained a necessary condition: if $R(M+1)$ continues R_M then (4.86) necessarily holds for a real number K_{M+1} such that $|K_{M+1}| \leq 1$.

Theorem 4.4.4 *Let R_M be a real non-negative Töplitz matrix. Let $R(M+1)$ be defined by (4.86), where K_{M+1} is a real number such that $|K_{M+1}| < 1$. Then $R(M+1)$ continues R_M.*

Proof Starting from $a(M)$ (obtained by applying the Levinson–Durbin algorithm to R_M) and from K_{M+1} we construct the vector $a(M+1)$ as follows:

$$a_k(M+1) = a_k(M) - K_{M+1}a_{M+1-k}(M), \quad (1 \leq k \leq M), \qquad (*)$$

$$a_{M+1}(M+1) = K_{M+1}.$$

The corresponding filter

$$E_{M+1}(z) = 1 - \sum_{i=1}^{M+1} a_i(M+1)z^i$$

is then linked to E_M by (4.83) and (4.84). It is strictly stable, since $|K_{M+1}| < 1$ (Lemma 4.4.1). We set $P_{M+1} = (1 - K_{M+1}^2)P_M$ (a quantity that is greater than zero since $P_M > 0$ and $|K_{M+1}| < 1$) and we construct an autoregressive WSS time series $\{X'_n\}_{n \in \mathbb{Z}}$ with spectral density

$$\frac{P_{M+1}}{2\pi} \frac{1}{|E_{M+1}(e^{-i\omega})|^2}.$$

We have to show that $\{X'_n\}_{n\in\mathbb{Z}}$ realises R_{M+1}. The one-step prediction filter with memory size $(M+1)$ of $\{X'_n\}_{n\in\mathbb{Z}}$ is $E'_{M+1}(z) = E_{M+1}(z)$ and the prediction error at the $(M+1)$th step is $P'_{M+1} = P_{M+1}$. Let us calculate the one-step prediction filter with memory size M of $\{X'_n\}_{n\in\mathbb{Z}}$, denoted

$$E'_M(z) = 1 - \sum_{i=1}^{M} a'_i(M)z^i.$$

By the Levinson–Durbin algorithm, and taking account of $E'_{M+1}(z) = E_{M+1}(z)$,

$$a_k(M+1) = a'_k(M) - K_{M+1}a'_{M+1-k}(M) \quad (1 \le k \le M), \tag{†}$$

$$a_{M+1}(M+1) = K_{M+1}.$$

We conclude from (†) and (∗) that

$$E'_M(z) = E_M(z).$$

Furthermore,

$$P'_M = \frac{P'_{M+1}}{1 - K^2_{M+1}}$$
$$= \frac{P_{M+1}}{1 - K^2_{M+1}} = P_M.$$

If we denote the covariance of $\{X'_n\}_{n\in\mathbb{Z}}$ by R', we see that the Levinson–Durbin algorithm produces the two calculations

$$R_M \leftrightarrow (E_M(z), P_M), \qquad R'_M \leftrightarrow (E_M(z), P_M),$$

where $E_M(z)$ is strictly stable and $P_M > 0$. As the algorithm is, under these latter conditions, invertible, we deduce that $R_M = R'_M$. □

4.4.2 Maximum Entropy Realization

Let $R_M = T(R(0), \ldots, R(M))$ be a positive definite real Töplitz matrix. We know that there exists at least one vector $(X_0, \ldots, X_M)$ that we can, furthermore, choose to be Gaussian, admitting R_M as its covariance matrix. But can we find a WSS time series $\{X_n\}_{n\in\mathbb{Z}}$ whose covariance function R_X satisfies

$$R_X(i) = R(i) \quad (0 \le i \le M) ?$$ (4.87)

This problem is that of the *realization* of a Töplitz matrix. It happens that it always has at least one solution that is an autoregressive process of order M, called the *maximal entropy* realization.

To show the existence of the autoregressive solution let us construct the WSS time series $\{X_n\}_{n\in\mathbb{Z}}$ satisfying

$$E_M(z)X_n = \epsilon_n,$$ (4.88)

where $\{\epsilon_n\}_{n\in\mathbb{Z}}$ is a discrete-time white noise with variance $P_M > 0$, where E_M and P_M are obtained by running the Levinson–Durbin algorithm with data R_M. As $R_M > 0$ we know that $P_M > 0$ and that E_M is strictly stable, so that (4.88) defines an autoregressive WSS time series $\{X_n\}_{n\in\mathbb{Z}}$ with covariance function R_X.

It remains to verify (4.87). We have seen that under the conditions: E_M is strictly stable and $P_M > 0$, the Levinson–Durbin algorithm is invertible. Thus we certainly have (4.87) because the Levinson–Durbin algorithm produces the calculations

$$(R(0), \ldots, R(M)) \leftrightarrow (E_M(z), P_M)$$

and

$$(R_X(0), \ldots, R_X(M)) \leftrightarrow (E_M(z), P_M).$$

The singular case (where R_M is not invertible) leads to a realization as a line spectrum. In fact, $\det R_M = 0$ implies that there exists a dependence relation between the $X_0, \ldots, X_M$. Thus there exists $1 \le N \le M$ such that

$$X_N = \sum_{i-1}^{N} a_i X_{N-i}$$ (4.89)

for some vector $a = (a_1, \ldots, a_N)^T$. Therefore if there exists a process $\{X_n\}_{n\in\mathbb{Z}}$ realizing R_M we shall necessarily have, taking account of (4.89) and the wide sense stationarity,

$$X_n = \sum_{i=1}^{n} a_i X_{n-i} \quad \text{for all } n \in \mathbb{Z},$$

which corresponds to a line spectrum.

Theorem 4.4.5 *Among all realizations of the positive definite real Töplitz matrix $R_M = T(R(0), \ldots, R(M))$, there exists an autoregressive realization of order M. This realization maximises the one-step prediction error.*

For this reason, one also calls, more accurately, this realization the MPE (maximal prediction error) realization.

Proof For a given second-order time series $\{Z_n\}_{n \in \mathbb{Z}}$, denote the Hilbert subspace generated by $Z_{n-1}, \ldots, Z_{n-M}$ by $H_{n-1}(Z; M)$, and the subspace generated by $Z_{n-1}, Z_{n-2}, \ldots$ by $H_{n-1}(Z; \infty)$. Let $R_M = T(R(0), \ldots, R(M))$ be a positive real Töplitz matrix. Let $\{Y_n\}_{n \in \mathbb{Z}}$ be the autoregressive WSS time series of order M realizing it. Let $\{X_n\}_{n \in \mathbb{Z}}$ be another time series realizing R_M. We have to prove that

$$E[|X_n - P(X_n|H_{n-1}(X))|^2] \leq E[|Y_n - P(Y_n|H_{n-1}(Y))|^2]$$

This inequality follows from the following simple observations. First,

$$E[|X_n - P(H_{n-1}(X; M))|^2] = E[|Y_n - P(Y_n|H_{n-1}(Y; M))|^2]$$

because $(X_n, X_{n-1}, \ldots, X_{n-M})$ and $(Y_n, Y_{n-1}, \ldots, Y_{n-M})$ have the same covariance matrix R_M; secondly,

$$E[|X_n - P(X_n|H_{n-1}(X, \infty))|^2] \leq E[|X_n - P(X_n|H_{n-1}(X, M))|^2]$$

since $H_{n-1}(X; \infty) \supseteq H_{n-1}(X; M))$; and thirdly

$$E[|Y_n - P(Y_n|H_{n-1}(X, \infty))|^2] = E[|Y_n - P(Y_n|H_{n-1}(X, M))|^2]$$

because the prediction with infinite memory and that with memory M are equal for any AR time series of order M.   ☐

Algebraic Aspects of the MPE **Realization**

Define $b = (1, -a_1, \ldots, -a_M)^T$. Since

$$E\left|X_n - \sum_{k=1}^{M} a_k X_{n-k}\right|^2 = b^T R_M b,$$

we see that the regression vector $a(M)$ of X_n on $X_{n-1}, \ldots, X_{n-M}$ is a solution of the optimisation problem

$$\min b^T R_M b \text{ on } \{b\,;\ u_0^T b = 1\}.$$

where $u_0 = (1, 0, \ldots, 0)^T$ (constraint $u_0^T b = 1$ guarantees that b is of the form $(1, -a_1, \ldots, -a_M)^T$). The minimum value of the criterion $b^T R_M b$ is then P_M, the one-step prediction error with memory M.

We use Lagrange's method to solve the above optimization problem: we minimise the function

$$F(b, \lambda) = b^T R_M b + \lambda(1 - u_0^T b)$$

under the condition $u_0^T = 1$, where λ is the Lagrange multiplier. By differentiating the criterion with respect to b and equating the gradient to zero we obtain

$$R_M b = \lambda u_0.$$

To determine λ we multiply this vector equality on the left by b^T, which gives, in view of the constraint $u_0^T b = 1$,

$$P_M = b^T R_M b = \lambda.$$

Therefore, the optimal vector b satisfies

$$R_M b = P_M u_0, \qquad u_0^T b = 1.$$

In particular, if R_M is invertible,

$$b = P_M R_M^{-1} u_0, \tag{4.90}$$

and multiplying this equality on the left by u_0^T,

$$u_0^T b = 1 = P_M u_0^T R_M^{-1} u.$$

Thus we have

$$P_M = \frac{1}{u_0^T R_M^{-1} u_0} = \frac{1}{(R_M^{-1})_{00}}. \tag{4.91}$$

Equation (4.90) and (4.91) give an alternative expression of the solution of the Yule–Walker equations.

4.4.3 Pisarenko's Realization

In the previous subsection, we have seen that a strictly positive Töplitz matrix R_M can be realized by a spectral density of the autoregressive type of order M. But, as we have already mentioned several times, there is an infinity of spectral measures realizing R_M. This subsection is devoted to the so-called *Pisarenko realization*, with spectral measure

$$\mu(d\omega) = \frac{\sigma^2}{2\pi} d\omega + \sum_{k=1}^{N} \alpha_k \, \delta_{\omega_k}(d\omega) \qquad (4.92)$$

corresponding to a WSS time series of the type *white noise plus sinusoids*

$$X_n = \sum_{k=1}^{N} U_k e^{i\omega_k n} + \epsilon_n, \qquad (4.93)$$

where the U_k's are centered complex random variables with respective variances $E|U_k^2| = \alpha_k$, mutually uncorrelated and uncorrelated with the white noise $\{\epsilon_n\}_{n\in\mathbb{Z}}$ of variance σ^2.

We first treat the degenerate case: $\det R_N = 0$. Then, for all $n \in \mathbb{Z}$, the variables $X_n, \ldots, X_{n-N}$ are linked by a linear relation:

$$a_0 X_n - a_1 X_{n-1} - \cdots - a_N X_{n-N} = 0.$$

If, furthermore, $\det R_{N-1} > 0$ there does not exist a linear relation between $X_{n-1}, \ldots, X_{n-N}$, and we can suppose that $a_0 = 1$ in the last equality:

$$X_n = \sum_{k=1}^{N} a_k X_{n-k}.$$

Defining

$$E(z) = 1 - \sum_{k=1}^{N} a_k z^k,$$

we have

$$E(z)X_n = 0,$$

that is, in terms of the spectral measure μ of $\{X_n\}_{n\in\mathbb{Z}}$,

$$|E(e^{-i\omega})|^2 \, \mu(d\omega) = 0,$$

which implies that μ is a pure line spectrum

$$\mu(d\omega) = \sum_{k=1}^{N} \alpha_k \, \epsilon_{\omega_k}(d\omega),$$

where the $e^{i\omega_k}$'s are the roots of E and ϵ_a is the Dirac measure at $a \in \mathbb{R}$. The condition $\det R_{N-1} > 0$ implies that the frequencies ω_k ($1 \le k \le N$) are all *effective*, i.e. $\alpha_k > 0$ ($1 \le k \le N$). Indeed, if this were not so, assuming for example that $\alpha_N = 0$, we would have

$$|E'(e^{-i\omega})|^2 \, \mu(d\omega) = 0,$$

where E' is the so-called annihilating filter for the pulsations $\omega_1, \ldots \omega_{N-1}$:

$$E'(z) = \prod_{k=1}^{N-1} (1 - e^{-i\omega_k}z) = 1 - \prod_{k=1}^{N-1} a'_k z.$$

This implies that

$$E'(z)X_n = 0,$$

that is

$$X_N = \sum_{k=1}^{N-1} a'_k X_{n-k}.$$

Now, this linear relation between $X_n, X_{n-1}, \ldots, X_{n-N+1}$ would imply that $\det R_{N-1} = 0$, a contradiction.

We now treat the case of a non-degenerate Töplitz matrix $R_M = T(R(0), \ldots, R(M))$. For all $\alpha > 0$ we set

$$f_N(\alpha) = \det(R_N - \alpha I_{N+1}),$$

where I_N is the unit $N \times N$ matrix. Since $f_M(0) > 0$, and thus, by the preceding remarks, $f_N(0) > 0$ for all N such that $0 \le N \le M$ (otherwise $(X_n, \ldots, X_{n-N})$, and therefore $(X_n, \ldots, X_{n-M})$ would be degenerate, implying that $f_M(0) = 0$). For all $\alpha \ge 0$, if $f_N(\alpha) \ge 0$ then $f_k(\alpha) \ge 0$, $0 \le k \le N$, since $R_k - \alpha I_{k+1}$ is a Töplitz submatrix of $R_N - \alpha I_{N+1}$. Similarly, if $f_N(\alpha) > 0$ then $f_k(\alpha) > 0$ for all k such that $0 \le k \le N$. Let us also note that if $\beta < \alpha$ and $f_N(\alpha) \ge 0$, then $f_N(\beta) \ge 0$. From these remarks we deduce the existence of a strictly positive number σ^2 which is the largest $\alpha > 0$ such that

$$f_0(\alpha) > 0, \quad \ldots, \quad f_{N-1}(\alpha) > 0, \quad f_N(\alpha) = \cdots = f_M(\alpha) = 0$$

for some $N < M$. Defining

$$\tilde{R}(n) = R(n) - \sigma^2 1_{\{n=0\}}$$

we therefore have

$$\det \tilde{R}_0 > 0, \ldots, \det \tilde{R}_{N-1} > 0, \det \tilde{R}_N = \cdots = \det \tilde{R}_M = 0.$$

The matrix $\tilde{R}_M$ is therefore realized by a spectral measure of the type

$$\tilde{\mu}(d\omega) = \sum_{k=1}^{N} \alpha_k \, \epsilon_{\omega_k}(d\omega)$$

where all the frequencies are effective. Since

$$\sigma^2 \, 1_{\{n=0\}} = \int_{-\pi}^{+\pi} e^{in\omega} \frac{\sigma^2}{2\pi} \, d\omega,$$

we deduce from this that R_M is realised by the spectral measure (4.92).

Algebraic Characterization of the Pisarenko Realization

We shall now give an algebraic characterization of the order N, the frequencies ω_k, and the noise variance σ^2. For all $\omega \in (-\pi, +\pi]$ and all $k \geq 0$ let us set

$$p_k(\omega) = (1, e^{+i\omega}, \ldots, e^{+ik\omega})^T.$$

$p_M(\omega_k)$ is called the *source vector* corresponding to the source frequencies ω_k, and we call the matrix

$$S_M = [p_M(\omega_1), \ldots, p_M(\omega_N)] = \begin{pmatrix} 1 & 1 & \ldots & 1 \\ e^{i\omega_1} & e^{i\omega_2} & \ldots & e^{i\omega_N} \\ e^{i2\omega_1} & e^{i2\omega_2} & \ldots & e^{i2\omega_N} \\ \vdots & \vdots & \vdots & \vdots \\ e^{iM\omega_1} & e^{iM\omega_2} & \ldots & e^{iM\omega_N} \end{pmatrix}$$

the *position matrix* of the sources, while

$$\mathrm{diag}\{\alpha_1, \ldots, \alpha_N\}$$

is the *power matrix* of the sources. The direct calculation of R_M, starting from the expression (4.93), gives

$$R_M = \sigma^2 I_{N+1} + \sum_{k=1}^{N} \alpha_k \, p_M(\omega_k) p_M(\omega_k)^{\dagger},$$

or, again,

$$R_M = \sigma^2 I + S \alpha S^{\dagger} \tag{4.94}$$

(the dagger symbol represents *transposition and conjugation*; in the expression on the right hand side we have suppressed the indices, as the dimension is referred to in the left hand side).

The vector subspace of $\mathbb{C}^{M+1}$ generated by the *source vectors* $(p(\omega_k), 1 \leq k \leq N)$ is called the *source subspace* $\mathcal{S}$ to. As the pulsations ω_k are all different, the dimension of the source subspace is exactly N. The *noise subspace* $\mathcal{B}$ is defined as the orthogonal complement of $\mathcal{S}$. Its dimension is thus $M - N + 1$.

By (4.94), for every vector b of the noise subspace $\mathcal{B}$ we have

$$R_M b = \sigma^2 b + S \alpha S^\dagger b = \sigma^2 b, \qquad (4.95)$$

since by the orthogonality of b and $\mathcal{S}$, $S^\dagger b = 0$. Thus b is an eigenvector of R_M with associated eigenvalue σ^2. Conversely, if b is an eigenvector of R_M associated with the eigenvalue σ^2, we have the equality (4.95), which implies $S \alpha S^\dagger b = 0$, and, *a fortiori*, $b^\dagger S \alpha S^\dagger b = 0$, and thus $S^\dagger b = 0$: b is orthogonal to $\mathcal{S}$. The noise subspace is thus the eigensubspace associated with the eigenvalue σ^2. Furthermore, the eigenvalue σ^2 of R_M is the smallest eigenvalue of R_M. To show this, let λ be another eigenvalue, and let x be an eigenvector associated with λ that is also normalized ($x^\dagger x = 1$). By (4.95) we have

$$x^\dagger \lambda x = \lambda = x^\dagger R_M x$$
$$= \sigma^2 + x^\dagger S \alpha S^\dagger x \geq \sigma^2.$$

We can now summarize the findings of the present section in the following theorem.

Theorem 4.4.6 *Let R_M be a positive definite Töplitz matrix. Let σ^2 be its smallest eigenvalue. Then there exists $N \leq M$, such that R_M is realized by the spectral measure (4.92) with N effective sources ω_k ($1 \leq k \leq N$). These sources are characterised by the requirement*

$$p(\omega_k) \perp \mathcal{B},$$

where $\mathcal{B}$ is the $(M - N + 1)$-dimensional eigensubspace associated with the eigenvalue σ^2 of R_M.

Observe that N is determined by the fact that $(M - N + 1)$ is the dimension of the eigensubspace corresponding to σ^2.

ME **Realization of Sinusoids in White Noise**

Let us be given a process $\{X_n\}_{n \in \mathbb{Z}}$ that is the sum of a white noise of variance σ^2 and N effective spectral lines; See (4.93). Its Töplitz matrix of order M is given by (4.94). We are going to exactly calculate the AR spectral density of order M realising R_M, that is to say, the maximum entropy realisation. We have seen that the corresponding spectral density is

$$f(\omega) = \frac{P_M}{2\pi} \frac{1}{\left| 1 - \sum_{k=1}^{M} a_k e^{-i\omega k} \right|^2},$$

where $b = (1, -a_1, \ldots, -a_M)^T$ is given by

$$b = P_M R_M^{-1} u_0, \quad P_M^{-1} = u_0^T R_M^{-1} u_0, \tag{4.96}$$

where $u_0 = (1, 0, \ldots, 0)^T \in \mathbb{R}^{M+1}$. The calculations can be carried out since we know the form of R_M. We shall make use of the following lemma:

Lemma 4.4.2 *(i) Let S and Q be two invertible square matrices. For any matrices G and H of appropriate dimensions (those for which the following identity is meaningful):*

$$(S + GQH)^{-1} = S^{-1} - S^{-1}G(HS^{-1}G + Q^{-1})^{-1}HS^{-1}.$$

(ii) If α is an invertible matrix of the same dimension as S, then

$$(\sigma^2 I + S\alpha S^\dagger)^{-1} = \frac{1}{\sigma^2}(I + S\beta S^\dagger)$$

where

$$\beta := -\left[\sigma^2 \alpha^{-1} + S^\dagger S \right]^{-1}.$$

Proof (a) It suffices to multiply the right-hand side of the above equality by $S + GQH$ and to verify that the result is the identity matrix. (b) is immediate from (a).   □

From (4.94) we deduce, using the above matrix inversion lemma,

$$R_M^{-1} = \frac{1}{\sigma^2}(I + S\beta S^\dagger),$$

$$\beta = -[\sigma^2 \alpha^{-1} + S^\dagger S]^{-1}.$$

Expression (4.96) gives, denoting $\mathbf{1} = (1, 1, \ldots, 1)^T \in \mathbb{R}^{M+1}$,

$$b = \frac{P_M}{\sigma^2}(I + S\beta S^\dagger)u_0 = \frac{P_M}{\sigma^2}(u_0 + S\beta\mathbf{1}), \tag{4.97}$$

where we have taken account of the fact $p(\omega)^\dagger u_0 = 1$, and thus

$$S^\dagger u_0 = \mathbf{1}.$$

Let

$$d_i := \sum_{j=1}^{N} \beta_{ij} \text{ and } \beta\mathbf{1} := \begin{pmatrix} d_1 \\ \vdots \\ d_N \end{pmatrix} = d,$$

Therefore (4.97) becomes

$$b = \frac{P_M}{\sigma^2}(u_0 + Sd)$$

and (4.96) can be written as

$$P_M^{-1} = u_0^T R_M^{-1} u_0 = \frac{1}{\sigma^2} u_0^T [I + S\beta S^\dagger] u_0$$

$$= \frac{1}{\sigma^2}[1 + \mathbf{1}^T \beta\mathbf{1}] = \frac{1}{\sigma^2}[1 + \mathbf{1}^T d] = \frac{1}{\sigma^2}\left(1 + \sum_{i=1}^{N} d_i\right),$$

that is to say,

$$P_M = \sigma^2\left(1 + \sum_{i=1}^{N} d_i\right)^{-1}.$$

Finally, the error filter vector b may be written

$$b = \frac{u_0 + Sd}{1 + \mathbf{1}^T d} = \frac{u_0 + \sum_{i=1}^{N} d_i p(\omega_i)}{1 + \sum_{i=1}^{N} d_i}.$$

Since

$$\left(1 - \sum_{k=1}^{N} a_k e^{-ik\omega}\right) = p(\omega)^\dagger b,$$

we have

$$f(\omega) = \frac{P_M}{2\pi} \frac{1}{|p(\omega)^\dagger b|^2} = \frac{\sigma^2\left(1 + \sum_{i=1}^{N} d_i\right)}{2\pi \left|p(\omega)^\dagger u_0 + \sum_{i=1}^{N} d_i p(\omega)^\dagger p(\omega_i)\right|^2}.$$

Now, $p(\omega)^\dagger u_0 = 1$, and, on the other hand,

$$p(\omega)^\dagger p(\omega_i) = W(\omega - \omega_i),$$

where

$$W(\omega) = \sum_{k=0}^{M} e^{-ik\omega} = \frac{1 - e^{-i\omega(M+1)}}{1 - e^{-i\omega}}$$

$$= e^{-i\omega M/2} \frac{\sin[\frac{1}{2}\omega(M+1)]}{\sin \omega/2} .$$

Finally,

$$f(\omega) = \frac{\sigma^2 \left(1 + \sum_{i=1}^{N} d_i\right)}{2\pi \left|1 + \sum_{i=1}^{N} d_i W(\omega - \omega_i)\right|^2} .$$

Concerning the localization of the sources, the important part is the denominator

$$G(\omega) = \left|1 + \sum_{i=1}^{N} d_i W(\omega - \omega_i)\right|^2 , \qquad (4.98)$$

which must be close to zero for $\omega = \omega_1, \ldots, \omega_N$, so that the spectral lines may be well brought out.

Example 4.4.1 Well separated sources. The sources are said to be *well separated* if, for all $i \neq j$ $(1 \leq i, j \leq N)$

$$|\omega_i - \omega_j| \geq \frac{2\pi}{M+1} .$$

In the case of the matrix $S^\dagger S$, whose general term is $W(\omega_i - \omega_j) = p(\omega_i)^\dagger p(\omega_j)$, has off-diagonal elements that are negligible with respect to the elements on the diagonal, these latter all being equal to $W(0) = M + 1$, we thus have the approximation

$$\beta = -[\sigma^2 \sigma^{-1} + S^\dagger S]^{-1}$$
$$\simeq -[\sigma^2 \alpha^{-1} + (M+1)I]^{-1}$$
$$= \text{diag}\{\alpha_1, \ldots, \alpha_N\}\text{diag}\{\beta_{11}, \ldots, \beta_{NN}\},$$

where

$$\beta_{ii} = -\left(\frac{\sigma^2}{\alpha_i} + M + 1\right)^{-1} = -\frac{\alpha_i}{\sigma^2 + (M+1)\alpha_i} .$$

As β is diagonal, $d_i \simeq \beta_{ii}$. Finally, on substituting into (4.98),

$$G(\omega_i) \simeq |1 + d_i W(0)|^2 = \left|1 - \frac{\alpha_i(M+1)}{\sigma^2 + (M+1)\alpha_i}\right|^2 ,$$

that is to say,

$$G(\omega_i) \simeq \left| \frac{1}{1 + \frac{\alpha_i}{\sigma^2}(M+1)} \right|^2. \qquad (4.99)$$

If the process to noise ratios α_i/σ^2 are large then the AR spectrum will have marked peaks at the source frequencies. We also observe that for fixed process to noise ratios the possibility of detecting frequencies increases with the order M.

4.5 Exercises

Exercise 4.5.1 Gaussian ARMA time series. Consider the discrete time stochastic process $\{X_n\}_{n\geq 0}$ defined by

$$X_{n+1} = aX_n + \varepsilon_{n+1}, \quad n \geq 0$$

where X_0 is a Gaussian random variable of mean 0 and variance c^2, and $\{\varepsilon_n\}_{n\geq 0}$ is a sequence of IID Gaussian variables of mean 0 and variance σ^2, and independent of X_0.
1. Show that for all $n \geq 1$, the vector $(X_0, \ldots, X_n)$ is a Gaussian vector.
2. Express X_n in terms of $X_0, \varepsilon_1, \ldots, \varepsilon_n$ (and a). Give the mean and variance of X_n.
3. Show that X_n converges in distribution to a centered Gaussian variable of mean 0 and variance γ^2 to be computed.
4. Suppose now that X_0 is Gaussian with mean 0, with variance γ^2 as computed in the previous question. Show that $\{X_n\}_{n\geq 0}$ is a strictly stationary time series.

Exercise 4.5.2 Time series as sampled WSS stochastic processes. Let $\{X(t)\}_{t\in\mathbb{R}}$ be a WSS stochastic process with the power spectral density f_X. Define, for $T > 0$, $X_n = X(nT)$. What is the power spectral measure of the WSS time series $\{X_n\}_{n\in\mathbb{Z}}$?

Exercise 4.5.3 Show that in Example 4.1.7, the sum $\sum_{n\in\mathbb{Z}} c_k \varepsilon_{n-k}$ is almost surely absolutely convergent.

Exercise 4.5.4 Herglotz via Bochner. Let $\{X_n\}_{n\in\mathbb{Z}}$ be a discrete-time WSS stochastic process with the covariance function C. Define

$$Y(t) = X_n \text{ for } t \in (n, n+1],$$

and

$$\tilde{X}(t) = Y(t - U),$$

where U is a random variable uniformly distributed on $[0, 1]$ and independent of $\{X_n\}_{n\in\mathbb{Z}}$.
(a) Prove that the stochastic process $\{\tilde{X}(t)\}_{t\in\mathbb{R}}$ is a WSS continuous-time stochastic process whose covariance function $\tilde{C}$ is a continuous function, linear between k and $k+1$ for all $k \in \mathbb{Z}$, and such that $\tilde{C}(k) = C(k)$ for all $k \in \mathbb{Z}$.

(b) Using (a), deduce from Bochner's theorem the existence part of Herglotz's theorem.

Exercise 4.5.5 Real WSS time series. Show that the power spectral measure of a *real* WSS time series is symmetric.

Exercise 4.5.6 Polya's theorem for time series. Let $f : \mathbb{Z} \to \mathbb{R}_+$ be a function satisfying the following properties:
(i) it is symmetric ($f(-n) = f(n)$),
(ii) it is convex on $\mathbb{N}$, and
(iii) it decreases to 0 as $n \uparrow \infty$.
    Show that f is the autocovariance function of some WSS time series.

Exercise 4.5.7 Is this a covariance function? Let the non-null terms of the function $f : \mathbb{Z} \to \mathbb{R}_+$ be $R(0) = 1$, $R(-2) = R(+2) = -\frac{1}{2}$, $R(-3) = R(+3) = -\frac{1}{4}$. Is f the autocovariance function of some WSS time series?

Exercise 4.5.8 The shift operator. Let $\{X_n\}_{n\in\mathbb{Z}}$ be a WSS time series. Show that the requirement $SX_n = X_{n-1}$ for all $n \in \mathbb{Z}$ uniquely defines a linear operator $S : H(X) \to H(X)$ that is an isometry. Show that for all $n \in \mathbb{Z}$, $P_n S = S P_{n+1}$ where P_n is the projection operator from $H(X)$ to $H(X; n)$.

Exercise 4.5.9 Product of uncorrelated WSS time series. Let $\{X_n\}_{n\in\mathbb{Z}}$ and $\{Y_n\}_{n\in\mathbb{Z}}$ be uncorrelated centered WSS time series with respective covariance functions R_X and R_Y, and respective power spectral measures μ_X and μ_Y. Show that $\{Z_n\}_{n\in\mathbb{Z}} := \{X_n Y_n\}_{n\in\mathbb{Z}}$ is a WSS time series and give its power spectral measure μ_Z in terms of μ_X and μ_Y.

Exercise 4.5.10 $H(X; 0)$. Let $\{c_n\}_{n\geq 0}$ be an absolutely summable sequence of complex numbers and let $\{X_n\}_{n\in\mathbb{Z}}$ be a WSS time series. Show that $\sum_{n\geq 0} c_n X_{-n} \in H_{\mathbb{C}}(X_n; n \leq 0)$.

Exercise 4.5.11 Regular, singular. Prove that a white noise is regular whereas a harmonic series is singular.

Exercise 4.5.12 Féjer's identity. Prove Féjer's identity true for all $\beta \in \mathbb{C}$, $\beta \neq 0$, and for all $z \in \mathbb{C}$ such that $|z| = 1$:

$$(z - \beta)\left(z - \frac{1}{\beta^*}\right) = -\frac{1}{\beta^*} z|z - \beta|^2.$$

Exercise 4.5.13 Prediction of a time series. (i) Find the representation in canonical ARMA form of the power spectral density

$$f(\omega) = \frac{3 - 2\cos\omega}{2 - \cos\omega}$$

(ii) Find the whitening filter and the generating filter of the ARMA process of Example 4.2.2.
    (iii) Find the one-step predictor and give the corresponding prediction gain.

Exercise 4.5.14 Two-step predictor. Let $\{X_n\}_{n\in\mathbb{Z}}$ be a centered ARMA time series with the power spectral density

$$f_X(\omega) = \frac{1}{26 - 10\cos\omega}$$

1. Detail its *canonical* representation $P(z)X_n = \varepsilon_n$ (that is: give the polynomial P and the variance σ^2 of the noise sequence).
2. What is the whitening (or analysis) filter of this time series? What is its generating (or synthesis) filter?
3. Give the one-step predictor of $\widehat{X}_{n/n-1}$ and the corresponding prediction gain.
4. Give the covariance function this time series.
5. Write X_n in terms of ε_n, ε_{n-1} and X_{n-2}. Deduce from this the two-step predictor $\widehat{X}_{n;n-2}$ (the projection of X_n onto $H(n, X_{n-2})$, the Hilbert subspace spanned by the random variables $X_{n-2}, X_{n-3}, \ldots$).

Exercise 4.5.15 m-step prediction.
(1) Consider a regular ARMA process in canonical form $P(z)X_n = Q(z)\varepsilon_n$ where in addition Q has no roots on the unit circle centered at the origin. Give an expression of the projection $\widehat{X}_{0,-m}$ of X_0 on $H(X; -m)$ ($m \geq 1$) as a convolutional filter applied to the time series itself, and give the transfer function of the corresponding filter in terms of P and Q.
(2) Apply the result of (1) to the autoregressive ARMA process with $P(z) = 1 - \frac{2}{3}z$.

Exercise 4.5.16 Covariance and power spectra of ARIMA series. Let $d \in (-\frac{1}{2}, +\frac{1}{2})$, and let P and Q be two polynomials of orders p and q respectively, with no common roots.
(i) Show that if P has no root on the unit circle, the unique solution of (4.65)

$$X_n = \sum_{n\in\mathbb{Z}} h_k(1 - z)^{-d}\varepsilon_n ,$$

where $\sum_{n\in\mathbb{Z}} h_k z^z$ is the Laurent expansion of $\frac{Q(z)}{P(z)}$ in a donut containing the unit circle centered at 0.
(ii) If P and Q have no root inside the closed unit disk centered at 0, show that

$$C_X(n) \sim Cn^{2d-1} \text{ as } n \uparrow \infty$$

for some $C \neq 0$ and that its power spectral density is such that

$$f_X(\omega) \sim \frac{\sigma^2}{2\pi}\omega^{-2d} \text{ as } \omega \downarrow 0 .$$

Exercise 4.5.17 Multivariate ARMA models. Prove in detail Theorem 4.2.6.

Exercise 4.5.18 Recursions for the whitening and generating filters. Show that the matrices h_k and $w_k, k \geq 0$, of Theorems 4.2.5 and 4.2.6 respectively can be calculated recursively as follows from the initial conditions $h_0 = w_0 = I$:

$$h_k = -\sum_{j=1}^{k} a_j h_{k-j} + b_k$$

and

$$w_k = -\sum_{j=1}^{k} w_j h_{k-j} + a_k$$

Exercise 4.5.19 Maximum entropy realization of the covariance of a HMC. Find the maximum entropy realization of order 4 of the two-state HMC of Example 4.1.4.

Chapter 5
Power Spectra of Point Processes

A simple point process on the positive half-line is, roughly speaking, a strictly increasing sequence $\{T_n\}_{n \in \mathbb{N}_+}$ of random variables taking their values in $[0, +\infty]$ and called the event times. This sequence is sometimes, especially in the signal processing litterature, represented by a train of Dirac pulses:

$$X(t) = \sum_{n \in \mathbb{N}_+; |T_n| < \infty} \delta(t - T_n),$$

where $t \to \delta(t)$ is the Dirac "pseudo-function" (the "impulse function at time 0"). Of course, $\{X(t)\}_{t \in \mathbb{R}_+}$ is not a *bona fide* stochastic process. One may give a meaning to this "pseudo-process" in terms of the theory of distributions, but we shall not do this and instead take the more rewarding point of view according to which a point process is a special case of random measure. The notion of power spectral measure of a stationary point process is slightly different from that of stationary stochastic processes. The corresponding mathematical object is the so-called Bartlett spectral measure. This notion is defined precisely and the spectral measures of various point processes and of their transformations (thinning, translation, clustering, jittering, etc.) are computed. The formulas obtained are potentially useful for the computation of the power spectral measures of stochastic processes structured by point processes, such as those arising in communications and in biology. We begin with a review of the indispensable background on point processes and of two fundamental models: the Poisson process and the renewal point process.

© Springer International Publishing Switzerland 2014
P. Brémaud, *Fourier Analysis and Stochastic Processes*, Universitext,
DOI 10.1007/978-3-319-09590-5_5

5.1 Review of Point Processes

5.1.1 The Distribution of a Point Process

We start with the basic definition of point process on euclidean spaces. The majority
of results concerning the general theory of point processes in the present section will
be presented without proofs, the reader being referred to the literature.[1]

In the sequel, $E = \mathbb{R}^m$ for some $m \geq 1$ and $\mathcal{B}(E) = \mathcal{B}(\mathbb{R}^m)$.

Definition 5.1.1 A *locally finite* measure on $(E, \mathcal{B}(E))$ is a measure giving finite
mass to the bounded Borel sets. If it takes its values in $\overline{\mathbb{N}}$, it is called a *point measure*.

A locally finite measure on $(E, \mathcal{B}(E))$ is in particular sigma-finite, since there
exists a sequence of bounded sets increasing to E.

Definition 5.1.2 Let $M(E)$ be the set of *locally finite* measures on $(E, \mathcal{B}(E))$
and let $\mathcal{M}(E)$ be the sigma–field on $M(E)$ generated by the collection of sets
$\{\mu \in M(E); \mu(F) \in C\}, F \in \mathcal{B}(E), C \in \mathcal{B}(\mathbb{R}_+)$ (see footnote[2]). The measurable space $(M(E), \mathcal{M}(E))$ is called the *canonical space of (locally finite) measures*
on $(E, \mathcal{B}(E))$.

Let now $(\Omega, \mathcal{F}, P)$ be some probability space.

Definition 5.1.3 A *random measure* on $(E, \mathcal{B}(E))$ is a measurable mapping

$$N : (\Omega, \mathcal{F}) \to (M(E), \mathcal{M}(E)).$$

It is called a *point process* on $(E, \mathcal{B}(E))$ if, in addition, $N(\omega)$ is a point measure for
all $\omega \in \Omega$.

If N is a random measure, $N(C)$ is a random variable for all $C \in \mathcal{B}(E)$, since the
mapping $\omega \to N(\omega)(C)$ is the composition of the measurable mappings $\omega \to N(\omega)$
and $\mu \to p_C(\mu) = \mu(C)$.

The measurability of $N(C)$ need not be verified for all $C \in \mathcal{B}(E)$, but only
for the class $\mathcal{I}$ of rectangles $\prod_{j=1}^{d}(a_j, b_j]$, where $[a_j, b_j] \subseteq \mathbb{R}$. We have: For
$N : \Omega \to M(E)$ to be measurable, it suffices that $N(I) : \Omega \to \overline{\mathbb{R}}_+$ be a random
variable for all $I \in \mathcal{I}$.

For the next definition, recall that a *singleton* is a set, denoted by $\{a\}$, consisting
of exactly one element $a \in E$.

Definition 5.1.4 A point measure μ on $(E, \mathcal{B}(E))$ is called *simple* if

$$\mu(\{a\}) = 0 \text{ or } 1 \text{ for all } a \in E.$$

A point process N on E is called *simple* if for all $\omega \in \Omega$, the point measure $N(\omega)$ is
simple.

---

[1] For instance: [Kallenberg], [Neveu], [Daley and Vere-Jones].

[2] In particular, for any $C \in \mathcal{B}(E)$, the mapping $p_C : \mu \to \mu(C)$ is measurable.

We leave to the reader the task of defining the notions of P-a.s. sigma-finite random measures, P-a.s. locally bounded random measures, and P-a.s. simple point processes. (For instance, a point process N on E is called P-a.s. simple if $N(\omega)$ is simple for all ω outside some P-negligible event.)

Events and Points

Let μ be a *locally finite* measure on $\mathbb{R}$ taking its values in $\overline{\mathbb{N}}$. The following facts are easy to prove:

(i) There exists a non-decreasing sequence $\{t_n\}_{n \in \mathbb{Z}}$ in $\overline{\mathbb{R}}$ such that for all measurable sets C,

$$\mu(C) = \sum_{n \in \mathbb{Z}} \varepsilon_{t_n}(C)$$

(where ε_a is the Dirac measure at a if $a \in \mathbb{R}$, and the null measure if $a = +\infty$ or $-\infty$), and such that

$$t_0 \leq 0 < t_1.$$

(ii) The mappings $\mu \to t_n = t_n(\mu)$ are measurable from $(M(E), \mathcal{M}(E))$ to $(\overline{\mathbb{R}}, \mathcal{B}(\overline{\mathbb{R}}))$.

(iii) If, in addition, μ is *simple*, then

$$|t_n| < \infty \Rightarrow t_n < t_{n+1}.$$

(In other words, $\{t_n\}_{n \in \mathbb{Z}}$ is strictly increasing on $\mathbb{R}$.)

If N is a point process on $\mathbb{R}$, we define

$$T_n(\omega) := t_n(N(\omega)).$$

In particular, T_n is a random variable since $T_n = t_n \circ N$ is the composition of two measurable functions. If N is *simple* and *locally bounded*, the sequence of random variables $\{T_n\}_{n \in \mathbb{Z}}$ is *strictly increasing* on $\mathbb{R}$. The sequence $\{T_n\}_{n \in \mathbb{Z}}$ is called the sequence of *event times* (or *points*) of N, and we have, for all $C \in \mathcal{B}(\mathbb{R})$,

$$N(C) = \sum_{n \in \mathbb{Z}} 1_C(T_n).$$

A similar representation is available for locally finite point processes on $E = \mathbb{R}^m$: Let Δ be an arbitrary element not in E. Let N be a locally finite point process on E. It can be represented as

$$N = \sum_{n \in \mathbb{N}} \varepsilon_{X_n}$$

where $\{X_n\}_{n\in\mathbb{N}}$ is a sequence of random variables with values in $E \cup \{\Delta\}$ endowed with the measurable space generated by $\mathcal{B}(E)$ and $\{\Delta\}$, and ε_a is the Dirac measure at a if $a \in E$, the null measure if $a = \Delta$. The Δ element plays the role of ∞. Note that it may occur that some of the values in the list $\{X_n\}_{n\in\mathbb{N}}$ are the same (multiple points).

Marked Point Processes

We now introduce marks, which are attributes of points. More precisely:

Definition 5.1.5 Let N and $\{X_n\}_{n\in\mathbb{N}}$ be as above. Let $(K, \mathcal{K})$ be some measurable space of the form $(\mathbb{R}^d, \mathcal{B}(\mathbb{R}^d))$, and let $\{Z_n\}_{n\in\mathbb{N}}$ be a random sequence with values in K. The sequence $\{(T_n, Z_n)\}_{n\in\mathbb{N}}$ is called a *marked point process* on E with marks in K, $\{Z_n\}_{n\in\mathbb{N}}$ is the *mark sequence*, and N is the *basic point process*.

We obtain from a marked point process a point process on $E \times K$, denoted by $\widetilde{N}$, by defining

$$\widetilde{N}(C \times L) := \sum_{n\in\mathbb{N}} 1_C(X_n) 1_L(Z_n),$$

or, equivalently

$$\widetilde{N} := \sum_{n\in\mathbb{N}} \varepsilon_{X_n}\varepsilon_{Z_n} = \sum_{n\in\mathbb{N}} \varepsilon_{(X_n, Z_n)},$$

where $\varepsilon_{(\Delta,z)}$ is the null measure on $E \times K$ for all $z \in K$.

Definition 5.1.6 Suppose moreover that $\{Z_n\}_{n\in\mathbb{N}}$ is IID and independent of N, with common probability distribution Q_Z. Then $\widetilde{N}$ is called a marked point process with *independent* IID *marks*.

Campbell's Theorem

Let μ be a measure on $(E, \mathcal{B}(E))$, and let $\varphi : (E, \mathcal{B}(E)) \to (\overline{\mathbb{R}}, \overline{\mathcal{B}})$ be a measurable function for which the integral $\int_E \varphi \, d\mu$ is well-defined. We shall occasionally denote this integral by $\mu(\varphi)$. Similarly, when $\varphi : E \to \overline{\mathbb{R}}$ is a measurable function and N is a point process, the notations

$$\int_E \varphi(x) N(dx), \qquad N(\varphi)$$

represent the same mathematical object. In the situation of Definition 5.1.5, observe that

$$\int_{E \times K} \varphi(x, z) \overline{N}(dx \times dz) = \sum_{n \in \mathbb{N}} \varphi(X_n, Z_n).$$

(By convention, such a sum extends only to those indices n such that $X_n \in E$, excluding the points "at infinity".) For a simple locally finite point processes N on $\mathbb{R}$ with sequence of points $\{T_n\}_{n \in \mathbb{Z}}$, if $\varphi : \mathbb{R} \to \overline{\mathbb{R}}$ is a non-negative measurable function,

$$N(\varphi) = \sum_{n \in \mathbb{Z}} \varphi(T_n).$$

(Here also, the sum extends only to those indices n such that $T_n \in \mathbb{R}$.) If $\varphi : (E, \mathcal{B}(E)) \to (\overline{\mathbb{R}}, \mathcal{B}(\overline{\mathbb{R}}))$ is a non-negative measurable function, then, $N(\varphi)$ is a random variable (Exercise 5.4.1).

Definition 5.1.7 Let N be a point process on E. Then

$$\nu(C) = E[N(C)]$$

defines a measure ν on E, called the *mean measure*, or *intensity measure* of N. N is called a *first-order* point process if for all bounded measurable sets $C \subset E$,

$$E[N(C)] < \infty. \tag{5.1}$$

A first-order point process N is locally finite, since (5.1) implies that $P(N(C) < \infty) = 1$.

Theorem 5.1.1 *Let N be a point process on E with intensity measure ν. Then, for all measurable functions $\varphi : E \to \mathbb{R}$ which are either non-negative or in $L^1_{\mathbb{C}}(\nu)$, the integral $N(\varphi)$ is well-defined (possibly infinite when φ is only assumed non-negative) and*

$$E[N(\varphi)] = \nu(\varphi). \tag{5.2}$$

In particular, if $\varphi \in L^1_{\mathbb{C}}(\nu)$, $N(\varphi)$ is almost-surely finite.

Formula (5.2) is known as *Campbell's formula*.

Proof Let $\varphi = \sum_{h=1}^{K} \alpha_h 1_{C_h}$ (a simple non-negative measurable function), where $K \in \mathbb{N}$, $\alpha_h \in \mathbb{R}_+$ and $C_1, \ldots, C_K$ are disjoint measurable subsets of E. We have

$$E[N(\varphi)] = E\left[\sum_{h=1}^{K} a_h N(C_h)\right] = \sum_{h=1}^{K} a_h \nu(C_h) = \nu(\varphi)$$

Let now φ be non-negative and let $\{\varphi_n\}_{n \in \mathbb{N}}$, be a non-decreasing sequence of simple non-negative measurable functions, with limit φ. Letting n go to ∞ in the equality

$E[N(\varphi_n)] = \nu(\varphi_n)$ yields the announced result, by monotone convergence. In the case where $\varphi \in L_C^1(\nu)$, since $E\left[N(\varphi^\pm)\right] = \nu(\varphi^\pm) < \infty$, the random variables $N(\varphi^\pm)$ are P-a.s. finite, and therefore $N(\varphi) = N(\varphi^+) - N(\varphi^-)$ is well-defined and finite, and

$$E[N(\varphi)] = E[N(\varphi^+)] - E[N(\varphi^-)] = \nu(\varphi^+) - \nu(\varphi^-) = \nu(\varphi). \qquad \square$$

Example 5.1.1 (Campbell's theorem for marked point processes with IID independent marks) Let $\widetilde{N}$ be as in Definition 5.1.6, that is a point process N on E, with independent IID marks $\{Z_n\}_{n\in\mathbb{N}}$. Let Q_Z be the common distribution of the marks. Denote by ν the intensity measure of the basic point process N. For all $C \in \mathcal{B}(E)$, $L \in \mathcal{K}$,

$$
\begin{aligned}
E\left[\sum_{n\in\mathbb{N}} 1_{C\times L}(X_n, Z_n)\right] &= \sum_{n\in\mathbb{N}} E\left[1_{C\times L}(X_n, Z_n)\right] \\
&= \sum_{n\in\mathbb{N}} E\left[1_C(X_n)1_L(Z_n)\right] \\
&= \sum_{n\in\mathbb{N}} E\left[1_C(X_n)\right] E\left[1_L(Z_n)\right] \\
&= \sum_{n\in\mathbb{N}} E\left[1_C(X_n)\right] Q_Z(L) = \nu(C)Q_Z(L) \\
&= E\left[\sum_{n\in\mathbb{N}} 1_C(X_n)\right] Q_Z(L) = \nu(C)Q_Z(L).
\end{aligned}
$$

Therefore the intensity measure of $\widetilde{N}$ is the product measure $\widetilde{\nu}(dx \times dz) = \nu(dx)\,Q_Z(dz)$. Campbell's theorem then reads as follows. If the measurable function $\varphi : \mathbb{R}^m \times K \to \mathbb{R}$ is non-negative or in $L_C^1(\nu \times Q_Z)$, then the sum

$$\sum_{n\in\mathbb{N}} \varphi(X_n, Z_n)$$

is P-a.s. well-defined (possibly infinite if φ is only assumed non-negative), and

$$E\left[\sum_{n\in\mathbb{N}} \varphi(X_n, Z_n)\right] = \int_E E\left[\varphi(x, Z)\right] \nu(dx), \qquad (5.3)$$

where Z is any K-valued random variable with distribution Q_Z.

Distribution of a Point Process

The *fidi* (finite-dimensional) distribution of a point process N is, by definition, the collection of distributions of the vectors $(N(A_1), \ldots, N(A_n))$, for all $m \in \mathbb{N}_+$, $A_1, \ldots, A_m \in \mathcal{B}(E)$.

Definition 5.1.8 Let N be a random measure on $(E, \mathcal{B}(E))$ and let P be a probability measure on $(\Omega, \mathcal{F})$. The probability P_N on $(M(E), \mathcal{M}(E))$ defined by

$$P_N = P \circ N^{-1}$$

is called the *probability distribution* of N.

We quote without proof the following fundamental result of the theory of point processes.

Theorem 5.1.2 *The distribution P_N of a random measure N on $(E, \mathcal{B}(E))$ is characterized by the distributions of the vectors $(N(I_1), \ldots, N(I_m))$ for all integers $m \geq 1$, all $I_1, \ldots, I_m \in \mathcal{I}$, the collection of sets of the form $I = \prod_{i=1}^{m}(a_i, b_i] \subset \mathbb{R}^m$.*

This result says in particular that the *fidi* distributions restricted to $\mathcal{I}$ caracterize the distribution of a *locally finite* point process.

Definition 5.1.9 Let N be a point process on E. The *Laplace functional* of N is the mapping associating to a non-negative measurable function $\varphi : E \to \mathbb{R}_+$ the non-negative real number

$$L(\varphi) = E\left[e^{-N(\varphi)}\right].$$

Theorem 5.1.3 *The Laplace functional of a point process N on E characterizes its distributions.*

Proof It suffices to observe that for all $K \geq 1$, all disjoint measurable sets $C_1, \ldots, C_K$ of $\mathcal{B}(E)$, the Laplace transform of the vector $(N(C_1), \ldots, N(C_K))$, that is the function

$$(t_1, \ldots, t_K) \in \mathbb{R}_+^K \to E\left[e^{-t_1 N(C_1) - \cdots - t_K N(C_K)}\right]$$

is of the form $E\left[e^{-N(\varphi)}\right]$ (Take $\varphi = t_1 1_{C_1} + \cdots + t_K 1_{C_K}$). $\qquad\square$

Given the random measure N on E, we define $\mathcal{F}^N$ to be the sigma-field generated by the random variables $N(C)$; $C \in \mathcal{B}(E)$. It is called the sigma-field *generated by the measure N*.

Definition 5.1.10 The family N_i, $i \in I$, of point processes on E, where I is an arbitrary index set, is said to be *independent* if the associated family $\mathcal{F}^{N_i}$, $i \in I$, is independent.

Theorem 5.1.4 *Let N_i, $i \in I$, be a collection of point processes on E where I is an arbitrary index set. If for any finite subset $J \subseteq I$, any collection φ_i, $i \in J$, of non-negative measurable functions from E to $\mathbb{R}$,*

$$E\left[e^{-\sum_{i \in J} N_i(\varphi_i)}\right] = \prod_{i \in J} E\left[e^{-N_i(\varphi_i)}\right], \qquad (5.4)$$

then N_i, $i \in I$, is an independent family of point processes.

Proof Exercise 5.4.3   □

5.1.2 Poisson Processes

Definition 5.1.11 Let ν be a sigma-finite measure on $\mathbb{R}^m$. The point process N on $\mathbb{R}^m$ is called a *Poisson process* on $\mathbb{R}^m$ with *mean measure ν* if

(i) For all finite families of mutually *disjoint* $C_1, \ldots, C_K \in \mathcal{B}(\mathbb{R}^m)$, the random variables $N(C_1), \ldots, N(C_K)$ are independent, and
(ii) For any set $C \in \mathcal{B}(\mathbb{R}^m)$,

$$P(N(C) = k) = e^{-\nu(C)} \frac{\nu(C)^k}{k!}, \; k \geq 0$$

$(= 0 \text{ if } \nu(C) = +\infty)$.

If ν is of the form

$$\nu(C) = \int_C \lambda(x) dx$$

for some non-negative measurable function $\lambda : \mathbb{R}^m \to \mathbb{R}_+$, the Poisson process N is said to admit the *intensity function $\lambda(x)$*. If in addition $\lambda(x) \equiv \lambda$, N is called a *homogeneous* Poisson process (HPP) on $\mathbb{R}^m$ with *intensity*, or *rate*, λ.

Theorem 5.1.5 *Let T be a Poisson random variable of mean θ. Let $\{Z_n\}_{n \geq 1}$ be an IID sequence of random vectors with values in $\mathbb{R}^m$ and common distribution Q. Assume that T is independent of $\{Z_n\}_{n \geq 1}$. Define the point process N on $\mathbb{R}^m$ by*

$$N(C) = \sum_{n=1}^{T} 1_C(Z_n), \text{ for all } C \in \mathcal{B}(\mathbb{R}^m).$$

Then, N is a Poisson process with mean measure $\nu = \theta \times Q$.

Proof What we want to prove can be stated in terms of characteristic functions: For for any finite family $C_1, \ldots, C_K$ of pairwise disjoint measurable sets of $\mathbb{R}^m$,

$$E[e^{i \sum_{j=1}^{K} u_j N(C_j)}] = \Pi_{j=1}^{K} \exp\left\{ \nu(C_j)(e^{iu_j} - 1) \right\}.$$

We have

$$\sum_{j=1}^{K} u_j N(C_j) = \sum_{j=1}^{K} u_j \left(\sum_{n=1}^{T} 1_{C_j}(Z_n) \right)$$

$$= \sum_{n=1}^{T} \left(\sum_{j=1}^{K} u_j 1_{C_j}(Z_n) \right) = \sum_{n=1}^{T} Y_n,$$

where

$$Y_n = \sum_{j=1}^{K} u_j 1_{C_j}(Z_n).$$

A standard computation yields

$$E[e^{i \sum_{n=1}^{T} Y_n}] = g_T(E[e^{iY_1}]),$$

where g_T is the generating function of T, here, since T is Poisson mean θ,

$$g_T(z) = \exp\left\{\theta(z-1)\right\}.$$

The random variable Y_1 takes the values $u_1, \ldots, u_K$ and 0 with the respective probabilities $Q(C_1), \ldots, Q(C_K)$ and $1 - \sum_{j=1}^{K} Q(C_j)$. Therefore

$$E[e^{iY_1}] = \sum_{j=1}^{K} e^{iu_j} Q(C_j) + 1 - \sum_{j=1}^{K} Q(C_j)$$

$$= 1 + \sum_{j=1}^{K} \left(e^{iu_j} - 1 \right) Q(C_j).$$

The conclusion then readily follows.   □

The above result is a special case of what we want to do. We want to construct a Poisson process on $\mathbb{R}^m$ with a mean measure ν that is *sigma-finite* (not just finite). We first observe that a locally finite measure ν on $\mathbb{R}^m$ can be decomposed as

$$\nu = \sum_{j=1}^{\infty} \theta_j \times Q_j,$$

where $\theta_j \geq 0$ and Q_j is a probability measure on $\mathbb{R}^m$, for all $j \geq 1$. We can construct independent Poisson processes N_j on $\mathbb{R}^m$ with respective mean measures $\theta_j \times Q_j$ and apply the following lemma.

Lemma 5.1.1 *Let ν be a locally finite measure on $\mathbb{R}^m$, of the form $\nu = \sum_{i=1}^{\infty} \nu_i$, where the ν_i's are locally finite measures on $\mathbb{R}^m$. Let N_i, $i \geq 1$, be a family of independent Poisson processes on $\mathbb{R}^m$ with respective mean measures ν_i, $i \geq 1$. Then the point process*

$$N = \sum_{j=1}^{\infty} N_j$$

is a Poisson process with mean measure ν.

Proof For mutually disjoint ν-bounded bounded measurable sets $C_1, \ldots, C_K$, and non-negative reals $t_1, \ldots, t_K$,

$$E\left[e^{-\sum_{\ell=1}^{K} t_\ell N(C_\ell)}\right] = E\left[e^{-\sum_{\ell=1}^{K} t_\ell \sum_{j=1}^{\infty} N_j(C_\ell)}\right]$$

$$= E\left[e^{-\lim_{n\uparrow\infty} \sum_{\ell=1}^{K} t_\ell \sum_{j=1}^{n} N_j(C_\ell)}\right]$$

$$= \lim_{n\uparrow\infty} E\left[e^{-\sum_{\ell=1}^{K} t_\ell \sum_{j=1}^{n} N_j(C_\ell)}\right],$$

by dominated convergence. (All the fuss about taking limits is because the product formula for expectation, as we know it, is for a *finite* number of terms in the product.) But

$$E\left[e^{-\sum_{\ell=1}^{K} t_\ell \sum_{j=1}^{n} N_j(C_\ell)}\right] = \prod_{j=1}^{n} E\left[e^{-\sum_{\ell=1}^{K} t_\ell N_j(C_\ell)}\right]$$

$$= \prod_{j=1}^{n} exp\left\{\sum_{\ell=1}^{K} \left(e^{-t_\ell} - 1\right) \nu_j(C_\ell)\right\}$$

$$= exp\left\{\sum_{\ell=1}^{K} \left(e^{-t_\ell} - 1\right) \sum_{j=1}^{n} \nu_j(C_\ell)\right\}.$$

Letting $n \uparrow \infty$, we obtain by dominated convergence that

$$E\left[e^{-\sum_{\ell=1}^{K} t_\ell N(C_\ell)}\right] = exp\left\{\sum_{\ell=1}^{K} \left(e^{-t_\ell} - 1\right) \nu(C_\ell)\right\}.$$

Therefore $N(C_1), \ldots, N(C_\ell)$ are independent Poisson random variables with respective means $\nu(C_1), \ldots, \nu(C_\ell)$.   □

Cox Processes

The following type of point processes on $\mathbb{R}^m$ is very common in applications. It is a "doubly stochastic" Poisson process, in the sense that it can be constructed in two steps. First one draws a random intensity function, that is a real non-negative measurable locally integrable stochastic process

$$\{\lambda(x)\}_{x \in \mathbb{R}^m}$$

and having done so, one generates a Poisson process N with the intensity function $\lambda(x)$. Formally,

Definition 5.1.12 Let $\mathcal{G} \supseteq \mathcal{F}^\lambda := \sigma(\lambda(x), \, x \in \mathbb{R}^m)$ where $\{\lambda(x)\}_{x \in \mathbb{R}^m}$ is a real non-negative measurable locally integrable stochastic process. A point process N on $\mathbb{R}^m$ such that given $\mathcal{G}$, N is a Poisson process on $\mathbb{R}^m$ with the intensity $\lambda(x)$, is called a *doubly stochastic Poisson process* (or *Cox process*) with respect to $\mathcal{G}$ with the stochastic intensity $\{\lambda(x)\}_{x \in \mathbb{R}^m}$.

In other words, N is a point process on $\mathbb{R}^m$ such that for all non-negative measurable functions $\varphi : \mathbb{R}^m \to \mathbb{R}$,

$$E\left[e^{-N(\varphi)} | \mathcal{G}\right] = \exp\left\{ \int_{\mathbb{R}^m} \left(e^{-\varphi(x)} - 1\right) \lambda(x) \, dx \right\} \tag{5.5}$$

Expectations concerning Cox processes are usually computed by first conditioning on $\mathcal{G}$.

Marked Poisson Process

A Poisson process on $\mathbb{R}^m$ with independent IID $\mathbb{R}^d$—valued marks can be viewed as a Poisson process on $\mathbb{R}^{m+d}$. More precisely, let

(α) N be a simple locally finite process on $\mathbb{R}^m$, with point sequence $\{X_n\}_{n \geq 1}$, and
(β) $\{Z_n\}_{n \geq 1}$ be a sequence of random vectors taking their values in $\mathbb{R}^d$.

The sequence $\{X_n, Z_n\}_{n \geq 1}$ is a *marked point process*, with the interpretation that Z_n is the *mark* associated with the *point* X_n. N is the *base* point process of the marked point process, and $\{Z_n\}_{n \geq 1}$ is the associated *sequence of marks*. We shall also say: "N is a simple and locally finite point process on $\mathbb{R}^m$ with marks $\{Z_n\}_{n \geq 1}$". If moreover

(1) N is a Poisson process with mean measure ν,
(2) $\{Z_n\}_{n \geq 1}$ is an IID sequence; and
(3) $\{Z_n\}_{n \geq 1}$ and N are independent,

we call the corresponding marked point process a Poisson processes on $\mathbb{R}$ *with independent* IID *marks*.

In this situation, we say: "N is a simple locally finite Poisson process on $\mathbb{R}^m$ with independent IID marks $\{Z_n\}_{n\geq 1}$".

We shall slightly generalize the model by allowing the mark distribution to depend on the location of the marked point. More precisely, we replace requirements (2) and (3) by

(2′) $\{Z_n\}_{n\geq 1}$ is, conditionally on $\mathcal{F}^N := \sigma(X_n, n \geq 1)$, an independent sequence;
(3′) given X_n, the random vector Z_n is independent of $(X_k; k \geq 1, k \neq n)$; and
(4′) for all $n \geq 1$ and all $L \in \mathcal{B}(\mathbb{R}^d)$,

$$P(Z_n \in C | X_n = x) = Q(x, L),$$

where $Q(\cdot, \cdot)$ is a stochastic kernel from $\mathbb{R}^m$ to $\mathbb{R}^d$, that is, Q is a function from $\mathbb{R}^m \times \mathcal{B}(\mathbb{R}^d)$ to $[0, 1]$ such that: for all $L \in \mathcal{B}(\mathbb{R}^d)$ the application $x \to Q(x, L)$ is measurable, and for all $x \in \mathbb{R}^m$, $Q(x, \cdot)$ is a probability measure on $\mathbb{R}^d$.

Theorem 5.1.6 *Let* $\{X_n, Z_n\}_{n\geq 1}$ *be as defined above by* (α) *and* (β), *and define the point process* $\tilde{N}$ *on* $\mathbb{R}^m \times \mathbb{R}^d$ *by*

$$\tilde{N}(A) = \sum_{n\geq 1} 1_A(X_n, Z_n)$$

for all measurable sets $A \in \mathcal{B}(\mathbb{R}^{m+d})$. *If conditions (1), (2′), (3′), and (4′) above are satisfied, then* $\tilde{N}$ *is a simple Poisson process with mean measure* $\tilde{\nu}$ *defined by*

$$\tilde{\nu}(C \times L) = \int\limits_{\mathbb{R}^m} Q(x, L)\,\nu(dx).$$

Proof In view of Theorem 5.1.3 we just have to show that the Laplace transform of $\tilde{N}$ has the appropriate form, that is for any non-negative measurable function $\tilde{\varphi} : \mathbb{R} \times \mathbb{R}^d \to \mathbb{R}$,

$$E\left[e^{-\tilde{N}(\tilde{\varphi})}\right] = exp\left\{ \int\limits_{\mathbb{R}^m}\int\limits_{\mathbb{R}^d} \left(e^{-\tilde{\varphi}(t,z)} - 1\right) \tilde{\nu}(dt \times dz) \right\}.$$

We have, by dominated convergence,

$$E\left[e^{-\tilde{N}(\tilde{\varphi})}\right] = E\left[e^{-\sum_{n\geq 1}\tilde{\varphi}(X_n, Z_n)}\right]$$

$$= \lim_{K\uparrow\infty} E\left[e^{-\sum_{n=1}^{+K}\tilde{\varphi}(X_n, Z_n)}\right].$$

For the time being fix a positive integer K. Then, using the assumptions $(2')$ and $(3')$,

$$E\left[e^{-\sum_{n=1}^{+K}\tilde{\varphi}(X_n,Z_n)}\right] = E\left[\prod_{n=1}^{+K}e^{-\tilde{\varphi}(X_n,Z_n)}\right]$$

$$= E\left[E^{\sigma(X_1,\dots,X_K)}\left[\prod_{n=1}^{+K}e^{-\tilde{\varphi}(X_n,Z_n)}\right]\right]$$

$$= E\left[\prod_{n=1}^{+K}\int_{\mathbb{R}^d}e^{-\tilde{\varphi}(X_n,z)}Q(X_n,dz)\right]$$

$$= E\left[e^{\sum_{n=1}^{+K}\ln\int_{\mathbb{R}^d}e^{-\tilde{\varphi}(X_n,z)}Q(X_n,dz)}\right]$$

$$= E\left[e^{-\sum_{n=1}^{+K}\psi(X_n)}\right],$$

where $\psi(x) = -\ln\int_{\mathbb{R}^d}e^{-\tilde{\varphi}(x,z)}Q(x,dz)$, a non-negative function. Letting $K \uparrow \infty$, we obtain, by dominated convergence,

$$E\left[e^{-\tilde{N}(\tilde{\varphi})}\right] = E\left[e^{-\sum_{n\geq 1}\psi(X_n)}\right] = E\left[e^{-N(\psi)}\right]$$

$$= exp\left\{\int_{\mathbb{R}^m}\left(e^{-\psi(x)}-1\right)\nu(dx)\right\}$$

$$= exp\left\{\int_{\mathbb{R}^m}\left[\int_{\mathbb{R}^d}e^{-\tilde{\varphi}(x,z)}Q(x,dz)-1\right]\nu(dx)\right\}$$

$$= exp\left\{\int_{\mathbb{R}^m}\left[\int_{\mathbb{R}^d}\left(e^{-\tilde{\varphi}(x,z)}-1\right)Q(x,dz)\right]\nu(dx)\right\}$$

$$= exp\left\{\int_{\mathbb{R}^m}\int_{\mathbb{R}^d}\left(e^{-\tilde{\varphi}(x,z)}-1\right)\tilde{\nu}(dx\times dz)\right\}. \qquad \square$$

Example 5.1.2 (The M/GI/ ∞ model, I) The model of this example is of interest in queuing theory, and also in traffic analysis in communications networks. We shall give here the queuing interpretation. Let

(a) A be HPP on $\mathbb{R}$ with intensity λ (the arrival process), and denote by $\{T_n\}_{n\in\mathbb{Z}}$ the ordered sequence of points of A, with the usual convention $T_0 \leq 0 < T_1$; and let

(b) $\{\sigma_n\}_{n\in\mathbb{Z}}$ be a sequence of random vectors taking their values in $\mathbb{R}_+$ with the probability distribution Q.

(c) $\{\sigma_n\}_{n\in\mathbb{Z}}$ and N are independent,

Here, T_n is interpreted as the time of arrival of the nth customer, and σ_n as the service time requested by this customer. Define the point process $\tilde{N}$ on $\mathbb{R} \times \mathbb{R}_+$ by

$$\tilde{N}(C) = \sum_{n\in\mathbb{Z}} 1_C(T_n, \sigma_n)$$

for all $A \in \mathcal{B}(\mathbb{R}) \otimes \mathcal{B}(\mathbb{R}_+)$. By Theorem 5.1.6, $\tilde{N}$ is a simple Poisson process with mean measure

$$\tilde{\nu}(dt \times dz) = \lambda dt \times Q(dz).$$

In the M/GI/∞ model,[3] a customer arriving at time T_n is immediately served, and therefore departs from the "system" at time $T_n + \sigma_n$. The number $X(t)$ of customers present in the system at time t is therefore given by the formula

$$X(t) = \sum_{n\in\mathbb{Z}} 1_{(-\infty, t]}(T_n) 1_{(t, \infty)}(T_n + \sigma_n).$$

(Indeed, the nth customer is in the system at time t if and only if she arrived at time $T_n \leq t$ and departed at time $T_n + \sigma_n > t$.) The *departure* process is, by definition, the point process D of departure times, that is

$$D(C) = \sum_{n\in\mathbb{Z}} 1_C(T_n + \sigma_n).$$

Assume that the service times have finite expectation: $E[\sigma_1] < \infty$. Then:

A. For all $t \in \mathbb{R}$, $X(t)$ is a Poisson random variable with mean $\lambda E[\sigma_1]$.
B. For all $t, \tau \in \mathbb{R}$, $\tau \geq 0$,

$$\text{cov}\,(X(t), X(t+\tau)) = \lambda \int_\tau^\infty P(\sigma_1 > y)\,dy.$$

C. The departure process is a homogeneous Poisson process with intensity λ.

Proof A. Observe that

$$X(t) = \tilde{N}(C(t)).$$

---

[3] The ∞ represents the number of servers. This model is sometimes called a "queuing" system, although in reality there is no queuing, since customers are served immediately upon arrival and without interruption. It is in fact a "pure delay" system.

where

$$C(t) := \{(s, z); \ s \le t, \ s + z > t\} \subset \mathbb{R} \times \mathbb{R}_+ .$$

In particular, $X(t)$ is a Poisson random variable with mean $\tilde{\nu}(C(t))$. We compute $\tilde{\nu}(C(t))$, using Fubini's theorem, as follows:

$$\tilde{\nu}(C(t)) = \int_{\mathbb{R}} \int_{\mathbb{R}_+} 1_{\{s+z>t\}} 1_{\{s \le t\}} \tilde{\nu}(ds \times dz)$$

$$= \int_{\mathbb{R}} \int_{\mathbb{R}_+} 1_{\{s+z>t\}} 1_{\{s \le t\}} \lambda \, ds \times Q(dz)$$

$$= \int_{\mathbb{R}} \left(\int_{\mathbb{R}_+} 1_{\{s+z>t\}} Q(dz) \right) 1_{\{s \le t\}} \lambda \, ds$$

$$= \lambda \int_{-\infty}^{t} Q((t - s, +\infty)) \, ds$$

$$= \lambda \int_{0}^{\infty} Q((s, +\infty)) \, ds = \lambda \int_{0}^{\infty} P(Z_1 > s) ds = \lambda E[Z_1].$$

Therefore $X(t)$ is Poisson, with mean $\lambda E[Z_1]$.

B. Take $\tau \ge 0$ and define the sets $C = C(t) \cap C(t + \tau)$, $A = C(t) - C$ and $B = C(t + \tau) - C$. In particular, $X(t) = \tilde{N}(A) + \tilde{N}(C)$ and $X(t + \tau) = \tilde{N}(B) + \tilde{N}(C)$. We have therefore

$$\text{cov}(X(t), X(t + \tau)) = \text{cov}(\tilde{N}(A) + \tilde{N}(C), \tilde{N}(B) + \tilde{N}(C))$$

$$= \text{cov}(\tilde{N}(A), \tilde{N}(B)) + \text{cov}(\tilde{N}(A), \tilde{N}(C))$$

$$+ \text{cov}(\tilde{N}(C), \tilde{N}(B)) + \text{cov}(\tilde{N}(C), \tilde{N}(C))$$

$$= \text{Var}(\tilde{N}(C)) = E\left[\tilde{N}(C)\right],$$

where we have taken into account that $\tilde{N}(A)$, $\tilde{N}(B)$, $\tilde{N}(C)$ are independent Poisson random variables of variances $E[\tilde{N}(A)]$, $E[\tilde{N}(B)]$, $E[\tilde{N}(C)]$. Therefore

$$\text{cov}\ (X(t), X(t + \tau)) = \tilde{\nu}(C).$$

Now, by Fubini, with computations quite similar to those performed in detail in Part A of the proof:

$$\tilde{\nu}(C) = \lambda \int_0^\infty P(\sigma_1 > y + \tau)\, dy = \lambda \int_\tau^\infty P(\sigma_1 > y)\, dy\,,$$

and therefore finally

$$\mathrm{cov}\,(X(t), X(t+\tau)) = \lambda \int_\tau^\infty P(\sigma_1 > y)\, dy\,.$$

C. Observe that $D(C) = \tilde{N}(\tilde{C})$ where $\tilde{C} := \{(t, z) \in \mathbb{R} \times \mathbb{R}_+;\ t + z \in C\}$, and that if $C_1, \ldots, C_K$ are disjoint measurable sets of $\mathbb{R}$, then $\tilde{C}_1, \ldots, \tilde{C}_K$ are disjoint measurable sets of $\mathbb{R} \times \mathbb{R}_+$. In particular, $D(C_1) = \tilde{N}(\tilde{C}_1), \ldots, D(C_K) = \tilde{N}(\tilde{C}_K)$ are independent Poisson random variables with means $\tilde{\nu}(\tilde{C}_1), \ldots, \tilde{\nu}(\tilde{C}_K)$. Since $\tilde{\nu}(\tilde{C}) = \lambda \times \ell(C)$ (the usual Fubini computation), we see that the departure process is a homogeneous Poisson process of intensity λ. □

Example 5.1.3 (The M/GI/ ∞ model, II) Consider the situation of Example 5.1.2. For easier notation, we write σ for σ_1. Suppose that $E[\sigma^2] < \infty$. Then the covariance function C is integrable, and the process $\{X(t)\}_{t\in\mathbb{R}}$ admits the power spectral density

$$f(\nu) = \frac{\lambda}{2\pi^2\nu^2} \left(1 - \mathrm{Re}\left\{E\left[e^{-2i\pi\nu\sigma}\right]\right\}\right).$$

Proof We have

$$\int_0^\infty C(\tau)\, d\tau = \lambda \int_0^\infty \int_0^\tau P(\sigma > t)\, dt\, d\tau$$

$$= \lambda E\left[\int_0^\infty \int_0^\infty 1_{\{\tau < t < \sigma\}}\, dt\, d\tau\right]$$

$$= \lambda E\left[\int_0^\infty (\sigma - \tau)^+\, d\tau\right] = \lambda E\left[\sigma^2\right] < \infty.$$

Therefore C is integrable and we can compute its Fourier transform. Denoting

$$g(\nu) := \int_0^\infty C(\tau) e^{-2i\pi\nu\tau}\, d\tau\,,$$

we have that

$$g(\nu) = \int\limits_0^\infty \left(\lambda \int\limits_\tau^\infty P(\sigma > t)\, dt \right) e^{-2i\pi\nu\tau}\, d\tau$$

$$= \lambda E \left[\int\limits_0^\infty \left(\int\limits_0^\infty 1_{\{\tau < t\}} e^{-2i\pi\nu\tau}\, d\tau \right) 1_{\{t < \sigma\}}\, dt \right]$$

$$= \lambda E \left[\int\limits_0^\infty \left(\int\limits_0^t e^{-2i\pi\nu\tau}\, d\tau \right) 1_{\{t < \sigma\}}\, dt \right]$$

$$= \lambda E \left[\int\limits_0^\sigma \frac{1}{2i\pi\nu} \left(1 - e^{-2i\pi\nu t} \right) dt \right]$$

$$= \frac{\lambda}{2i\pi\nu} E\left[\sigma\right] - \frac{\lambda}{2i\pi\nu} E \left[\int\limits_0^\sigma e^{-2i\pi\nu t}\, dt \right]$$

$$= \frac{\lambda}{2i\pi\nu} E\left[\sigma\right] + \frac{\lambda}{4\pi^2\nu^2} \left(1 - E\left[e^{-2i\pi\nu\sigma} \right] \right).$$

Taking into account the fact that C is a real symmetric function, we have that $f(\nu) = 2\mathrm{Re}\,\{g(\nu)\}$, which is the announced result. □

The Covariance Formula

Let ν be a measure on $(\mathbb{R}^m, \mathcal{B}(\mathbb{R}^m))$. We recall the standard notation

$$\nu(\varphi) := \int\limits_{\mathbb{R}^m} \varphi(x)\nu(dx).$$

For the special case of a point process N on $\mathbb{R}^m$, the corresponding integral reduces to the sum

$$N(\varphi) = \int\limits_{\mathbb{R}^m} \varphi(x)N(dx) = \sum_{n\in\mathbb{Z}} \varphi(X_n),$$

with the usual convention that this sum excludes the points "at infinity" ($X_n = \Delta$).

Theorem 5.1.7 *Let N be a Poisson process on $\mathbb{R}^m$ with mean measure ν. Let $\varphi : \mathbb{R}^m \to \mathbb{R}$ be a ν-integrable measurable function. Then $N(\varphi)$ is a well-defined* integrable *random variable, and*

$$E[N(\varphi)] = \nu(\varphi). \tag{5.6}$$

Let $\varphi, \psi : \mathbb{R}^m \to \overline{\mathbb{R}}$ *be two* ν-*integrable measurable functions such that moreover* $|\varphi|^2$ *and* $|\psi|^2$ *are* ν-*integrable. Then* $N(\varphi)$ *and* $N(\psi)$ *are well-defined* square-integrable *random variables and*

$$cov\,(N(\varphi), N(\psi)) = \nu(\varphi\psi). \tag{5.7}$$

Proof The first part is just a special case of Campbell's formula (Theorem 5.1.1).

For the second moments, first suppose that φ and ψ are *simple* non-negative measurable functions, that is, of the form $\sum_{h=1}^{K} \alpha_h 1_{A_h}$, where $K \in \mathbb{N}$, $\alpha_h \in \mathbb{R}_+$ and A_h is a measurable subset of $\mathbb{R}^m$, $1 \le h \le K$. It may be assumed that

$$\varphi = \sum_{h=1}^{K} a_h 1_{C_h}, \quad \psi = \sum_{h=1}^{K} b_h 1_{C_h}$$

where $C_1, \ldots, C_K$ are *disjoint*. In particular, $\varphi(x)\psi(x) = \sum_{h=1}^{K} a_h b_h 1_{C_h}(x)$.

Using the facts that if $i \ne j$, $N(C_i)$ and $N(C_j)$ are independent, and that the variance of a Poisson random variable with mean θ is θ,

$$E[N(\varphi)N(\psi)] = \sum_{\substack{h,l=1}}^{K} a_h b_l E[N(C_h)N(C_l)]$$

$$= \sum_{\substack{h,l=1 \\ h \ne l}}^{K} a_h b_l E[N(C_h)N(C_l)] + \sum_{l=1}^{K} a_l b_l E[N(C_l)^2]$$

$$= \sum_{\substack{h,l=1 \\ h \ne l}}^{K} a_h b_l E[N(C_h)]E[N(C_l)] + \sum_{l=1}^{K} a_l b_l E[N(C_l)^2]$$

$$= \sum_{\substack{h,l=1 \\ h \ne l}}^{K} a_h b_l \nu(C_h)\nu(C_l) + \sum_{l=1}^{k} a_l b_l [\nu(C_l) + \nu(C_l)^2]$$

$$= \sum_{h,l=1}^{k} a_h b_l \nu(C_h)\nu(C_l) + \sum_{l=1}^{k} a_l b_l \nu(C_l)$$

$$= \nu(\varphi)\nu(\psi) + \nu(\varphi\psi).$$

Let now φ, ψ be as in the statement of the theorem and let $\{\varphi_n\}_{n\ge1}$, $\{\psi_n\}_{n\ge1}$ be non-decreasing sequences of simple non-negative functions, with respective limits φ and ψ. Letting n go to ∞ in the equalitiy

$$E[N(\varphi_n)N(\psi_n)] = \nu(\varphi_n\psi_n) + \nu(\varphi_n)\nu(\psi_n)$$

yields the announced result, by monotone convergence. We have that

$$E[N(\varphi)] = E[N(\varphi^+)] - E[N(\varphi^-)]$$
$$= \nu(\varphi^+) - \nu(\varphi^-) = \nu(\varphi).$$

Also by the result in the non-negative case,

$$E\left[N(|\varphi|)^2\right] = \nu(|\varphi|^2) + \nu(|\varphi|)^2 < \infty.$$

Therefore, since $|N(\varphi)| \leq N(|\varphi|)$, $N(\varphi)$ is a square integrable variable, as well as $N(\psi)$ for the same reasons. Therefore, by Schwarz's inequality, $E[N(\varphi)N(\psi)]$ is well-defined. We have

$$E[N(\varphi)N(\psi)] = E\left[\left(N(\varphi^+) - E[N(\varphi^-)]\right)\left(N(\psi^+) - E[N(\psi^-)]\right)\right]$$
$$= E[N(\varphi^+)N(\psi^+)] + E[N(\varphi^-)N(\psi^-)]$$
$$\quad - E[N(\varphi^+)N(\psi^-)] - E[N(\varphi^-)N(\psi^+)]$$
$$= \left(\nu(\varphi^+\psi^+) + \nu(\varphi^+)\nu(\psi^+)\right) + \left(\nu(\varphi^-\psi^-) + \nu(\varphi^-)\nu(\psi^-)\right)$$
$$\quad - \left(\nu(\varphi^+\psi^-) + \nu(\varphi^+)\nu(\psi^-)\right) - \left(\nu(\varphi^-\psi^+) + \nu(\varphi^-)\nu(\psi^+)\right)$$
$$= \nu(\varphi\psi) + \nu(\varphi)\nu(\psi),$$

from which (5.7) follows. □

Formulas (5.6) and (5.7) immediately extend to *complex* functions $\varphi, \psi : \mathbb{R}^m \to \mathbb{C}$ that are ν-integrable measurable functions such that moreover $|\varphi|^2$ and $|\psi|^2$ are ν-integrable. Only Formula (5.7) has to be slightly modified to

$$\text{cov}\,(N(\varphi), N(\psi)) = \nu(\varphi\psi^*).$$

Extension to Marked Poisson Processes

Consider the situation in Theorem 5.1.6. Let Q be the common probability distribution of the marks Z_n. We consider for a function $\tilde{\varphi} : \mathbb{R}^m \times \mathbb{R}^d \to \mathbb{R}$, sums of the type

$$\tilde{N}(\tilde{\varphi}) := \sum_{n \geq 1} \tilde{\varphi}(X_n, Z_n). \tag{5.8}$$

Note that, denoting by $Z_1(x)$ any random vector of $\mathbb{R}^d$ with the distribution $Q(x, dz)$,

$$\tilde{\nu}(\tilde{\varphi}) = \int_{\mathbb{R}^m} \int_{\mathbb{R}^d} \tilde{\varphi}(x, z) Q(x, dz) \nu(dx)$$

$$= \int_{\mathbb{R}^m} E\left[\tilde{\varphi}(x, Z_1(x))\right] \nu(dx),$$

whenever the quantities involved have a meaning. Using this observation, the formulas obtained in the previous subsection can be applied in terms of marked point processes. The "corollaries" below do not have to be proven, since they are just *reformulations* of previous results, namely Theorems 5.1.7 and 5.1.8.

Let $0 < p < \infty$. Recall that a measurable function $\tilde{\varphi} : \mathbb{R}^m \times \mathbb{R}^d \to \mathbb{R}$ (resp. $\to \mathbb{C}$) is said to be in $L^p_{\mathbb{R}}(\tilde{\nu})$ (resp. $L^p_{\mathbb{C}}(\tilde{\nu})$) if

$$\int_{\mathbb{R}^m} \int_{\mathbb{R}^d} |\tilde{\varphi}(x, z)|^p \, \nu(dx) \, Q(x, dz) < \infty.$$

Corollary 5.1.1 *Suppose that $\tilde{\varphi} \in L^1_{\mathbb{C}}(\tilde{\nu})$. Then the sum (5.8) is well-defined, and moreover*

$$E\left[\sum_{n \geq 1} \tilde{\varphi}(X_n, Z_n)\right] = \int_{\mathbb{R}^m} E\left[\tilde{\varphi}(x, Z_1(x))\right] \nu(dx).$$

Let $\tilde{\varphi}, \tilde{\psi} : \mathbb{R} \times \mathbb{R}^m \to \mathbb{C}$ be two measurable functions in $L^1_{\mathbb{C}}(\tilde{\nu}) \cap L^2_{\mathbb{C}}(\tilde{\nu})$. Then

$$cov\left(\sum_{n \geq 1} \tilde{\varphi}(X_n, Z_n), \sum_{n \geq 1} \tilde{\psi}(X_n, Z_n)\right) = \int_{\mathbb{R}^m} E\left[\tilde{\varphi}(x, Z_1(x))\tilde{\psi}(x, Z_1(x))^*\right] \nu(dx).$$

Poissonian Shot Noises

Definition 5.1.13 Let N is a simple and locally finite point process on $\mathbb{R}^m$ with marks $\{Z_n\}_{n \geq 1}$. Let $h : \mathbb{R}^m \times \mathbb{R}^d \to \mathbb{C}$ be a measurable function. The complex-valued spatial stochastic process $\{X(y)\}_{y \in \mathbb{R}^m}$ given by

$$X(y) = \sum_{n \geq 1} h(y - X_n, Z_n) \tag{5.9}$$

(where it is assumed that the right-hand side is well-defined, for instance when h takes real non-negative values), is called a *spatial shot noise with random impulse response*. If N is a simple and locally finite Poisson process on $\mathbb{R}^m$ with independent IID marks $\{Z_n\}_{n \geq 1}$, it is called a *Poisson spatial shot noise with random impulse response and independent* iid *marks*.

The following result is a direct application of Theorem 5.1.6.

Corollary 5.1.2 *Consider the Poisson spatial shot noise with random impulse response and independent* IID *marks of Definition 5.1.13. Supppose that for all* $y \in \mathbb{R}^m$,

$$\int_{\mathbb{R}^m} E\left[|h(y - x, Z_1)|\right] \nu(dx) < \infty$$

and

$$\int_{\mathbb{R}^m} E\left[|h(y - x, Z_1)|^2\right] \nu(dx) < \infty.$$

Then the complex-valued spatial stochastic process $\{X(y)\}_{y \in \mathbb{R}^m}$ *given by (5.9) is well-defined, and for any* $y, \xi \in \mathbb{R}^m$, *we have*

$$E[X(y)] = \int_{\mathbb{R}^m} E[h(y - x, Z_1)] \nu(dx)$$

and

$$cov(X(y + \xi), X(y)) = \int_{\mathbb{R}^m} E\left[h(y - x, Z_1)h^*(y + \xi - x, Z_1)\right] \nu(dt).$$

In the case where the base point process N is an HPP with intensity λ, we find that

$$E[X(y)] = \lambda \int_{\mathbb{R}^m} E[h(x, Z_1)] dx$$

and

$$cov(X(y + \xi), X(y)) = \lambda \int_{\mathbb{R}^m} E\left[h(+x, Z_1)h^*(\xi + x, Z_1)\right] \lambda dx.$$

Observe that these quantities do not depend on $y \in \mathbb{R}^m$. The process $\{X(y)\}_{y \in \mathbb{R}^m}$ is for that reason called a *wide-sense stationary process*.

Example 5.1.4 The process $\{X(t)\}_{t \in \mathbb{R}}$ of Example 5.1.2 is a special case of shot noise: Take $h(t, \sigma) = 1_{[0,\sigma]}(t)$.

The Exponential Formula

Theorem 5.1.8 *Let N be a Poisson process on $\mathbb{R}^m$ with mean measure ν. Let $\varphi : \mathbb{R}^m \to \overline{\mathbb{R}}$ be a non-negative measurable function. Then,*

$$E[e^{-N(\varphi)}] = exp\left\{\nu(e^{-\varphi} - 1)\right\} \tag{5.10}$$

and

$$E[e^{N(\varphi)}] = exp\left\{\nu(e^{\varphi} - 1)\right\} \tag{5.11}$$

Proof We prove (5.10), the proof of (5.11) being quite similar. Suppose that φ is simple and non-negative, that is, $\varphi = \sum_{h=1}^{K} a_h 1_{C_h}$ where $C_1, \ldots, C_K$ are mutually disjoint measurable subsets of $\mathbb{R}^m$. Then

$$E[e^{-N(\varphi)}] = E\left[e^{-\sum_{h=1}^{K} a_h N(C_h))}\right] = E\left[\prod_{h=1}^{K} e^{-a_h N(C_h)}\right]$$

$$= \prod_{h=1}^{K} E\left[e^{-a_h N(C_h)}\right] = \prod_{h=1}^{K} exp\left\{(e^{-a_h} - 1)\nu(C_h)\right\}$$

$$= exp\left\{\sum_{h=1}^{k}(e^{-a_h} - 1)\nu(C_h)\right\} = exp\left\{\nu(e^{-\varphi} - 1)\right\}.$$

The formula is therefore true for non-negative simple functions. Take now a non-decreasing sequence $\{\varphi_n\}_{n\geq 1}$ of such functions converging to φ. We have for all $n \geq 1$

$$E[e^{-N(\varphi_n)}] = exp\left\{\nu(e^{-\varphi_n} - 1)\right\}.$$

By monotone convergence, the limit as n tends to ∞ of $N(\varphi_n)$ is $N(\varphi)$. Consequently, by dominated convergence, the limit of the left-hand side is $E[e^{-N(\varphi)}]$. The function $g_n = -(e^{-\varphi_n} - 1)$ is a non-negative function increasing to $g = -(e^{-\varphi} - 1)$, and therefore, by monotone convergence, $\nu(e^{-\varphi_n} - 1) = -\nu(g_n)$ converges to $\nu(e^{-\varphi} - 1) = -\nu(g)$, which in turn implies that the right-hand side of the last displayed equality tends to $exp\left\{\nu(e^{-\varphi} - 1)\right\}$ as n tends to ∞. □

Corollary 5.1.3 *Let $\tilde{\varphi}$ be a non-negative function from $\mathbb{R}^m \times \mathbb{R}^d$ to $\mathbb{R}$. Then,*

$$E\left[e^{-\sum_{n\geq 1} \tilde{\varphi}(X_n, Z_n)}\right] = exp\left\{\int_{\mathbb{R}^m} E\left[e^{-\tilde{\varphi}(x, Z_1(x))} - 1\right]\nu(dx)\right\}$$

Corollary 5.1.4 *Suppose that $\tilde{\varphi} \in L^1_{\mathbb{R}}(\tilde{\nu})$. (In particular $\sum_{n\geq 1} \tilde{\varphi}(X_n, Z_n)$ is a well-defined a.s. finite random variable). Then, for all $u \in \mathbb{R}$,*

$$E\left[e^{iu\sum_{n\geq 1} \tilde{\varphi}(X_n, Z_n)}\right] = exp\left\{\int_{\mathbb{R}^m} E\left[e^{iu\tilde{\varphi}(x, Z_1(x))} - 1\right]\nu(dx)\right\}.$$

According to Theorem 5.1.8, the Laplace functional of a Poisson process N on $\mathbb{R}^m$ with the mean measure ν is

$$L_N(\varphi) = exp\left\{\nu\left(e^{-\varphi} - 1\right)\right\}.$$

Theorem 5.1.9 *Let N_i, $i \in I$, be an arbitrary collection of simple point processes on $\mathbb{R}^m$. If for any finite subset $J \subset I$ and any collection φ_i, $i \in J$, of non-negative measurable functions from $\mathbb{R}^m$ to $\mathbb{R}$,*

$$E\left[e^{-\sum_{i \in J} N_i(\varphi_i)}\right] = \prod_{i \in J} exp\left\{\int_{\mathbb{R}^m} \left(e^{-\varphi_i(x)} - 1\right)\nu_i(dx)\right\} \qquad (5.12)$$

where ν_i, $i \in I$, is a collection of sigma-finite measures on $\mathbb{R}^m$, then N_i, $i \in I$, is a family of independent Poisson processes with respective mean measures ν_i, $i \in I$.

Proof Taking all the φ_i identically null except the first one, we have

$$E\left[e^{-N_1(\varphi_1)}\right] = exp\left\{\int_{\mathbb{R}^m} \left(e^{-\varphi_1(x)} - 1\right)\nu_1(dx)\right\},$$

and therefore N_1 is a Poisson process with mean measure ν_1. Similarly, for any $i \in J$, N_i is a Poisson process with mean measure ν_i. The independence follows from Theorem 5.1.4. $\qquad\square$

5.1.3 Renewal Processes

Consider an IID sequence $\{S_n\}_{n\geq 1}$ of *non-negative* random variables with common cumulative distribution function

$$F(x) = P(S_n \leq x).$$

These random variables can take the values 0 and ∞. The cumulative distribution function F is called *defective* when $F(\infty) = P(S_1 < \infty) < 1$, and *proper* when $F(\infty) = P(S_1 < \infty) = 1$.

We henceforth eliminate the case where $P(S_1 = 0) = 1$.

Definition 5.1.14 The above sequence is called the *inter-renewal sequence*. The associated *renewal sequence* $\{T_n\}_{n\geq 1}$ is defined as follows:
$T_n = T_{n-1} + S_{n-1}, n \geq 2$, and

T_1 is a non-negative FINITE random variable, called the *initial delay*, and assumed independent of the inter-renewal sequence. When $T_1 = 0$, the renewal sequence is called *undelayed*. (Here we momentarily abandon the convention $T_1 > 0$.)

Time T_n is called a *renewal time*, or an *event*. The stochastic process $\{N(t)\}_{t\geq0}$ defined by

$$N(t) = \sum_{n\geq1} 1_{\{T_n \leq t\}}$$

is the *counting process* of the renewal sequence; $N(t)$ counts the number of events in the closed interval $[0, t]$. Clearly the random function $t \to N(t)$ is almost surely right-continuous with limits on the left for each $t > 0$, namely $N(t-)$ is the number of renewal times in $[0, t)$. Note that his convention differs from the one adopted for Poisson processes on the positive haf-line, in that we count the point at 0 when there is one.

Definition 5.1.15 The function $R : \mathbb{R}_+ \to \overline{\mathbb{R}}_+$ defined by

$$R(t) = E[N(t)],$$

where $\{N(t)\}_{t\geq0}$ is the counting process of the UNDELAYED renewal sequence, is called the *renewal function*.

The renewal function R defined on $[0, \infty)$ takes finite values, is non-decreasing and right-continuous (Exercise 5.4.2). It therefore defines a locally finite measure, still denoted by R and called the *renewal measure*, by

$$R((a, b]) = R(b) - R(a).$$

Example 5.1.5 The inter-renewal sequence consists of IID exponential random variables with mean $\lambda^{-1} > 0$:

$$P(S_n \leq x) = 1 - e^{-\lambda x} \quad (x \geq 0)$$

and therefore, in the undelayed case, the renewal process $\{T_n\}_{n\geq1}$ is a *homogeneous Poisson process* of intensity λ. In this case

$$R(t) = 1 + \lambda t.$$

The renewal measure is therefore

$$R(dt) = \varepsilon_0(dt) + \lambda\, dt,$$

where ε_0 is the Dirac measure at 0.

It will be convenient to express the renewal function in terms of the common cumulative distribution function of the random variables S_n. For this, observe that in the undelayed case, $T_n = S_1 + \cdots + S_{n-1}$ is the sum of $n-1$ independent random variables with common cumulative distribution function F, and therefore

$$P(T_n \le t) = F^{*(n-1)}(t), \tag{5.13}$$

where F^{*n} is the nth fold convolution of F, defined recursively by

$$F^{*1}(t) = 1_{[0,\infty)}(t), \quad F^{*n}(t) = \int_{[0,t]} F^{*(n-1)}(t-s)\,dF(s). \tag{5.14}$$

Definition 5.1.16 A renewal process (delayed or not) is called *recurrent* when $P(S_1 < \infty) = 1$ (F is proper), and *transient* when $P(S_1 < \infty) < 1$ (F is defective).

In Exercise 5.4.2, it is shown that recurrence is equivalent to $P(N(\infty) = \infty) = 1$ which is in turn equivalent to $R(\infty) := \lim_{t \uparrow \infty} R(t) = \infty$.

Definition 5.1.17 Let $F : \mathbb{R}_+ \to \mathbb{R}_+$ be a *generalized cumulative distribution function* on $\mathbb{R}_+$, that is $F(x) = c\,G(x)$ where $c \in (0, \infty)$ and G is the cumulative distribution function of a proper non-negative real random variable, and in particular $G(\infty) = 1$. The basic object of renewal theory is the *renewal equation*

$$f = g + f * F,$$

that is, by definition of the convolution symbol $*$,

$$f(t) = g(t) + \int_{[0,t]} f(t-s)\,dF(s), \qquad t \ge 0, \tag{5.15}$$

where $g : \mathbb{R}_+ \to \mathbb{R}$ is a function called the *data*.

If $F(\infty) = 1$ one refers to the renewal equation as a *proper* renewal equation, or just a renewal equation. If $F(\infty) < 1$ the renewal equation is called *defective* and if $F(\infty) > 1$ it is called *excessive*.

Example 5.1.6 Writing the renewal function

$$R(t) = E[N(t)]$$
$$= E\Big[\sum_{n\ge1} 1_{\{T_n \le t\}}\Big] = \sum_{n\ge1} P(T_n \le t),$$

we find the expression, since $P(T_1 \leq t) = 1$ in the undelayed case:

$$R(t) = 1 + \sum_{n=2}^{\infty} F^{*n}(t) = 1 + F * (1 + \sum_{n=2}^{\infty} F^{*n}).$$

Therefore

$$R = 1 + R * F.$$

An expression of the solution of the renewal equation in terms of the renewal function is easy to obtain. Recall the following definition: A function $g : \mathbb{R}_+ \to \mathbb{R}$ is called *locally bounded* if for all $a \geq 0$, $\sup_{t \in [0,a]} |g(t)| < \infty$.

Theorem 5.1.10 When $F(\infty) \leq 1$, the renewal equation (5.15) where the (measurable) data function $g : \mathbb{R}_+ \to \mathbb{R}$ is locally bounded admits a unique locally bounded solution $f : \mathbb{R}_+ \to \mathbb{R}$ given by $f = g * R$, that is

$$f(t) = \int_{[0,t]} g(t - s) dR(s). \tag{5.16}$$

Proof The function f as in (5.16) is indeed locally bounded since g is locally bounded and $R(t)$ is finite for all t. Also

$$\begin{aligned}
f * F &= (g * R) * F \\
&= g * (R * F) \\
&= g * (R - 1) \\
&= g * R - g = f - g.
\end{aligned}$$

Therefore f is indeed solution of the renewal equation. Let f_1 be another locally bounded solution, and define $h = f - f_1$. This is a locally bounded solution which verifies $h = h * F$. By iteration,

$$h = h * F^{*n}.$$

Therefore for all $t \geq 0$

$$|h(t)| \leq \left(\sup_{s \in [0,t]} |h(s)| \right) F^{*n}(t).$$

Since $R(t) = \sum_{n \geq 1} F^{*n}(t) < \infty$, we have $\lim_{n \to \infty} F^{*n}(t) = 0$, which implies, in view of the last displayed inequality, $|h(t)| = 0$. $\square$

Regenerative Processes

Let $(E, \mathcal{E})$ be an arbitrary measurable space.

Definition 5.1.18 Let $\{X(t)\}_{t\geq 0}$, be a measurable E-valued stochastic process and let $\{T_n\}_{n\geq 1}$, be a proper recurrent renewal process, possibly delayed (recall however that the initial delay T_1 is then assumed finite). The process $\{X(t)\}_{t\geq 0}$ is said to be *regenerative* with respect to $\{T_n\}_{n\geq 1}$ if:

(a) for each $n \geq 1$, the distributions of the stochastic process $\{X(t + T_1)\}_{t\geq 0}$ and $\{X(t + T_n)\}_{t\geq 0}$ are identical, and:
(b) for each $n \geq 1$, the process $\{X(t + T_n)\}_{t\geq 0}$, is independent of $T_1, \ldots, T_n$.

The times T_n are called *regeneration times* of the regenerative process.

Example 5.1.7 (Continuous-time HMC) Let $\{X(t)\}_{t\geq 0}$, be a recurrent continuous-time homogeneous Markov chain taking its values in the state space $E = \mathbb{N}$. Suppose that it starts from state 0 at time $t = 0$. By the strong Markov property, $\{X(t)\}_{t\geq 0}$ is regenerative with respect to the sequence $\{T_n\}_{n\geq 1}$ where T_n is the nth time of visit of the chain in state 0.

Regenerative processes are the main sources of renewal equations. For instance:

Theorem 5.1.11 Let $\{X(t)\}_{t\geq 0}$ and $\{T_n\}_{n\geq 1}$ be as in Definition 5.1.18 except for the additional assumption $T_1 \equiv 0$ (undelayed renewal process) and let $A \in \mathcal{E}$. The function

$$f(t) = P(X(t) \in A)$$

satisfies the renewal equation with data

$$g(t) = P(X(t) \in A, t < T_2).$$

Proof Write:

$$1_{\{X(t)\in A\}} = 1_{\{X(t)\in A\}}1_{\{t<T_2\}} + 1_{\{X(t)\in A\}}1_{\{t\geq T_2\}}.$$

But, if $t \geq T_2$, $X(t) = \tilde{X}(t - T_2)$ where $\tilde{X}(t) = X(t - T_2)$, so that

$$1_{\{X(t)\in A\}} = 1_{\{X(t)\in A\}}1_{\{t<T_2\}} + 1_{\{\tilde{X}(t-T_2)\in A\}}1_{\{t\geq T_2\}}$$

Taking expectations and using the hypothesis that $\{\tilde{X}(t)\}_{t\geq 0}$ is independent of T_2 and has the same distribution as $\{X(t)\}_{t\geq 0}$, one obtains the announced result.   □

More generally, and with the same proof, the function $f(t) = E[\varphi(X(t))]$ where $\varphi : \mathbb{R} \to E$ is measurable and non-negative verifies the renewal equation with data $E[\varphi(X(t))1_{\{t<T_2\}}]$.

Stationarization

Very often, the processes of regenerative type (renewal processes, regenerative processes, semi-Markov processes) are described with initial conditions that do not imply stationarity. A fundamental issue is that of giving the initial conditions that make these processes stationary. For instance, by a proper choice of the initial delay, a delayed renewal process can be made stationary, in a sense to be made precise. We consider a delayed renewal process,

$$T_1 = S_0, \ T_2 = S_0 + S_1, \ldots, \ T_{n+1} = S_0 + \cdots + S_n$$

where $0 \leq T_0 < \infty$. We call G the cumulative distribution function of T_0, and suppose that $P(S_1 < \infty) = 1$ (the renewal process is proper) and, as usual, exclude trivialities by imposing the condition $P(S_1 = 0) < 1$. For all $t \geq 0$ define

$$S_0(t) = T_{N(t)+1} - t, \ S_n(t) = T_{N(t)+n+1} - T_{N(t)+n} \ (n \geq 1).$$

In particular, $S_n(0) = S_n$ for all $n \geq 0$.

Definition 5.1.19 The delayed renewal process is called *stationary* if the distribution of the sequence

$$S_0(t), \ S_1(t), \ S_2(t), \ldots$$

is independent of time $t \geq 0$.

For a delayed renewal process to be stationary, it is necessary and sufficient that $E[S_1] < \infty$ and that

$$P(T_1 \leq x) = F_0(x),$$

where

$$F_0(x) = \frac{1}{E[S_1]} \int_0^x (1 - F(y)) \, dy. \tag{5.17}$$

(F_0 is called the stationary forward recurrence time distribution.)

More generally, we can define a stationary renewal process on the line. It consists of a point process on $\mathbb{R}$ with the time sequence $\{T_n\}_{n \in \mathbb{Z}}$ such that the sequence of interarrival times not containing 0, that is $\{T_{n+1} - T_n\}_{\substack{n \in \mathbb{Z} \\ n \neq 1}}$ is IID with common CDF F and independent of the times T_0 and T_1 (We now revert to the convention $T_0 \leq 0 < T_1$) and such that

$$P(T_1 > v, -T_0 > w) = \lambda \int_{v+w}^{\infty} (1 - F(u)) \, du$$

Example 5.1.8 (Switching process) This is a process of the form

$$X(t) = Z_n \text{ for } t \in (T_n, T_{n+1}]$$

where the sequence $\{T_n\}_{n \in \mathbb{Z}}$ is the stationary renewal process on $\mathbb{R}$ and $\{Z_n\}_{n \in \mathbb{Z}}$ is an IID sequence independent of this renewal process. The process $\{X(t)\}_{t \in \mathbb{R}}$ is then stationary.

Let $\mathbf{P} = \{p_{ij}\}_{i,j \in E}$, be a stochastic matrix, assumed irreducible and positive recurrent with stationary distribution denoted by π. For all $i, j \in E$, let G_{ij} be the cumulative distribution function of some random variable taking its values in $(0, \infty)$, with mean m_{ij}. Let $\{X_n\}_{n \in \mathbb{Z}}$, be a stationary Markov chain with transition matrix $\mathbf{P}$, defined on some probability space with a probability P^0, and let $\{S_n\}_{n \in \mathbb{Z}}$, be a sequence of random variables defined on the same probability space, conditionally independent given $\{X_n\}_{n \in \mathbb{Z}}$. Assume moreover that S_n is conditionnally independent of X_k, $k \neq n, n+1$, and that conditionally on $X_n = i$ and $X_{n+1} = j$, S_n is distributed according to the CDF $G_{ij}(t)$.

We can now define a sequence $\{T_n\}_{n \in \mathbb{Z}}$ by

$$T_0 = 0, \ T_{n+1} - T_n = S_n \quad (n \in \mathbb{Z})$$

and a stochastic process $\{X(t)\}_{t \in \mathbb{R}}$, called a *semi-Markov process* by

$$X(t) = X_n, \ \text{for } T_n \le t < T_{n+1} \quad (t \in \mathbb{R}). \tag{5.18}$$

Observe that the number of T_n's in a finite interval is almost-surely finite due to the fact that $\{X_n\}_{n \in \mathbb{Z}}$ is recurrent. It will be assumed that

$$E^0[T_1] = \sum_{i \in E} \pi(i) \sum_{j \in E} p_{ij} m_{ij} < \infty$$

and that

$$E^0[N(0, t]] < \infty, \quad (t \in \mathbb{R}_+).$$

The process $\{X(t)\}_{t \in \mathbb{R}}$ is not stationary under P^0. However, if one changes the distribution of $(X_0, X_1, -T_0, T_1)$ is an appropriate way, one obtains a stationary version. Therefore, we now work under a probability P such that the sequences $\{X_n\}_{n \in \mathbb{Z}}$ and $\{T_n\}_{n \in \mathbb{Z}}$ have the following probabilistic structure:

(a) Conditionally on $X_1 = j$, the sequence

$$S_1, X_2, S_2, X_3, S_3, \ldots \tag{*}$$

has the same distribution under P or P^0.

(b) Conditionally on $X_0 = i$, the sequence

$$X_{-1}, S_{-1}, X_{-2}, S_{-2}, \ldots \tag{†}$$

has the same distribution under P and P^0.

(c) Conditionally on $X_0 = i, X_1 = j, -T_0 > x, T_1 > y$, the sequences (*) and (†) are independent.

(d) Moreover

$$P(X_0 = i, \; X_1 = j, \; -T_0 > x, \; T_1 > y) = \lambda\pi(i)p_{ij}\int_{x+y}^{\infty}(1 - G_{ij}(t))dt.$$

The process $\{X(t)\}_{t\in\mathbb{R}}$ is defined as before by (5.18). It then turns out that it is stationary (under this new probability P).[4]

Example 5.1.9 Consider the semi-Markov stationary process $\{X(t)\}_{t\in\mathbb{R}}$ described above. Let $f : E \to \mathbb{R}$ be some function such that the process $\{Y(t)\}_{t\in\mathbb{R}}$, where $Y(t) := f(X(t))$, is of the second order (and therefore WSS) with covariance function C_Y. We have, for $\tau \geq 0$:

$$C_Y(\tau) = E\left[f(X(0))f(X(\tau))\right]$$
$$= E\left[f(X(0))^2 1_{\{\tau < T_1\}}\right] + E\left[f(X(0))f(X(\tau))1_{\{\tau \geq T_1\}}\right] = A + B$$

where

$$A := E\left[f(X(0))^2 1_{\{\tau < T_1\}}\right] = \sum_{i\in E}\sum_{j\in E}\pi(i)f(i)^2 p_{ij}G_{ij}((\tau, \infty))$$

and, defining for $t \geq 0$, $\widetilde{X}(t) = X(t + T_1)$ $(t \geq 0)$,

$$B = E\left[f(X(0))f(X(\tau))1_{\{\tau \geq T_1\}}\right]$$
$$= \sum_{i\in E}\sum_{j\in E}E\left[f(X(0))1_{\{X_0=i\}}1_{\{X_1=j\}}f(\widetilde{X}(\tau - T_1))1_{\{\tau \geq T_1\}}\right]$$
$$= \sum_{i\in E}\sum_{j\in E}f(i)E\left[1_{\{X_0=i\}}1_{\{X_1=j\}}E_{0,j}\left[\int_{(0,\tau]}f(X(\tau - t))\lambda(1 - G_{ij}(t))\,dt\right]\right]$$
$$= \sum_{i\in E}\sum_{j\in E}f(i)\pi(i)p_{ij}\int_{(0,\tau]}\lambda(1 - G_{ij}(t))E_{0,j}\left[f(X(\tau - t))\right]dt.$$

It remains to obtain expressions for the quantities of the type $E_{0,k}\left[f(X(t))\right]$. We have

$$E_{0,k}\left[f(X(t))\right] = E_{0,k}\left[f(X(t))1_{\{\tau < T_1\}}\right] + E_{0,k}\left[f(X(t))1_{\{t \geq T_1\}}\right] = C + D$$

where

---

[4] See for instance, Baccelli and Brémaud, *Elements of Queuing Theory*, Springer, New York, 1994 (2nd edition 2003).

$$C = f(k) \sum_{\ell \in E} p_{k\ell} G_{k\ell}((t, \infty))$$

and

$$D = \sum_{\ell \in E} E_{0,k} \left[f(\widetilde{X}(t - T_1)) 1_{\{X_1 = \ell\}} 1_{\{t \ge T_1\}} \right]$$

$$= \sum_{\ell \in E} p_{k\ell} \int_{[0,t]} E_{0,\ell} \left[f(X(t - s)) \right] G_{k\ell}(ds).$$

Therefore, denoting

$$Z_k(t) = E_{0,k} \left[f(X(t)) \right]$$

$$z_k(t) = f(k) \sum_{\ell \in E} p_{k\ell} G_{k\ell}((t, \infty))$$

$$F_{k\ell}(t) = p_{k\ell} G_{k\ell}(t)$$

we obtain the multivariate renewal equations

$$Z_k(t) = z_k(t) + \sum_{\ell} \int_{[0,t]} Z_\ell(t - s) \, dF_{k\ell}(s) \qquad (k \in E). \tag{5.19}$$

Going back to the first lines of computation

$$C_Y(\tau) = \sum_{i \in E} \sum_{j \in E} \pi(i) f(i)^2 p_{ij} G_{ij}((\tau, \infty))$$

$$+ \sum_{i \in E} \sum_{j \in E} f(i)\pi(i) p_{ij} \int_{(0,\tau]} \lambda(1 - G_{ij}(t)) Z_j(\tau - t) \, dt. \tag{5.20}$$

5.2 Power Spectral Measure

5.2.1 Covariance Measure and Bartlett Spectrum

A point process N on $\mathbb{R}$ with sequence of event times $\{T_n\}_{n \in \mathbb{Z}}$ can be represented—at least symbolically—by the "random Dirac comb"

$$X(t) := \sum_{n \in \mathbb{Z}} \delta(t - T_n), \tag{5.21}$$

the sum therein extending, according to the usual convention, to all event times at finite distance. The right-hand side is certainly not a *bona fide* stochastic process since the delta function is not a function in the ordinary sense, but a distribution. In particular, one cannot define for the random Dirac comb associated with a stationary process a power spectral measure in the sense of Bochner's theorem. The natural extension of the notion of power spectral density that is suitable for stationary point processes and that is discussed in the present chapter is the so-called *Bartlett spectral measure*.

Recall that a point process N on $\mathbb{R}^m$ such that for all bounded Borel sets $C \subset \mathbb{R}^m$, $E[N(C)] < \infty$ is called a *first-order* point process. The measure ν on $\mathbb{R}^m$ defined by $\nu(C) = E[N(C)]$ is then a locally finite measure called the *intensity measure* or *first moment measure* of N.

Definition 5.2.1 Let N be a *simple* point process on $\mathbb{R}^m$. It is called a *second-order* point process if for all bounded Borel sets $C \subset \mathbb{R}^m$,

$$E\left[N(C)^2\right] < \infty. \tag{5.22}$$

In particular, a second-order point process is also a first order point process.

For a second-order point process N, by Caratheodory's extension theorem (Theorem A.1.9), the *second moment* measure M_2 on $\mathbb{R}^m \times \mathbb{R}^m$ is well and uniquely defined by the formula

$$M_2(A \times B) = E[N(A)N(B)] \, (A, B \in \mathcal{B}(\mathbb{R}^m))$$

(Exercise 5.4.18). This measure is locally finite. Indeed, if D is a bounded measurable set of $\mathbb{R}^{2m}$, it is contained in a "square" $C \times C$, where C is a bounded measurable subset of $\mathbb{R}^m$, and therefore $M_2(D) \leq M_2(C \times C) = E\left[N(C)^2\right] < \infty$.

Denote by $N \times N$ the point process on $\mathbb{R}^m \times \mathbb{R}^m$ whose points are the ordered pairs (X_n, X_k), $n, k \in \mathbb{N}$. Observing that $(N \times N)(A \times B) = N(A)N(B)$, we see that M_2 is the intensity measure of $N \times N$, and therefore, by Campbell's theorem, for all non-negative measurable functions $g : \mathbb{R}^m \times \mathbb{R}^m \to \mathbb{R}$

$$E\left[\sum_{n \in \mathbb{N}} \sum_{k \in \mathbb{N}} g(X_n, X_k)\right] = \int_{\mathbb{R}^m} \int_{\mathbb{R}^m} g(t, s) \, M_2(dt \times ds).$$

This formula extends to all measurable functions $g : \mathbb{R}^m \times \mathbb{R}^m \to \mathbb{C}$ such that

$$\int_{\mathbb{R}^m} \int_{\mathbb{R}^m} |g(t, s)| \, M_2(dt \times ds) < \infty.$$

Let $L_N^2(M_2)$ be the collection of measurable funtions $\varphi : \mathbb{R} \to \mathbb{C}$ such that

$$\int_{\mathbb{R}^m} \int_{\mathbb{R}^m} |\varphi(t)\varphi(s)| \, M_2(dt \times ds) < \infty. \tag{5.23}$$

or, equivalently,

$$E\left[N(|\varphi|)^2\right] < \infty,$$

and in particular $E\left[N(|\varphi|)\right] < \infty$ and $\varphi \in L^1_{\mathbb{C}}(\nu)$. If $\varphi, \psi \in L^2_N(M_2)$,

$$E\left[\left(\int_{\mathbb{R}^m} \varphi(t)\, N\,(dt)\right)\left(\int_{\mathbb{R}^m} \psi(t)\, N\,(dt)\right)^*\right] = \int_{\mathbb{R}^m} \int_{\mathbb{R}^m} \varphi(t)\,\psi^*(s)\, M_2\,(dt \times ds).$$

$$\tag{5.24}$$

Also, $L^2_N(M_2)$ is a vector space that contains all bounded functions with compact support. In fact, we have the following inclusion:

$$L^2_N(M_2) \subseteq L^1_{\mathbb{C}}(\nu) \cap L^2_{\mathbb{C}}(\nu). \tag{5.25}$$

Proof The defining inequality $E\left[N(|\varphi|)^2\right] < \infty$ of $L^2_N(M_2)$ implies in particular that $E\left[N(|\varphi|)\right] < \infty$ and, therekore, by Campbell's formula, $\varphi \in L^1_{\mathbb{C}}(\nu)$. On the other hand, since

$$\sum_{n \in \mathbb{Z}} |\varphi(X_n)|^2 \leq \left(\sum_{n \in \mathbb{Z}} |\varphi(X_n)|\right)^2,$$

we have that is $\varphi \in L^2_N(M_2)$, then $E\left[\int_{\mathbb{R}^m} |\varphi(t)|^2\, N(dt)\right] < \infty$, that is, by Campbell's formula, $\int_{\mathbb{R}^m} |\varphi(t)|^2\, \nu(dt) < \infty < \infty$. $\qquad\square$

Example 5.2.1 (Second moment measure of a Poisson process) In view of Campbell's theorem, for a Poisson process N on $\mathbb{R}^m$ with mean measure ν to be a first-order point process, it is necessary and sufficient that ν be locally finite. In this case, N is also a second order point process since a Poisson variable has a variance equal to its mean. By standard computations

$$M_2(A \times B) = \operatorname{cov}(N(A), N(B)) + E[N(A)]E[N(B)] = \nu(A \cap B) + \nu(A)\nu(B)$$

and by the covariance formula (5.7), for all $\varphi, \psi \in L^1_{\mathbb{R}}(\nu) \cap L^2_{\mathbb{R}}(\nu)$,

$$E\left[N(\varphi)N(\psi)^*\right] = \int_{\mathbb{R}} \varphi(t)\,\psi^*(t)\,\nu(dt) + \left(\int_{\mathbb{R}} \varphi(t)\,\nu(dt)\right)\left(\int_{\mathbb{R}} \psi^*(t)\,\nu(dt)\right).$$

Also, in this case, $L^2_N(M_2) = L^1_{\mathbb{R}}(\nu) \cap L^2_{\mathbb{R}}(\nu)$ (Exercise 5.4.19).

Definition 5.2.2 A *wide-sense stationary point process* is, by definition, a second-order point process N such that for all measurable sets $C \subseteq \mathbb{R}^m$ and all $t \in \mathbb{R}^m$,

292   5   Power Spectra of Point Processes

$$E\,[N(C+t)] = E\,[N(C)]\,,$$

and such that for all bounded measurable sets A, $B \subset \mathbb{R}^m$ and all $t \in \mathbb{R}^m$,

$$E\,[N(A+t)N(B+t)] = E\,[N(A)N(B)]\,.$$

In particular, for all non-negative measurable functions ϕ, $\psi{:}\mathbb{R}{\to}\mathbb{R}$, the quantity

$$E\left[\left(\int_{\mathbb{R}} \phi(\tau+t)\,N(dt)\right)\left(\int_{\mathbb{R}} \psi(\tau+t)\,N(dt)\right)\right]$$

is independent of $\tau \in \mathbb{R}$ (Exercise 5.4.20).

Example 5.2.2 (Regular grid, I) Consider the point process on $\mathbb{R}^2$ whose points form a regular (T_1, T_2)-grid on $\mathbb{R}^2$ with random origin, that is

$$N = \left\{(n_1 T_1 + U_1, n_2 T_1 + U_2)\,, (n_1, n_2) \in \mathbb{Z}^2\right\}$$

where $T_1 > 0$, $T_2 > 0$, and U_1, U_2 are independent random variables uniformly distributed on $[0, T_1]$, $[0, T_2]$ respectively. The point process is obviously WSS (in fact: strictly stationary) with average intensity $\lambda = 1/(T_1 T_2)$ (Exercise 5.4.16).

Let N be a WSS point process. Its intensity measure ν is invariant under translation, and therefore (Theorem A.1.12)

$$\nu\,(C) = \lambda\,\ell^m\,(C)$$

(where ℓ^m is the Lebesgue measure on $\mathbb{R}^m$) for some $\lambda \in \mathbb{R}_+$, called the *intensity*. In particular $L^p_{\mathbb{C}}(\nu) = L^p_{\mathbb{C}}(\mathbb{R}^m)$ and therefore, the inclusion (5.25) reads in the stationary case

$$L^2_N(M_2) \subseteq L^1_{\mathbb{C}}(\mathbb{R}^m) \cap L^2_{\mathbb{C}}(\mathbb{R}^m)\,.$$

By stationarity again, for all Borel sets A, B in $\mathbb{R}^m$, all $t \in \mathbb{R}^m$

$$M_2\,((A+t) \times (B+t)) = M_2\,(A \times B)\,.$$

It follows from Theorem A.1.20 that there exists a unique *locally finite* measure σ such that for all φ, $\psi \in L^2_N(M_2)$,

$$\int_{\mathbb{R}^m}\int_{\mathbb{R}^m} \varphi\,(t)\,\psi^*\,(s)\,M_2\,(dt \times ds) = \int_{\mathbb{R}^m}\left(\int_{\mathbb{R}^m} \varphi\,(t)\,\psi^*\,(s+t)\,dt\right)\sigma\,(ds)\,. \quad (5.26)$$

Since for all φ, $\psi \in L^1_{\mathbb{C}}(\mathbb{R}^m)$,

$$E\left[N(\varphi)\right] E\left[N(\psi)\right]^* = \left(\lambda \int_{\mathbb{R}^m} \varphi(t)\,dt\right)\left(\lambda \int_{\mathbb{R}^m} \psi^*(s)\,ds\right)$$

$$= \lambda^2 \int_{\mathbb{R}^m}\left(\int_{\mathbb{R}^m} \varphi(t)\,\psi^*(t+s)\,dt\right)ds\,,$$

we have from (5.24) and (5.26) that for all $\varphi, \psi \in L_N^2(M_2)$,

$$\mathrm{cov}\,(N(\varphi), N(\psi)) = \int_{\mathbb{R}^m}\left(\int_{\mathbb{R}^m} \varphi(t)\,\psi^*(t+s)\,dt\right)\Gamma_N(ds) \qquad (5.27)$$

where the locally finite measure

$$\Gamma_N := \sigma - \lambda^2 \ell^m$$

is called the *covariance measure* of the stationary second-order point process N.

Example 5.2.3 (Covariance measure of the homogeneous Poisson process) Let N be a homogeneous Poisson process on the line with intensity $\lambda > 0$. By the covariance formula (5.7),

$$\mathrm{cov}\,(N(\varphi), N(\psi)) = \lambda \int_{\mathbb{R}} \varphi(t)\,\psi^*(t)\,dt\,.$$

If we denote by ε_0 the Dirac measure at 0, we have for the right-hand side of the last display the expression

$$\lambda \int_{\mathbb{R}}\left(\int_{\mathbb{R}} \varphi(t)\,\psi^*(t+s)\,dt\right)\varepsilon_0(ds)\,,$$

and therefore, comparing with the right-hand side of (5.27),

$$\Gamma_N = \lambda \varepsilon_0\,.$$

Example 5.2.4 (Covariance of the renewal process) Let N be a stationary renewal point process with renewal function R. For any measurable non-negative functions $\varphi, \psi : \mathbb{R}_+ \to \mathbb{R}$,

$$E\left[\left(\int_{\mathbb{R}} \varphi(t)\, N(dt)\right)\left(\int_{\mathbb{R}} \psi(t)\, N(dt)\right)\right]$$

$$= E\left[\sum_{n\in\mathbb{Z}} \varphi(T_n)\left(\sum_{m\geq n} \psi(T_n + (T_{n+m} - T_n))\right)\right.$$

$$\left. + \sum_{n\in\mathbb{Z}} \psi(T_n)\left(\sum_{m>1} \varphi(T_n + (T_{n+m} - T_n))\right)\right]$$

$$= \sum_{n\in\mathbb{Z}} E\left[\varphi(T_n)\left(\sum_{m\geq 1} \psi(T_n + (T_{n+m} - T_n))\right)\right]$$

$$+ \sum_{n\in\mathbb{Z}} E\left[\psi(T_n)\left(\sum_{m>1} \varphi(T_n + (T_{n+m} - T_n))\right)\right]$$

Call A and B the sums in the right-hand side. Since $\{(T_{n+m} - T_n)\}_{m\geq 1}$ is an undelayed renewal process independent of T_n and with the same interval distribution as the original renewal process, we have (applying twice Campbell's formula):

$$A = \sum_{n\in\mathbb{Z}} E\left[\varphi(T_n) E\left[\sum_{m\geq 1} \psi(T_n + (T_{n+m} - T_n)) \mid T_n\right]\right]$$

$$= \sum_{n\in\mathbb{Z}} E\left[\varphi(T_n) \int_{[0,\infty)} \psi(t)\, R(dt + T_n)\right]$$

$$= E\left[\sum_{n\in\mathbb{Z}} \varphi(T_n) \int_{[0,\infty)} \psi(t)\, R(dt + T_n)\right]$$

$$= E\left[\int_{\mathbb{R}} \varphi(s)\left(\int_{[0,\infty)} \psi(t)\, R(dt + s)\right)\lambda\, ds\right]$$

$$= E\left[\int_{\mathbb{R}} \varphi(s)\left(\int_{[s,+\infty)} \psi(s + t)\, R(dt)\right)\lambda\, ds\right].$$

Similarly

$$B = E\left[\int_{\mathbb{R}} \psi(s)\left(\int_{(s,+\infty)} \varphi(s + t)\, R(dt)\right)\lambda\, ds\right]$$

Finally

$$
E\left[\left(\int_{\mathbb{R}} \varphi(t)\, N(dt)\right)\left(\int_{\mathbb{R}} \psi(t)\, N(dt)\right)\right] = E\left[\int_{\mathbb{R}} \varphi(s)\left(\int_{\mathbb{R}} \psi(s+t)\, R(dt)\right)\lambda\, ds\right],
$$

from which it follows that $\sigma = \lambda R$ and the covariance measure is in this case:

$$
\Gamma_N(dt) = \lambda(R(dt) - \lambda\, dt). \tag{5.28}
$$

In the particular case of the homogeneous Poisson process with intensity λ, since the renewal measure is the Dirac measure at 0, denoted by ε_0, plus λ times the Lebesgue measure, we find that

$$
\Gamma_N(dt) = \lambda\,\varepsilon_0(dt)
$$

as we already know from Example 5.2.3.

The link with the usual notion of covariance function of a *bona fide* wide-sense stationary (WSS) stochastic process $\{X(t)\}_{t\in\mathbb{R}^m}$ is the following. Let C_X be the covariance function of such process. Then, for all $\varphi, \psi \in L_{\mathbb{C}}^1(\mathbb{R}^m)$,

$$
\mathrm{cov}\left(\int_{\mathbb{R}^m} \varphi(t)\, X(t)\, dt, \int_{\mathbb{R}^m} \psi(s)\, X(s)\, ds\right) = \int_{\mathbb{R}^m}\left(\int_{\mathbb{R}^m} \varphi(t)\, \psi^*(t+s)\, dt\right) C_X(s)\, ds.
$$

The formal analogy with formula (5.27) is clear.

The Bartlett Spectrum

The power spectral measure μ_X of a wide-sense stationary stochastic process $\{X(t)\}_{t\in\mathbb{R}^m}$ is caracterized by the following condition: For all $\varphi, \psi \in L_{\mathbb{C}}^1(\mathbb{R}^m)$

$$
\mathrm{cov}\left(\int_{\mathbb{R}^m} \varphi(t)\, X(t)\, dt, \int_{\mathbb{R}^m} \psi(s)\, X(s)\, ds\right) = \int_{\mathbb{R}^m} \widehat{\varphi}(\nu)\, \widehat{\psi}^*(\nu)\, \mu_X(d\nu),
$$

or equivalently: For all $\varphi \in L_{\mathbb{C}}^1(\mathbb{R}^m)$

$$
\mathrm{Var}\left(\int_{\mathbb{R}^m} \varphi(t)\, X(t)\, dt\right) = \int_{\mathbb{R}^m} |\widehat{\varphi}(\nu)|^2\, \mu_X(d\nu).
$$

The last definition is fit for an extension to the case of point processes.

Definition 5.2.3 Let N be a simple wide-sense stationary point process on $\mathbb{R}^m$ with intensity λ. A measure μ_N on $\mathbb{R}^m$ is called the *Bartlett spectral measure* of N if it is the unique locally finite measure μ_N such that

$$\mathrm{Var}\left(\int_{\mathbb{R}^m} \varphi(t)\, N(dt)\right) = \int_{\mathbb{R}^m} |\widehat{\varphi}(\nu)|^2 \, \mu_N(d\nu) \tag{5.29}$$

for all $\varphi \in \mathcal{B}_N$, where $\mathcal{B}_N \subseteq L_N^2(M^2)$ is a vector space of functions called the *test function space*, a function therein being called a *test function*.

By polarization of (5.29), we have that for all $\varphi, \psi \in \mathcal{B}_N$,

$$\mathrm{cov}\left(N(\varphi),\, N(\psi)\right) = \int_{\mathbb{R}^m} \widehat{\varphi}(\nu)\widehat{\psi}^*(\nu)\mu_N(d\nu). \tag{5.30}$$

The extent of the test function space $\mathcal{B}_N$ is given in each situation. It may be not the largest one, but it should in any case contain a class of functions rich enough to guarantee uniqueness of the measure μ_N. By this, the following is meant. If the locally finite measures μ_1 and μ_2 are such that

$$\int_{\mathbb{R}^m} |\widehat{\varphi}(\nu)|^2 \, \mu_1(d\nu) = \int_{\mathbb{R}^m} |\widehat{\varphi}(\nu)|^2 \, \mu_2(d\nu)$$

for all $\varphi \in \mathcal{B}_N$, then $\mu_1 \equiv \mu_2$.

Note that $\mathcal{B}_N \subseteq L^1_{\mathbb{C}}(\mathbb{R}^m) \cap L^2_{\mathbb{C}}(\mathbb{R}^m)$ since, as we observed earlier, $L^2_N(M^2) \subseteq L^1_{\mathbb{C}}(\mathbb{R}^m) \cap L^2_{\mathbb{C}}(\mathbb{R}^m)$. In particular the Fourier transform of any $\varphi \in \mathcal{B}_N$ is well-defined. The existence and uniqueness of the Bartlett spectral measure is proven in Theorem 5.2.1 where it will be shown that $\mathcal{B}_N$ contains at least the functions that are, together with their Fourier transform, $O\left(1/|t|^2\right)$ as $|t| \to \infty$.

The definition of the Bartlett spectral measure immediately yields the spectral measure of the homogeneous Poisson process:

Example 5.2.5 (Poisson impulsive white noise) For a Poisson process, as we have seen in Example 5.2.4 (and as could be proven more directly!), the covariance function is λ times the Dirac measure at the origin, and therefore its spectral measure is λ times the Lebesgue measure, that is, it admits a power spectral density that is a constant:

$$f_N(\nu) \equiv \lambda.$$

For this reason the random Dirac comb (5.21) is sometimes called the *impulsive Poisson white noise*.

The Bochner spectral measure of a shot noise based on a WSS point process whose Bartlett spectral measure is given follows at once from the definition:

Example 5.2.6 (Shot noise) Let N be a WSS point process with spectral measure μ_N and space of test functions $\mathcal{B}_N$. For all $h \in \mathcal{B}_N$, all $u, v \in \mathbb{R}^m$, define

$$\varphi(t) = h(u - t), \quad \psi(t) = h(v - t).$$

Applying (5.30), we obtain

$$\operatorname{cov}\left(\int_{\mathbb{R}} h(u - t) N(dt), \int_{\mathbb{R}} h(v - t) N(dt) \right) = \int_{\mathbb{R}} |\widehat{h}(\nu)|^2 e^{2i\pi\nu(v-u)} \mu_N(d\nu).$$

In the univariate case ($m = 1$), the point process is represented by its sequence of event times $\{T_n\}_{n\in\mathbb{Z}}$, and

$$X(t) = \sum_{n\in\mathbb{Z}} h(t - T_n)$$

is the stochastic process obtained by passing the Dirac comb $\sum_n \delta(t - T_n)$ through the filter with impulse response h, the Bochner spectral measure of the output $\{X(t)\}_{t\in\mathbb{R}}$ is

$$\mu_X(d\nu) = |\widehat{h}(\nu)|^2 \mu_N(d\nu)$$

which indeed corresponds to the filtering formula for *bona fide* WSS stochastic processes if we assimilate the Dirac comb $\sum \delta(t - T_n)$ to a *bona fide* WSS stochastic process (which of course it is not) with spectral measure μ_N. The spectral measure μ_N is however *not* a finite measure as it would be for ordinary WSS stochastic processes.

One of the issues of interest is to determine $\mathcal{B}_N$ as large as possible, if not the largest. Here are two fundamental examples:

Example 5.2.7 (Regular grid, II) Consider the regular grid of Example 5.2.2. We shall argue that its Bartlett spectral measure is

$$\mu_N = \frac{1}{T_1^2 T_2^2} \sum_{(n_1,n_2)\neq(0,0)} \varepsilon_{(\frac{n_1}{T_1}, \frac{n_2}{T_2})}, \tag{5.31}$$

and that we can take for test function space

$$\mathcal{B}_N := \left\{ \varphi \, ; \, \varphi \in L^1_{\mathbb{C}}(\mathbb{R}^2) \text{ and } \sum_{n_1,n_2\in\mathbb{Z}} \left| \widehat{\varphi}\left(\frac{n_1}{T_1}, \frac{n_2}{T_2} \right) \right| < \infty \right\}.$$

Proof The conditions defining $\mathcal{B}_N$ guarantee that we can apply the weak Poisson formula. More precisely, both sides of the following equality

$$\sum_{n_1,n_2\in\mathbb{Z}} \varphi\left(u_1 + n_1T_1, u_2 + n_2T_2\right) = \frac{1}{T_1T_2} \sum_{n_1,n_2\in\mathbb{Z}} \widehat{\varphi}\left(\frac{n_1}{T_1}, \frac{n_2}{T_2}\right) e^{2i\pi\left(\frac{n_1}{T_1}u_1 + \frac{n_2}{T_2}u_2\right)}$$

$$(*)$$

are well-defined, and the equality holds for almost-all $(u_1, u_2) \in \mathbb{R}^{2m}$ with respect to the Lebesgue measure (this is an easy adaptation to the multidimensional case of Theorem 1.1.20). Note that condition

$$\sum_{n_1,n_2\in\mathbb{Z}} \left|\widehat{\varphi}\left(\frac{n_1}{T_1}, \frac{n_2}{T_2}\right)\right| < \infty$$

implies $(\ell^1_C(\mathbb{Z}^2) \subset \ell^2_C(\mathbb{Z}^2))$

$$\sum_{n_1,n_2\in\mathbb{Z}} \left|\widehat{\varphi}\left(\frac{n_1}{T_1}u_1, \frac{n_2}{T_2}u_2\right)\right|^2 < \infty.$$

By (*)

$$\int_{\mathbb{R}^2} \varphi(t)\, N(dt) = \sum_{n_1,n_2\in\mathbb{Z}} \varphi\left(U_1 + n_1T_1, U_2 + n_2T_2\right)$$

$$= \frac{1}{T_1T_2} \sum_{n_1,n_2\in\mathbb{Z}} \widehat{\varphi}\left(\frac{n_1}{T_1}, \frac{n_2}{T_2}\right) e^{2i\pi\left(\frac{n_1}{T_1}U_1 + \frac{n_2}{T_2}U_2\right)}$$

and therefore

$$E\left[\left|\int_{\mathbb{R}^2} \varphi(t)\, N(dt)\right|^2\right]$$

$$= \frac{1}{T_1^2T_2^2} E\left[\sum_{n_1,n_2\in\mathbb{Z}}\sum_{k_1,k_2\in\mathbb{Z}} \widehat{\varphi}\left(\frac{n_1}{T_1}, \frac{n_2}{T_2}\right) \widehat{\varphi}^*\left(\frac{k_1}{T_1}, \frac{k_2}{T_2}\right) e^{2i\pi\left(\frac{n_1-k_1}{T_1}U_1 + \frac{n_2-k_2}{T_2}U_2\right)}\right]$$

$$= \frac{1}{T_1^2T_2^2} \sum_{n_1,n_2\in\mathbb{Z}} \left|\widehat{\varphi}\left(\frac{n_1}{T_1}, \frac{n_2}{T_2}\right)\right|^2.$$

Also

$$E\left[\int_{\mathbb{R}^2} \varphi(t) N(dt)\right] = \sum_{n_1,n_2 \in \mathbb{Z}} E\left[\varphi(U_1 + n_1 T_1, U_2 + n_2 T_2)\right]$$

$$= \frac{1}{T_1 T_2} \int_0^{T_1} \int_0^{T_2} \varphi(u_1 + n_1 T_1, u_2 + n_2 T_2)\, du$$

$$= \frac{1}{T_1 T_2} \int_{\mathbb{R}^2} \varphi(t)\, dt = \frac{1}{T_1 T_2} \widehat{\varphi}(0,0).$$

Therefore

$$\operatorname{Var}\left(\int_{\mathbb{R}^2} \varphi(t) N(dt)\right) = \frac{1}{T_1^2 T_2^2} \sum_{n_1,n_2 \in \mathbb{Z}} \left|\widehat{\varphi}\left(\frac{n_1}{T_1}, \frac{n_2}{T_2}\right)\right|^2 - \frac{1}{T_1^2 T_2^2} |\widehat{\varphi}(0,0)|^2$$

$$= \frac{1}{T_1^2 T_2^2} \sum_{(n_1,n_2) \neq (0,0)} \left|\widehat{\varphi}\left(\frac{n_1}{T_1}, \frac{n_2}{T_2}\right)\right|^2$$

$$= \int_{\mathbb{R}} \int_{\mathbb{R}} |\widehat{\varphi}(\nu_1, \nu_2)|^2\, \mu_N(d\nu_1 \times d\nu_2),$$

where μ_N is given by (5.31). That $\mathcal{B}_N$ is a test function space (guaranteeing uniqueness of the spectral measure, as explained after Definition 5.2.3) is the object of Exercise 5.4.17. □

Example 5.2.8 (The Bartlett spectral measure of a WSS Cox process) Let N be a Cox point process (see Sect. 5.1.2) on $\mathbb{R}^m$ with stochastic intensity $\{\lambda(t)\}_{t \in \mathbb{R}^m}$. We suppose that $\{\lambda(t)\}_{t \in \mathbb{R}^m}$ is a WSS process with mean λ and Bochner spectral measure μ_λ. The Bartlett spectral measure of N is then

$$\mu_N(d\nu) = \mu_\lambda(d\nu) + \lambda d\nu,$$

and one may take for test function space $\mathcal{B}_N = L^1_{\mathbb{C}}(\mathbb{R}^m) \cap L^2_{\mathbb{C}}(\mathbb{R}^m)$. Even more, in this case $\mathcal{B}_N \equiv L^2_N(M_2)$

Proof Denote by $\mathcal{F}^\lambda_\infty$ the sigma-field generated by the stochastic process $\{\lambda(t)\}_{t \in \mathbb{R}}$. As $\varphi \in L^2_{\mathbb{C}}(\mathbb{R}^m)$, we have that

$$E\left[\int_{\mathbb{R}^m} |\varphi(t)|^2 \lambda(t) dt\right] = \lambda \int_{\mathbb{R}^m} |\varphi(t)|^2\, dt < \infty$$

and therefore, almost surely,

$$\int_{\mathbb{R}^m} |\varphi(t)|^2 \lambda(t)dt \; < \; \infty.$$

Similarly , since $\varphi \in L^1_{\mathbb{C}}(\mathbb{R}^m)$, almost surely

$$\int_{\mathbb{R}^m} |\varphi(t)| \lambda(t)dt \; < \; \infty.$$

By the covariance formulas for Poisson processes (Theorem 5.1.7),

$$E\left[\left(\int_{\mathbb{R}^m} |\varphi(t)| N(dt)\right)^2 \Big| \mathcal{F}^\lambda_\infty\right] = \int_{\mathbb{R}^m} |\varphi(t)|^2 \lambda(t)dt + \left(\int_{\mathbb{R}^m} |\varphi(t)| \lambda(t)dt\right)^2,$$

(5.32)

and therefore,

$$E\left[\left(\int_{\mathbb{R}^m} |\varphi(t)| N(dt)\right)^2\right] = \lambda \int_{\mathbb{R}^m} |\varphi(t)|^2 \, dt + E\left[\left(\int_{\mathbb{R}^m} |\varphi(t)| \lambda(t)dt\right)^2\right]$$

$$\geq \lambda \int_{\mathbb{R}^m} |\varphi(t)|^2 \, dt + \left(E\left[\int_{\mathbb{R}^m} |\varphi(t)| \lambda(t)dt\right]\right)^2$$

$$= \lambda \int_{\mathbb{R}^m} |\varphi(t)|^2 \, dt + \left(\lambda \int_{\mathbb{R}^m} |\varphi(t)| \, dt\right)^2.$$

Therefore if $\varphi \in L^2_N(M_2)$, then $\varphi \in L^1_{\mathbb{C}}(\mathbb{R}^m) \cap L^2_{\mathbb{C}}(\mathbb{R}^m)$. Also, from the first equality of the previous display and the fact that if $\varphi \in L^1_{\mathbb{C}}(\mathbb{R}^m)$, then $E\left[\left(\int_{\mathbb{R}^m} |\varphi(t)| \lambda(t)dt\right)^2\right] < \infty$ (this is because $\{\lambda(t)\}_{t\in\mathbb{R}}$ is WSS), we see that if $\varphi \in L^1_{\mathbb{C}}(\mathbb{R}^m) \cap L^2_{\mathbb{C}}(\mathbb{R}^m)$, then $\varphi \in L^2_N(M_2)$. Therefore in the case of Cox processes with a WSS conditional intensity, $\varphi \in L^1_{\mathbb{C}}(\mathbb{R}^m) \cap L^2_{\mathbb{C}}(\mathbb{R}^m) \Leftrightarrow \varphi \in L^2_N(M_2)$. By the *conditional variance formula*,

$$\mathrm{Var}\left(\int_{\mathbb{R}^m} \varphi(t)N(dt)\right) = E\left[\mathrm{Var}\left(\int_{\mathbb{R}^m} \varphi(t)N(dt)\Big| \mathcal{F}^\lambda_\infty\right)\right]$$

$$+ \mathrm{Var}\left(E\left[\int_{\mathbb{R}^m} \varphi(t)N(dt)\Big| \mathcal{F}^\lambda_\infty\right]\right)$$

$$= E\left[\int_{\mathbb{R}^m} \varphi(t)^2 \lambda(t)dt\right] + \text{Var}\left(\int_{\mathbb{R}^m} \varphi(t)\lambda(t)dt\right)$$

$$= \lambda \int_{\mathbb{R}^m} \varphi(t)^2 dt + \text{Var}\left(\int_{\mathbb{R}^m} \varphi(t)\lambda(t)dt\right).$$

By definition of the Bochner spectral measure, for $\varphi \in L^1_{\mathbb{C}}(\mathbb{R}^m)$,

$$\text{Var}\left(\int_{\mathbb{R}^m} \varphi(t)\lambda(t)dt\right) = \int_{\mathbb{R}^m} |\widehat{\varphi}(\nu)|^2 \mu_\lambda(d\nu),$$

and by the Plancherel-Parseval formula

$$\int_{\mathbb{R}^m} \varphi^2(t)dt = \int_{\mathbb{R}^m} |\widehat{\varphi}(\nu)|^2 d\nu.$$

Therefore

$$\text{Var}\left(\int_{\mathbb{R}^m} \varphi(t)N(dt)\right) = \int_{\mathbb{R}^m} |\widehat{\varphi}(\nu)|^2 (\mu_\lambda(d\nu) + \lambda d\nu),$$

which proves the announced result. $\qquad\square$

Existence and Uniqueness of the Spectral Measure

Let N be a wide-sense stationary point process with intensity $\lambda \in (0, \infty)$. In particular, if $f : \mathbb{R}^m \to \mathbb{R}$ is a bounded function with compact support, $E\left[N(f)^2\right] < \infty$. We already know that the covariance measure σ is locally finite. In fact,

Lemma 5.2.1 *Let N be as above. Then,*

$$\sup_{t \in \mathbb{R}^m} \sigma(K + t) < \infty \qquad (5.33)$$

for all compact sets $K \subset \mathbb{R}^m$.

Proof Select $\varphi : \mathbb{R}^m \to \mathbb{R}$, a non-negative function with compact support such that

$$\varphi * \check{\varphi} \geq 1_K.$$

(where $\check{\varphi}(t) = \varphi(-t)$). Then, τ_t being the shift operator $s \to s - t$, we have by (5.26),

$$\sigma\left(K+t\right) \leq \int_{\mathbb{R}^m} \left(\varphi * \breve{\varphi}\right)\left(s-t\right)\sigma\left(ds\right)$$

$$= \int_{\mathbb{R}^m} \left(\left(\varphi \circ \tau_t\right) * \breve{\varphi}\right)\left(s\right)\sigma\left(ds\right)$$

$$= E\left[N\left(\varphi \circ \tau_t\right) N\left(\varphi\right)\right].$$

Therefore, by Schwarz's inequality

$$\sigma\left(K+t\right) \leq E\left[N\left(\varphi\right)^2\right],$$

a finite quantity since φ is bounded with compact support and N is a second order point process. $\qquad\square$

Theorem 5.2.1 *Let N be as above. There exists a unique non-negative locally finite measure $\widehat{\sigma}$ on $(\mathbb{R}^m, \mathcal{B}(\mathbb{R}^m))$ such that, if f is continuous and both f and its Fourier transform $\widehat{f}$ are (as functions of the argument $x \in \mathbb{R}^m$) $O\left(1/||x||^{m+1}\right)$ as $||x|| \to \infty$, then*

$$\int_{\mathbb{R}^m} f\left(\nu\right)\widehat{\sigma}\left(d\nu\right) = \int_{\mathbb{R}^m} \widehat{f}\left(t\right)\sigma\left(dt\right), \tag{5.34}$$

and, if g satisfies the same conditions as f,

$$E\left[N\left(f\right)N\left(g\right)\right] = \lambda \int_{\mathbb{R}^m} \widehat{f}\left(\nu\right)\overset{\smallsmile}{\widehat{g}}\left(\nu\right)\widehat{\sigma}\left(d\nu\right). \tag{5.35}$$

Proof We do the case $\mathbb{R}^m = \mathbb{R}$ (the general case follows the same lines of argument).[5] We shall use the following fact: if $f : \mathbb{R} \to \mathbb{R}$ is a non-negative function that is non-increasing on $[1, \infty)$ and non-decreasing on $(-\infty, 0]$, and such that

$$\sum_{n\in\mathbb{Z}} f\left(n\right) < \infty,$$

we have that

$$\sup_{t\in\mathbb{R}} \int_{\mathbb{R}} f\left(t+s\right)\sigma\left(ds\right) < \infty.$$

Indeed,

---

[5] This proof is borrowed from [Neveu]. See also [Daley and Vere-Jones]. The latter has also the equivalent of the Cramér–Khinchin decomposition.

$$\sup_{t \in \mathbb{R}} \int_{\mathbb{R}} f(t+s)\,\sigma\,(ds) \le \sum_{n \ge 0} f(n)\,\sigma\,([n, n+1) - t) + \sum_{n \le -1} f(n)\,\sigma\,([n-1, n) - t)$$

$$\le \sum_{n \in \mathbb{Z}} f(n)\left(\sup_{l \in \mathbb{Z}, t \in \mathbb{R}} \sigma\,([l, l+1) - t)\right) \le C \sum_{n \in \mathbb{Z}} f(n)\,.$$

where $C < \infty$, by application of (5.33) to the compact set $[0, 1]$. Consider now for $a > 0$ the function

$$h_a(t) = \frac{1}{1 + 4\pi^2 a^2 t^2}$$

whose Fourier transform is

$$\widehat{h_a}(\nu) = \frac{1}{2a} e^{-a|\nu|}\,.$$

Observe that

$$h_{a/2} * h_{a/2} = \frac{2}{a} h_a \qquad\qquad (5.36)$$

since in the Fourier domain, $(\widehat{h_{a/2}})^2 = \frac{2}{a}\widehat{h_a}$. Also, observe that it is equivalent to say that f is $O\left(1/|t|^2\right)$ as $|t| \to \infty$ or that f is bounded by a multiple of h_1.

By Lemma 5.2.1, for all $a > 0$, $\sigma_a(dt) = h_a(t)\,\sigma(dt)$ is a finite measure on $(\mathbb{R}, \mathcal{B}(\mathbb{R}))$. Its Fourier transform

$$\widehat{\sigma_a}(\nu) = \int_{\mathbb{R}} e^{-2i\pi\nu t} \sigma_a(dt)$$

is therefore bounded and continuous. Using (5.36), we have

$$\frac{a}{2} E\left[\left|\int_{\mathbb{R}} h_{a/2}(t)\, e^{-2i\pi\nu t} N\,(dt)\right|^2\right] = \int_{\mathbb{R}} h_a(t)\, e^{-2i\pi\nu t} \sigma\,(dt) = \widehat{\sigma_a}(\nu)$$

and therefore $\widehat{\sigma_a}(\nu) \ge 0$ for all $a > 0$, $\nu \in \mathbb{R}$. Therefore, the measure

$$\widehat{\sigma_a}(\nu)\, h_1(\nu)\, d\nu$$

is a non-negative finite measure on $(\mathbb{R}, \mathcal{B}(\mathbb{R}))$. Its characteristic function (with t as argument) is

$$\int_{\mathbb{R}} e^{-2i\pi\nu t} \widehat{\sigma}_a(\nu) h_1(\nu) d\nu = \int_{\mathbb{R}} e^{-2i\pi\nu t} \left(\int_{\mathbb{R}} h_a(s) e^{-2i\pi\nu s} \sigma(ds) \right) h_1(\nu) d\nu$$

$$= \int_{\mathbb{R}} \left(\int_{\mathbb{R}} e^{-2i\pi\nu(t+s)} h_1(\nu) d\nu \right) h_a(s) \sigma(ds)$$

$$= \int_{\mathbb{R}} \widehat{h}_1(t+s) h_a(s) \sigma(ds). \tag{5.37}$$

(Fubini's theorem is applicable because $h_1(\nu) h_a(s)$ is $\ell \times \sigma$-integrable). Therefore

$$\lim_{a\downarrow 0} \int_{\mathbb{R}} e^{-2i\pi\nu t} \widehat{\sigma}_a(\nu) h_1(\nu) d\nu = \lim_{a\downarrow 0} \int_{\mathbb{R}} \widehat{h}_1(t+s) h_a(s) \sigma(ds)$$

$$= \int_{\mathbb{R}} \widehat{h}_1(t+s) \sigma(ds) < \infty$$

(by dominated convergence since $\widehat{h}_1(t+s) h_a(s) \le \widehat{h}_1(t+s)$, a σ-integrable function by Lemma 5.2.1). The limit is continuous in t (by dominated convergence again).

Therefore, by Paul Lévy's theorem for finite measures (Theorem 2.3.5), the measure $\widehat{\sigma}_a(\nu) h_1(\nu) d\nu$ converges weakly as $a \downarrow 0$ to a finite measure which we may always write as

$$h_1(\nu) \widehat{\sigma}(d\nu)$$

since $h_1(\nu) > 0$, for all $\nu \in \mathbb{R}$. Weak convergence means that for any bounded and continuous function $\varphi : \mathbb{R} \to \mathbb{R}$

$$\lim_{a\downarrow 0} \int_{\mathbb{R}} \varphi(\nu) h_1(\nu) \widehat{\sigma}_a(\nu) d\nu = \int_{\mathbb{R}} \varphi(\nu) h_1(\nu) \widehat{\sigma}(\nu) d\nu$$

In particular, if f is continuous and bounded by a multiple of h_1

$$\lim_{a\downarrow 0} \int_{\mathbb{R}} f(\nu) \widehat{\sigma}_a(\nu) d\nu = \int_{\mathbb{R}} f(\nu) \widehat{\sigma}(\nu) d\nu$$

On the other hand, by a computation similar to that in (5.37),

$$\int_{\mathbb{R}} f(\nu) \widehat{\sigma}_a(\nu) d\nu = \int_{\mathbb{R}} \widehat{f}(t) h_a(t) \sigma(dt),$$

and therefore

$$\lim_{a\downarrow 0} \int_{\mathbb{R}} f(\nu)\,\widehat{\sigma}_a(\nu)\,d\nu = \lim_{a\downarrow 0} \int_{\mathbb{R}} \widehat{f}(t)\,h_a(t)\,\sigma(dt) = \int_{\mathbb{R}} \widehat{f}(t)\,\sigma(dt)$$

if $\widehat{f}$ is bounded by a multiple of h_1 (this implying that $\int_{\mathbb{R}} |\widehat{f}|(t)\sigma(dt) < \infty$). This proves (5.34).

If f and g are $O\left(1/|t|^2\right)$ as well as their Fourier transform, so is $f * \check{g}$, and therefore

$$E\left[N(f)\,N(g)\right] = \int_{\mathbb{R}^m} (f * \check{g})\,d\sigma = \int_{\mathbb{R}^m} \widehat{f}\,\overset{\times}{\widehat{g}}\,d\widehat{\sigma}$$

It remains to prove uniqueness. In view of (5.34), we have to show that if two locally finite measures agree on all functions which are, together with their Fourier transforms, of the order of x^{-2} at infinity, these measures are identical. We leave this as an exercise (Hint: Approximate interval indicator functions by functions of the order of x^{-2} at infinity, together with their Fourier transforms). $\qquad\square$

Example 5.2.9 (Bartlett spectrum of renewal processes) We now compute the Bartlett spectrum of a stationary renewal point process with intensity λ. The renewal distribution F is assumed non-lattice (the support of the distribution of S_1 is not a regular grid, that is, not of the form $\{na\, ; n \in \mathbb{N}\}$ where $a > 0$). Define

$$\widehat{F}(2i\pi\nu) = \int_{\mathbb{R}_+} e^{-2i\pi\nu t}\,dF(t).$$

Note that, since F is non-lattice, $\widehat{F}(\nu) \neq 1$, except for $\nu = 0$. As we saw in Example 5.2.4, the covariance measure is given by the formula

$$\Gamma(dx) = \lambda R(dx) - \lambda^2 \ell(dx).$$

The renewal measure R is the sum of a Dirac measure at 0, ε_0, and of a symmetric measure R' given by, for $C \subset (0, \infty)$ by

$$R'(C) = \sum_{n \geq 1} F^{*n}(C),$$

where $F^{*n}(C) = \int_C dF^{*n}(x)$. We henceforth consider the case where R' admits a density u with respect to the Lebesgue measure and make the assumption that

$$\int_0^\infty |u(t) - \lambda|\,dt < \infty.$$

Then

$$\Gamma(C) = \lambda\varepsilon(C) + \lambda R'(C) - \lambda^2 \ell(C)$$
$$= \lambda\varepsilon(C) + \lambda \int_C (u(t) - \lambda)\, dt.$$

The Fourier transform of the Dirac measure at 0 is the constant 1. It therefore remains to compute the Fourier transform of the function $t \to u(t) - \lambda$.

Define

$$\widehat{g}(\nu) = \int_0^\infty e^{-2i\pi\nu t}(u(t) - \lambda)\, dt$$

Taking into account the symmetry of u, we have

$$\int_{\mathbb{R}} e^{-2i\pi\nu t}(u(t) - \lambda)\, dt = \widehat{g}(\nu) + \widehat{g}^*(\nu)$$

We will prove below that

$$\widehat{g}(\nu) = \frac{\widehat{F}(2i\pi\nu)}{1 - \widehat{F}(2i\pi\nu)} + \frac{1}{2i\pi\nu}. \tag{5.38}$$

Combining the above results, we see that the Bartlett spectrum of N admits the density

$$f_N(\nu) = \lambda\left(1 + Re\left(\frac{\widehat{F}(2i\pi\nu)}{1 - \widehat{F}(2i\pi\nu)}\right)\right)$$

Proof For $\theta > 0$, we have

$$\int_0^\infty e^{-(\theta+2i\pi\nu)t}(u(t) - \lambda)\, dt = \sum_{n\geq 1}\int_0^\infty e^{-(\theta+2i\pi\nu)t} F^{*n}(dt) - \int_0^\infty e^{-(\theta+2i\pi\nu)t}\lambda\, dt$$

$$= \sum_{n\geq 1}\widehat{F}(\theta + 2i\pi\nu)^n - \frac{\lambda}{\theta + 2i\pi\nu}$$

$$= \frac{\widehat{F}(\theta + 2i\pi\nu)}{1 - \widehat{F}(\theta + 2i\pi\nu)} - \frac{\lambda}{\theta + 2i\pi\nu}$$

Letting θ tend to 0 in the first term of the above equality, we obtain by dominated convergence $\int_0^\infty e^{-2i\pi\nu t}(u(t) - \lambda)dt$. For $\nu \neq 0$, letting θ tend to 0 in $\widehat{F}(\theta + 2i\pi\nu)$, we obtain $\widehat{F}(2i\pi\nu)$, again by dominated convergence. $\square$

5.2.2 Transformed Point Processes

A Universal Covariance Formula

Given the Bartlett spectral measure of a WSS point process, what is the spectral measure of WSS point processes or stochastic processes obtained by various transformations—such as thinning, translation, filtering and clustering—of the initial point process? This can be done using a universal formula from which all desired power spectral measures can be derived by specializing the abstract elements therein to the situation of interest.[6]

Let N a simple locally finite point process on $\mathbb{R}^m$ with points sequence $\{X_n\}_{n\geq 1}$. Let $\{Z_n\}_{n\geq 1}$ be an IID mark sequence taking its values in the measurable space $(K, \mathcal{K})$, with probability distribution Q, and independent of the basic point process N. Assume that N is a second-order stationary point process with Bartlett spectral measure μ_N and test function space $\mathcal{B}_N$. We shall obtain a formula for covariances of the type

$$\mathrm{cov}\left(\sum_{n\geq 1} \varphi(X_n, Z_n), \sum_{n\geq 1} \psi(X_n, Z_n)\right).$$

Let p be a positive integer, and let Z be a random element with values in $(K, \mathcal{K})$, with distribution Q. $L_{\mathbb{C}}^p(\ell \times Q)$ is the set of measurable functions $\varphi : \mathbb{R}^m \times K \to \mathbb{C}$ such that

$$\int_{\mathbb{R}^m} E\left[|\varphi(t, Z)|^p\right] dt < \infty.$$

In particular, $\varphi(t, Z) \in L_{\mathbb{C}}^p(P)$ for almost all $t \in \mathbb{R}$ with respect to the Lebesgue measure.

Let $\varphi : \mathbb{R}^m \times K \to \mathbb{R}$ be a measurable function such that

$$\varphi \in L_{\mathbb{C}}^1(\ell \times Q) \tag{5.39}$$

In particular, $\varphi(t, Z) \in L_{\mathbb{C}}^1(P)$ for almost all $t \in \mathbb{R}$ (with respect to the Lebesgue measure) and we can define for almost all t

$$\overline{\varphi}(t) := E\left[\varphi(t, Z)\right].$$

---

[6] For applications of the formulas of the present subsection to communications systems, see [Ridolfi].

It also follows from assumption (5.39) that $\overline{\varphi} \in L_{\mathbb{C}}^1(\mathbb{R}^m)$ and that for Q-almost all $z \in K$, $\varphi(\cdot, z) \in L_{\mathbb{C}}^1(\mathbb{R}^m)$. Therefore we can define the Fourier transforms of these two functions, and we will denote them by $\widehat{\overline{\varphi}}$ and $\widehat{\varphi}(\cdot, z)$ respectively. Suppose moreover that

$$\varphi \in L_{\mathbb{C}}^2(\ell \times Q). \tag{5.40}$$

Note that condition (5.40) implies that $\int_{\mathbb{R}^m} |E[\varphi(t, Z)]|^2 \, dt < \infty$, that is $\overline{\varphi} \in L_{\mathbb{C}}^2(\mathbb{R}^m)$, and for Q-almost all $z \in K$, $\varphi(\cdot, z) \in L_{\mathbb{C}}^2(\mathbb{R}^m)$. Observe that

$$\widehat{\overline{\varphi}}(\nu) = E[\widehat{\varphi}(\nu, Z)] := \widehat{\overline{\varphi}}(\nu).$$

Finally, suppose that

$$\overline{\varphi} \in \mathcal{B}_N. \tag{5.41}$$

Theorem 5.2.2 *Let N and $\{Z_n\}_{n \geq 1}$ be as above, and $\varphi, \psi : \mathbb{R}^m \times K \to \mathbb{R}$ satisfy conditions (5.39)–(5.41). Then*

$$\mathrm{cov}\left(\sum_{n \geq 1} \varphi(X_n, Z_n), \sum_{n \geq 1} \psi(X_n, Z_n)\right) = \int_{\mathbb{R}^m} E[\widehat{\varphi}(\nu, Z)] E[\widehat{\psi}(\nu, Z)^*] \mu_N(d\nu)$$

$$+ \lambda \int_{\mathbb{R}^m} \mathrm{cov}\left(\widehat{\varphi}(\nu, Z), \widehat{\psi}^*(\nu, Z)\right) d\nu, \tag{5.42}$$

where Z is any K-valued random variable with distribution Q.

Proof Formally:

$$E\left[\left(\sum_{n \geq 1} \varphi(X_n, Z_n)\right)\left(\sum_{n \geq 1} \psi(X_n, Z_n)\right)\right]$$

$$= E\left[\sum_{n,k \geq 1, n \neq k} \varphi(X_n, Z_n)\psi(X_k, Z_k)\right] + E\left[\sum_{n \geq 1} \varphi(X_n, Z_n)\psi^*(X_n, Z_n)\right]$$

$$= E\left[\sum_{n,k \geq 1, n \neq k} \overline{\varphi}(X_n)\overline{\psi}^*(X_k)\right] + E\left[\sum_{n \geq 1} \varphi(X_n, Z_n)\psi^*(X_n, Z_n)\right].$$

This is equal to

$$a - b + c := E\left[\left(\sum_{n\geq 1}\overline{\varphi}(X_n)\right)\left(\sum_{k\geq 1}\overline{\psi}^*(X_k)\right)\right]$$
$$- E\left[\sum_{n\geq 1}\overline{\varphi}(X_n)\overline{\psi}^*(X_n)\right]$$
$$+ E\left[\sum_{n\geq 1}\varphi(X_n, Z_n)\psi^*(X_n, Z_n)\right].$$

The above formal computations are justified because all three terms are, when φ and ψ are replaced by their absolute values, finite. This follows from Schwarz's inequality, and the facts that (for a and b) $\overline{\varphi}$ and $\overline{\psi}$ are in $L_N^2(M_2) \subseteq L_{\mathbb{C}}^1(\mathbb{R}^m) \cap L_{\mathbb{C}}^2(\mathbb{R}^m)$; and for c because of condition (5.40). Since $E\left[\sum_{n\geq 1}\varphi(X_n, Z_n)\right] = E\left[\sum_{n\geq 1}\overline{\varphi}(X_n)\right]$,

$$\mathrm{cov}\left(\sum_{n\geq 1}\varphi(X_n, Z_n), \sum_{n\geq 1}\psi(X_n, Z_n)\right) = \mathrm{cov}\left(\sum_{n\geq 1}\overline{\varphi}(X_n), \sum_{n\geq 1}\overline{\psi}(X_n)\right)$$
$$- E\left[\sum_{n\geq 1}\overline{\varphi}(X_n)\overline{\psi}^*(X_n)\right]$$
$$+ E\left[\sum_{n\geq 1}\varphi(X_n, Z)\psi^*(X_n, Z)\right].$$

Denote by $A - B + C$ the right-hand side of the above equation. By definition of the Bartlett spectral measure and by hypothesis (5.41),

$$A = \int_{\mathbb{R}^m} \widehat{\overline{\varphi}}(\nu)\widehat{\overline{\psi}}^*(\nu)\mu_N(d\nu).$$

By Campbell's formula,

$$B = \lambda \int_{\mathbb{R}^m} \overline{\varphi}(t)\overline{\psi}(t)^* dt, \quad C = \lambda \int_{\mathbb{R}^m} E\left[\varphi(t, Z)\psi(t, Z)^*\right] dt.$$

By the Plancherel–Parseval identity,

$$B = \lambda \int_{\mathbb{R}^m} \widehat{\overline{\varphi}}(\nu)\widehat{\overline{\psi}}(\nu)^* d\nu = \lambda \int_{\mathbb{R}^m} \widehat{\overline{\varphi}}(\nu)\widehat{\overline{\psi}}(\nu)^* d\nu = \lambda \int_{\mathbb{R}^m} E\left[\widehat{\varphi}(\nu, Z)\right] E\left[\widehat{\psi}(\nu, Z)^*\right] d\nu,$$

and

$$C = \lambda E \left[\int_{\mathbb{R}^m} \widehat{\varphi}(\nu, Z) \widehat{\psi}(\nu, Z)^* d\nu \right],$$

and the result (5.42) follows.   □

Formula (5.42) is rather versatile as we shall now see.

Thinning

Let N be a WSS stationary point process on $\mathbb{R}^m$ with spectral measure μ_N and test function space $\mathcal{B}_N$. Let $\{Z_n\}_{n \in \mathbb{Z}}$ be an IID sequence of random variables independent of N, taking their values in $\{0, 1\}$ and such that $P(Z_1 = 1) = \alpha$. Define the point process N_α by

$$N_\alpha(C) = \sum_{n \geq 1} Z_n 1_{\{X_n \in C\}}, \quad (C \in \mathcal{B}(\mathbb{R}^m)).$$

Thus, N_α is obtained by thinning N, a point of N being retained to be a point of N_α with probability α. The thinned point process N_α admits the spectral measure

$$\mu_{N_\alpha} := \alpha^2 \mu_N + \lambda \alpha (1 - \alpha) \ell$$

and that we may take $\mathcal{B}_{N_\alpha} := \varphi \in L^1_{\mathbb{C}}(\mathbb{R}^m) \cap L^2_{\mathbb{C}}(\mathbb{R}^m) \cap \mathcal{B}_N$ for test function space. To prove this we must show that for any function $\tilde{\varphi} \in \mathcal{B}_{N_\alpha}$,

$$\operatorname{Var} \int_{\mathbb{R}^m} \tilde{\varphi}(x) \, N_\alpha(dx) = \int_{\mathbb{R}^m} \left| \widehat{\tilde{\varphi}}(\nu) \right| \mu_{N_\alpha}(d\nu).$$

Now

$$\int_{\mathbb{R}^m} \tilde{\varphi}(x) \, N_\alpha(dx) = \sum_{n \geq 1} Z_n \tilde{\varphi}(X_n),$$

and therefore, the result follows from Theorem 5.2.1, by applying formula (5.42) with $\varphi(x, z) = \psi(x, z) = z\tilde{\varphi}(x)$ with $\tilde{\varphi} \in \mathcal{B}_N$ (Exercise 5.4.21).

Filtering (Shot Noise)

The power spectral measure of a shot noise with random impulse function based on a general WSS point process is obtained as a particular case of formula (5.2.1).

Corollary 5.2.1 *Consider the situation in Theorem 5.2.2 and let $h : \mathbb{R}^m \times K \to \mathbb{R}$ satisfy conditions (5.39)–(5.41). Define the* shot noise $\{X(t)\}_{t \in \mathbb{R}^m}$ *by*

$$X(t) = \sum_{n \geq 1} h(t - X_n, Z_n).$$

Then

$$E[X(t)] = \lambda \int_{\mathbb{R}^m} \overline{h}(t) dt,$$

and

$$\text{cov}(X(u), X(v)) = \int_{\mathbb{R}^m} e^{2i\pi\langle v, u-v \rangle} \mu_X(dv),$$

where (filtering formula for WSS *point processes)*

$$\mu_X(dv) = \left| E\left[\widehat{h}(v, Z)\right] \right|^2 \mu_N(dv) + \lambda \text{Var}\left(\widehat{h}(v, Z)\right) dv. \tag{5.43}$$

Proof It suffices to apply the fundamental isometry formula to $\varphi(t, z) = h(u - t, z)$, $\psi(t, z) = h(v - t, z)$ to obtain

$$\text{cov}(X(u), X(v)) = \int_{\mathbb{R}^m} \left| \overleftrightarrow{h}(v) \right|^2 e^{-2i\pi\langle v, u-v \rangle} \mu_N(dv)$$

$$+ \lambda \int_{\mathbb{R}^m} \text{Var}\left(\widehat{h}(v, Z)\right) e^{-2i\pi\langle v, u-v \rangle} dv. \qquad \square$$

Jittering

Knowing the Bartlett spectrum μ_N of a WSS point process N, what is the Bartlett spectrum $\mu_{\widetilde{N}}$ of the point process obtained by independent and identically distributed displacements of the points of N? The following result is the answer.

Corollary 5.2.2 *Consider the marked point process of Theorem 5.2.1, with $K = \mathbb{R}^m$. A point process $\widetilde{N}$ is defined by*

$$\widetilde{N} = \{X_n + Z_n, n \geq 1\}.$$

Then, calling λ the intensity of N, and μ_N the Bartlett spectral measure of $\widetilde{N}$,

$$\mu_{\widetilde{N}}(dv) = |\psi_Z(v)|^2 \mu_N(dv) + \lambda\left(1 - |\psi_Z(v)|^2\right) dv, \tag{5.44}$$

where

$$\psi_Z(\nu) = E\left[e^{2i\pi<\nu,Z>}\right]. \qquad (5.45)$$

is the characteristic function of the random displacements distributed as Q. We can take for test function space

$$B_{\tilde{N}} = \left\{\tilde{\varphi}; \ E\left[\tilde{\varphi}(t+Z)\right] \in B_N \text{ and } \tilde{\varphi} \in L^1_{\mathbb{C}}(\mathbb{R}^m) \cap L^2_{\mathbb{C}}(\mathbb{R}^m)\right\}. \qquad (5.46)$$

Proof Define $\varphi(t,z) = \tilde{\varphi}(t+z)$. Conditions (5.39) and (5.40) for the function φ are equivalent to conditions $\tilde{\varphi} \in L^1_{\mathbb{C}}(\mathbb{R}^m)$ and $\tilde{\varphi} \in L^2_{\mathbb{C}}(\mathbb{R}^m)$ respectively, since for any $p \geq 0$,

$$E\left[\int_{\mathbb{R}^m} |\varphi(t,Z)|^p \, dt\right] = \int_{\mathbb{R}^m} |\tilde{\varphi}(t)|^p \, dt.$$

Condition (5.41) for the function φ is satisfied by the *ad hoc* definition of $B_{\tilde{N}}$. We may therefore apply Theorem 5.2.1. We have

$$\widehat{\varphi}(\nu,z) = e^{2i\pi<\nu,z>}\widehat{\widetilde{\varphi}}(\nu),$$
$$\widehat{\overline{\varphi}}(\nu) = \overline{\widehat{\varphi}}(\nu) = \psi_Z(\nu)\widehat{\widetilde{\varphi}}(\nu),$$
$$\mathrm{cov}\left(\widehat{\varphi}(\nu,Z), \widehat{\varphi}(\nu,Z)^*\right) = \left(1 - |\psi_Z(\nu)|^2\right)\left|\widehat{\widetilde{\varphi}}(\nu)\right|^2.$$

Also,

$$\mathrm{Var}\left(\int_{\mathbb{R}^m} \tilde{\varphi}(t)\tilde{N}(dt)\right) = \mathrm{Var}\left(\sum_{n\geq 1} \tilde{\varphi}(T_n + Z_n)\right),$$

and therefore, applying formula (5.42),

$$\mathrm{Var}\left(\int_{\mathbb{R}^m} \tilde{\varphi}(t)\tilde{N}(dt)\right) = \int_{\mathbb{R}^m} \left|\widehat{\widetilde{\varphi}}(\nu)\right|^2 \mu_{\tilde{N}}(d\nu),$$

where $\mu_{\tilde{N}}$ is given by (5.44).   □

Example 5.2.10 (Jittered regular grid) We consider the case where N is the grid process of Example 5.2.7. We can take for test functions the functions in

$$B_{\tilde{N}} = \left\{\tilde{\varphi}; \ \sum_{n_1,n_2\in\mathbb{Z}} \left|\widehat{\widetilde{\varphi}}(\frac{n_1}{T_1}, \frac{n_2}{T_2})\right| < \infty \text{ and } \tilde{\varphi} \in L^1_{\mathbb{C}}(\mathbb{R}^2) \cap L^2_{\mathbb{C}}(\mathbb{R}^2)\right\}$$

Proof Indeed, observing that

$$\left| E\left[\widehat{\tilde{\varphi}(\cdot + Z)} \right](\nu) \right| = \left| \widehat{\tilde{\varphi}}(\nu) \right| ,$$

we see that condition $E\left[\tilde{\varphi}(t + Z) \right] \in \mathcal{B}_N$ is equivalent to $\sum_{n_1, n_2 \in \mathbb{Z}} \left| \widehat{\tilde{\varphi}}(\frac{n_1}{T_1}, \frac{n_2}{T_2}) \right| < \infty.$ □

When the grid is one-dimensional, this type of point processes arise in communications, where the information contained in the sequence $\{Z_n\}_{n \in \mathbb{Z}}$ is transmitted via the point process $\tilde{N}$, or a shot noise based on $\tilde{N}$.

Example 5.2.11 (Jittered Cox process) We consider the case where N is the Cox process of Example 5.2.8. We can take for test functions the functions in

$$B_{\tilde{N}} = \left\{ \tilde{\varphi} ; \; \tilde{\varphi} \in L^1_{\mathbb{C}}(\mathbb{R}^m) \cap L^2_{\mathbb{C}}(\mathbb{R}^m) \right\} .$$

Indeed condition $E\left[\tilde{\varphi}(t + Z) \right] \in \mathcal{B}_N$, that is, in this particular case, $E\left[\tilde{\varphi}(t + Z) \right] \in L^1_{\mathbb{C}}(\mathbb{R}^m) \cap L^2_{\mathbb{C}}(\mathbb{R}^m)$, is exactly $\tilde{\varphi} \in L^1_{\mathbb{C}}(\mathbb{R}^m) \cap L^2_{\mathbb{C}}(\mathbb{R}^m).$

Clustering

Let N be a wide-sense stationary point process on $\mathbb{R}^m$ with intensity $\lambda > 0$ represented by its point sequence $\{X_n\}_{n \geq 1}$, with power spectral measure μ_N and test function space $\mathcal{B}_N$. Let $\{Z_n\}_{n \geq 1}$ be an IID collection of point processes on $\mathbb{R}^m$, independent of N. Let Z be a point process on $\mathbb{R}^m$ with the same distribution as the common distribution of the Z_n's. Define

$$\psi_Z(\nu) := E\left[\int_{\mathbb{R}^m} e^{2i\pi\langle \nu, t \rangle} Z(dt) \right]$$

The function ψ_Z is well defined under the assumption

$$E\left[Z(\mathbb{R}^m) \right] < \infty.$$

(In particular, Z is almost surely a finite point process.) We now define two point process $\tilde{N}$ and $\widehat{N}$, on $\mathbb{R}^m$ by

$$\tilde{N}(C) = N(C) + \sum_{n \geq 1} Z_n(C - X_n),$$

$$\widehat{N}(C) = \sum_{n \geq 1} Z_n(C - X_n),$$

with the objective of computing their Bartlett spectral measures, starting with the
first one. Formally

$$
\mathrm{Var}\left(\int_{\mathbb{R}^m}\varphi(t)\tilde{N}(dt)\right) = \mathrm{Var}\left(\sum_{n\geq 1}\left\{\varphi(X_n) + \int_{\mathbb{R}^m}\varphi(X_n+s)\,Z_n(ds)\right\}\right)
$$

$$
= \mathrm{Var}\left(\sum_{n\geq 1}\varphi(X_n,Z_n)\right),
$$

where $\varphi(x,z) = \varphi(x) + \int_{\mathbb{R}^m}\varphi(x+s)\,z(ds)$. We have

$$
E\left[\varphi(x,Z)\right] = \varphi(x) + E\left[\int_{\mathbb{R}^m}\varphi(x+s)\,Z(ds)\right]
$$

$$
\widehat{\varphi}(\nu,z) = \widehat{\varphi}(\nu) + \int_{\mathbb{R}^m}\left(\int_{\mathbb{R}^m}\varphi(t+s)\,z(ds)\right)e^{-2i\pi\langle\nu,t\rangle}dt
$$

$$
= \widehat{\varphi}(\nu) + \int_{\mathbb{R}^m}\left(\int_{\mathbb{R}^m}\varphi(t+s)\,e^{-2i\pi\langle\nu,t\rangle}dt\right)z(ds)
$$

$$
= \widehat{\varphi}(\nu) + \int_{\mathbb{R}^m}\widehat{\varphi}(\nu)\,e^{2i\pi\langle\nu,s\rangle}z(ds) = \widehat{\varphi}(\nu)\left(1+\int_{\mathbb{R}^m}e^{2i\pi\langle\nu,s\rangle}z(ds)\right)
$$

Note that the exchange of order of integration is not a problem if z is a finite point
measure, in particular if z is replaced by its random version Z. Also

$$
\widehat{\widehat{\varphi}}(\nu) = \widehat{\varphi}(\nu)\left(1+\psi_Z(\nu)\right)
$$

Applying formally Theorem 5.2.1, we obtain

$$
\mathrm{Var}\left(\sum_{n\geq 1}\varphi(X_n,Z_n)\right) = \int_{\mathbb{R}^m}|\widehat{\varphi}(\nu)|^2\,|1+\psi_Z(\nu)|^2\,\mu_N(d\nu)
$$

$$
+ \lambda\int_{\mathbb{R}^m}|\widehat{\varphi}(\nu)|^2\,\mathrm{Var}\left(1+\int_{\mathbb{R}^m}e^{2i\pi\langle\nu,s\rangle}Z(ds)\right)d\nu.
$$

Observe that

$$\mathrm{Var}\left(1 + \int_{\mathbb{R}^m} e^{2i\pi\langle v,s \rangle} Z(ds)\right) = \mathrm{Var}\left(\int_{\mathbb{R}^m} e^{2i\pi\langle v,s \rangle} Z(ds)\right)$$

to obtain

$$\mathrm{Var}\left(\sum_{n \in \mathbb{N}} \varphi(X_n)\right) = \int_{\mathbb{R}^m} |\widehat{\varphi}(v)|^2 \, \mu_{\widetilde{N}}(dv)$$

where

$$\mu_{\widetilde{N}}(dv) = |1 + \psi_Z(v)|^2 \, \mu_N(dv) + \lambda \mathrm{Var}\left(\int_{\mathbb{R}^m} e^{2i\pi\langle v,s \rangle} Z(ds)\right) dv.$$

is the Bartlett spectral measure of $\widetilde{N}$. Similar computations lead to

$$\mu_{\widehat{N}}(dv) = |\psi_Z(v)|^2 \, \mu_N(dv) + \lambda \mathrm{Var}\left(\int_{\mathbb{R}^m} e^{2i\pi\langle v,s \rangle} Z(ds)\right) dv.$$

Multivariate Point Process

Attention will be restricted to bivariate point processes, the extension to an arbitrary number of dimensions being immediate. A bivariate point process on $\mathbb{R}^m$ is just a pair (N_1, N_2) of point processes on $\mathbb{R}^m$. It is called a second-order bivariate point process if both N_1 and N_2 are second-order point processes. It is called WSS if both N_1 and N_2 are WSS and if moreover they are *jointly* WSS, that is if for all bounded sets $A, B \in \mathcal{B}(\mathbb{R}^m)$ and all $t \in \mathbb{R}^m$

$$E[N_1(A+t)N_2(B+t)] = E[N_1(A)N_2(B)].$$

In this situation, we shall say that N_1 and N_2 admit the *cross-spectral measure* μ_{N_1,N_2} if the latter is a sigma-finite signed measure and if for all $\varphi_1 \in \mathcal{B}_{N_1}$, $\varphi_2 \in \mathcal{B}_{N_2}$

$$\mathrm{cov}\,(N_1(\varphi_1), N_2(\varphi_2)) = \int_{\mathbb{R}^m} \widehat{\varphi_1}(v)\widehat{\varphi_2}(v)^* \, \mu_{N_1,N_2}(dv).$$

Example 5.2.12 (Bivariate WSS Cox processes) Let N_1 and N_2 be WSS Cox processes on $\mathbb{R}$ whose respective stochastic intensities $\{\lambda_1(t)\}_{t \in \mathbb{R}}$ and $\{\lambda_2(t)\}_{t \in \mathbb{R}}$ are jointly stationary WSS stochastic processes with respective spectral measure μ_{λ_1} and μ_{λ_2}, and cross-spectral measure $\mu_{\lambda_1,\lambda_2}$. For $\varphi_1, \varphi_2 \in L^1_{\mathbb{C}}(\mathbb{R}) \cap L^2_{\mathbb{C}}(\mathbb{R})$ standard computations show that

$$E\left[N_1(\varphi_1)N_2(\varphi_2)^*\right] = E\left[E\left[N_1(\varphi_1)N_2(\varphi_2)^*|\mathcal{F}^{\lambda_1,\lambda_2}\right]\right]$$
$$= E\left[E\left[N_1(\varphi_1)|\mathcal{F}^{\lambda_1,\lambda_2}\right]E\left[N_1(\varphi_1)^*|\mathcal{F}^{\lambda_1,\lambda_2}\right]\right]$$
$$= E\left[\int_{\mathbb{R}} \varphi_1(t)\lambda_1(t)\,dt \times \int_{\mathbb{R}} \varphi_2(t)^*\lambda_2(t)\,dt\right]$$

and

$$E\left[N_1(\varphi_1)\right]E\left[N_2(\varphi_2)^*\right] = E\left[\int_{\mathbb{R}} \varphi_1(t)\lambda_1(t)\,dt\right] \times E\left[\int_{\mathbb{R}} \varphi_2(t)^*\lambda_2(t)\,dt\right].$$

Therefore

$$\mathrm{cov}\,(N_1(\varphi_1), N_2(\varphi_2)) = \mathrm{cov}\left(\int_{\mathbb{R}} \varphi_1(t)\lambda_1(t)\,dt, \int_{\mathbb{R}} \varphi_2(t)\lambda_2(t)\,dt\right)$$
$$= \int_{\mathbb{R}^m} \widehat{\varphi_1}(\nu)\widehat{\varphi_2}(\nu)^* \mu_{\lambda_1,\lambda_2}(d\nu).$$

Therefore

$$\mu_{N_1,N_2} = \mu_{\lambda_1,\lambda_2}.$$

The universal covariance formula can be used to compute cross-spectral measures between a point process and its transform by the above operations of thinning, clustering, translation or jittering, as well as between the transforms themselves. For instance:

Example 5.2.13 (Cross-spectral measure of a point process and its jittered version) The obtention of the cross-spectral measure of a WSS point process $N := N_1$ with its jittered version N_2 of Corollary 5.2.2 requires to compute

$$\mathrm{cov}\left(\sum_{n\in\mathbb{Z}} \varphi(X_n), \sum_{n\in\mathbb{Z}} \psi(X_n + Z_n)\right)$$

Formula (5.42) reduces in this case to

$$\mathrm{cov}\left(\sum_{n\geq1} \varphi(X_n), \sum_{n\geq1} \psi(X_n + Z_n)\right) = \int_{\mathbb{R}^m} \widehat{\varphi}(\nu)E\left[\widehat{\psi}(\nu+Z)^*\right]\mu_N(d\nu).$$

But

$$\widehat{\psi}(\nu + Z) = \int_{\mathbb{R}^m} \psi(t + Z)\, e^{-2i\pi\nu t}\, dt$$

$$= \int_{\mathbb{R}^m} \psi(t)\, e^{-2i\pi\nu(t-Z)}\, dt = \widehat{\psi}(\nu) E\left[e^{+2i\pi\nu Z}\right]$$

where the expectation is with respect to Z a random variable with the common probability distribution of the marks. Finally

$$\mathrm{cov}\left(\sum_{n\in\mathbb{Z}} \varphi(X_n), \sum_{n\in\mathbb{Z}} \psi(X_n + Z_n)\right) = \int_{\mathbb{R}^m} \widehat{\varphi}(\nu)\widehat{\psi}(\nu)^* E\left[e^{-2i\pi\nu Z}\right] \mu_N(d\nu),$$

and therefore

$$\mu_{N_1,N_2}(d\nu) = E\left[e^{-2i\pi\nu Z}\right] \mu_N(d\nu).$$

See Exercise 5.4.23 for more examples.

5.2.3 Pulse Amplitude Modulation

The following signal model, which mixes discrete time and continuous time, is omnipresent in communications theory. The so-called *pulse amplitude modulated* (PAM) signal is a stochastic process of the form

$$Z(t) = \sum_{n\in\mathbb{Z}} a_n g(t - nT),$$

where g is a complex valued function called the *base pulse* and $\{a_n\}_{n\in\mathbb{Z}}$ is a WSS discrete-time stochastic process, *uniformly bounded*, with mean m_a and covariance function C_a. In addition, we assume that

(H1) $\sum_{k\in\mathbb{Z}} |C_a(k)| < \infty$,

(H2) $g \in L^1_{\mathbb{C}}(\mathbb{R}) \cap L^2_{\mathbb{C}}(\mathbb{R})$

(H3) $\sum_{n\neq 0} |\widehat{g}(\frac{n}{T})| < \infty$ (which implies, since $\ell^1_{\mathbb{C}}(\mathbb{Z}) \subset \ell^2_{\mathbb{C}}(\mathbb{Z})$, that $\sum_{n\neq 0} |\widehat{g}(\frac{n}{T})|^2$).

Finally, we assume that

(H4) the absolute value time series $\{|a_n|\}_{n\in\mathbb{Z}}$ and the absolute value function $|g|$ satisfy the same conditions as the original quantities. (This is an *ad hoc* condition that justifies the computations below. It can be replaced by others in specific cases.)

In a communications theory context, a stochastic process such as $\{Z(t)\}_{t\in\mathbb{R}}$ is sometimes called a *cyclostationary* stochastic process with period T. It is not in general wide-sense stationary. However, if we define

$$Y(t) = Z(t - U) = \sum_{n \in \mathbb{Z}} a_n g(t - nT - U),$$

where U is a random variable uniformly distributed on $[0, T]$ and independent of $\{Z(t)\}_{t \in \mathbb{R}}$, the stochastic process $\{Y(t)\}_{t \in \mathbb{R}}$ is WSS.

Proof We have

$$E[Y(t)] = \sum_{n \in \mathbb{Z}} E[a_n] E[g(t - nT - U)]$$

$$= m_a \sum_{n \in \mathbb{Z}} \int_0^T g(t - nT - u) \frac{du}{T},$$

where we used the independence of U and $\{a_n\}_{n \in \mathbb{Z}}$, and the uniform distribution of U over $[0, T]$. Therefore

$$E[Y(t)] = \frac{m_a}{T} \int_{\mathbb{R}} g(t) \, dt.$$

We now calculate the covariance function. We begin by evaluating

$$E[Y(t + \tau) Y(t)^*] = E\left[\sum_{k \in \mathbb{Z}} \sum_{n \in \mathbb{Z}} a_n a_k^* g(t + \tau - nT - U) g^*(t - kT - U) \right]$$

$$= E\left[\sum_{n \in \mathbb{Z}} \sum_{j \in \mathbb{Z}} a_{n+j} a_n^* g(t + \tau - (n + j)T - U) g^*(t - nT - U) \right]$$

$$= \sum_{n \in \mathbb{Z}} \sum_{j \in \mathbb{Z}} E[a_{n+j} a_n^*] E[g(t + \tau - (n + j)T - U) g^*(t - nT - U)],$$

where we have taken into account the independence of U and $\{a_n\}_{n \in \mathbb{Z}}$. (Assumption H4 is there to allow the Fubini argument.) Since U is uniformly distributed on $[0, T]$ we have

$$E[g(t+\tau-(n+j)T-U)g^*(t-nT-U)] = \int_0^T g(t+\tau-(n+j)T-u)g^*(t-nT-u) \frac{du}{T}.$$

Furthermore,

$$E[a_{n+j} a_n^*] = C_a(j) + |m_a|^2.$$

Therefore

$$E[Y(t+\tau)Y(t)^*] = \sum_{j \in \mathbb{Z}} (C_a(j) + |m_a|^2)$$

$$\times \left\{ \sum_n \int_0^T g(t+\tau-(n+j)T-u)g^*(t-nT-u)\frac{du}{T} \right\}$$

$$= \sum_{j \in \mathbb{Z}} (C_a(j) + |m_a|^2) \int_{\mathbb{R}} g(t+\tau-jT-u)g^*(t-u)\frac{du}{T} .$$

Thus we see that $E[Y(t+\tau)Y(t)^*]$ does not depend upon t, and consequently $\{Y(t)\}$ is WSS. Its covariance function is

$$C_Y(\tau) = \sum_{j \in \mathbb{Z}} (C_a(j) + |m_a|^2) \int_{\mathbb{R}} g(\tau-jT-u)g^*(-u)\frac{du}{T}$$

$$- \frac{|m_a|^2}{T^2} \left| \int_{\mathbb{R}} g(s)\,ds \right|^2 . \qquad \square$$

We now compute the power spectral measure of $\{Y(t)\}_{t \in \mathbb{R}}$. For this, we write its covariance function as

$$C_Y(\tau) = C_1(\tau) + C_2(\tau),$$

where

$$C_1(\tau) = \sum_{j \in \mathbb{Z}} C_a(j) \int_{\mathbb{R}} g(\tau-jT-u)g^*(-u)\frac{du}{T}$$

and

$$C_2(\tau) = \frac{|m_a|^2}{T^2} \left\{ T \sum_{j \in \mathbb{Z}} \int_{\mathbb{R}} g(\tau-jT-u)g^*(-u)\,du - \left| \int_{\mathbb{R}} g(s)\,ds \right|^2 \right\}.$$

By the Plancherel–Parseval identity:

$$\int_{\mathbb{R}} g(\tau-jT-u)g^*(-u)\,du = \int_{\mathbb{R}} |\widehat{g}(\nu)|^2 e^{-2i\pi\nu jT} e^{2i\pi\nu\tau}\,d\nu,$$

and therefore, taking into account hypothesis (H1) which guarantees the absolute convergence and boundedness of $\sum C_a(j)e^{-2i\pi\nu jT}$,

$$C_1(\tau) = \int_{\mathbb{R}} \left(\frac{1}{T} \sum_{j \in \mathbb{Z}} C_a(j) e^{-2i\pi\nu jT} \right) |\widehat{g}(\nu)|^2 e^{2i\pi\nu\tau} \, d\nu,$$

that is to say,

$$C_1(\tau) = \int_{\mathbb{R}} f_1(\nu) e^{2i\pi\nu\tau} \, d\nu,$$

where

$$f_1(\nu) = \left(\frac{1}{T} \sum_{j \in \mathbb{Z}} C_a(j) e^{-2i\pi\nu jT} \right) |\widehat{g}(\nu)|^2$$

is the continuous part of the spectral measure of $\{Y(t)\}_{t\in\mathbb{R}}$. The discontinuous part corresponds to C_2. In order to calculate the latter we consider the function

$$p(t) = \int_{\mathbb{R}} g(t - u) g^*(-u) \, du$$

whose Fourier transform is $\widehat{p}(\nu) = |\widehat{g}(\nu)|^2$. By the weak Poisson summation formula (use Hypothesis H3), we obtain that, almost-everywhere,

$$T \sum_{j \in \mathbb{Z}} p(\tau - jT) = \sum_{n \in \mathbb{Z}} \widehat{p}\left(\frac{n}{T} \right) e^{2i\pi\frac{n}{T}\tau}.$$

Thus, taking into account the equality

$$\widehat{p}(0) = |\widehat{g}(0)|^2 = \left| \int_{\mathbb{R}} g(t) \, dt \right|^2$$

we have the expression

$$C_2(\tau) = \frac{|m_a|^2}{T^2} \sum_{n \neq 0} \left| \widehat{g}\left(\frac{n}{T} \right) \right|^2 e^{2i\pi(n/T)\tau},$$

that is to say

$$C_2(\tau) = \int_{\mathbb{R}} e^{2i\pi\nu\tau} \mu_2(d\nu),$$

where μ_2 is the discontinuous part of the spectrum with spectral lines at the frequencies n/T $(n \neq 0)$,

$$\mu_2(d\nu) = \sum_{n \neq 0} \frac{|m_a|^2}{T^2} \left| \widehat{g}\left(\frac{n}{T}\right) \right|^2 \varepsilon_{n/T}(d\nu),$$

where ε_a is the Dirac measure at the point a. Finally, we obtain the Bennett–Rice formula

$$\mu_Y(d\nu) = \left(\frac{1}{T} \sum_{j \in \mathbb{Z}} C_a(j) e^{-2i\pi\nu jT} \right) |\widehat{g}(\nu)|^2 \, d\nu + \sum_{n \neq 0} \frac{|m_a|^2}{T^2} \left| \widehat{g}\left(\frac{n}{T}\right) \right|^2 \varepsilon_{n/T}(d\nu).$$

The function

$$\frac{1}{2\pi} \sum_{j \in \mathbb{Z}} C_a(j) e^{-ij\omega} = f_a(\omega)$$

is the PSD of the second-order stationary time series $\{a_n\}_{n \in \mathbb{Z}}$, and it has period 2π. Therefore

$$\mu_Y(d\nu) = \frac{2\pi}{T} f_a(2\pi\nu T) |\widehat{g}(\nu)|^2 \, d\nu + \sum_{n \neq 0} \frac{|m_a|^2}{T^2} \left| \widehat{g}\left(\frac{n}{T}\right) \right|^2 \varepsilon_{n/T}(d\nu), \quad (5.47)$$

where $\nu \to f_a(2\pi\nu T)$ is a periodic function with period $1/T$.

If we use the symbolism of the Dirac pseudo-function, it is tempting to say that the process $\{Y(t)\}_{t \in \mathbb{R}}$ is obtained by filtering the *modulated impulse train*

$$X(t) = \sum_{n \in \mathbb{Z}} a_n \delta(t - nT - U) \tag{5.48}$$

by the filter with impulse response g. Thus if $\{X(t)\}_{t \in \mathbb{R}}$ were a genuine WSS stochastic process with the power spectral measure μ_X, the fundamental filtering formula would yield

$$\mu_Y(d\nu) = |\widehat{g}(\nu)|^2 \mu_X(d\nu)$$

On the other hand, in (5.47), we observe that the spectral measure μ_Y is obtained by multiplying the measure

$$\mu(d\nu) = \frac{2\pi}{T} f_a(2\pi\nu T) \, d\nu + \frac{|m_a|^2}{T^2} \sum_{n \neq 0} \varepsilon_{n/T}(d\nu)$$

by the function $|\widehat{g}|^2$. We are therefore naturally enclined to interpret this measure μ as the power spectral measure μ_X of the modulated impulse train. It will be noted that such μ is in general not a power spectral measure, as $\mu(\mathbb{R})$ may be infinite. But this is no surprise since the modulated impulse train is not a WSS stochastic process.

5.3 Random Sampling

5.3.1 Spectral Measure of the Sample Brush

Randomly sampling a continuous-time stochastic process $\{X(t)\}_{t\in\mathbb{R}}$—the *sampled process*—at the times $\{T_n\}_{n\in\mathbb{Z}}$ of a point process N—the *sampler*—yields a sequence of random samples —the *sample sequence*—$\{X(T_n)\}_{n\in\mathbb{Z}}$. The pseudo-process

$$Y(t) = \sum_{n\in\mathbb{Z}} X(T_n)\delta(t - T_n)$$

where $\delta(t)$ is the Dirac pseudo-function, is called the *sample comb*.

We now formulate random sampling in the spatial case. Here the sampled process is a WSS stochastic process $\{X(t)\}_{t\in\mathbb{R}^m}$ with mean m_X, covariance function C_X, power spectral measure μ_X and Cramér-Khinchin spectral decomposition Z_X:

$$X(t) = \int_{\mathbb{R}^m} e^{2i\pi\langle\nu,t\rangle} Z_X(d\nu) + m_X$$

where the integral thereof is a Doob integral. Recall that for all functions $g \in L^2_{\mathbb{C}}(\mu_X)$, the Doob integral $\int_{\mathbb{R}^m} g(\nu)Z_X(d\nu)$ is a well-defined element of $L^2_{\mathbb{C}}(P)$. Moreover

$$E\left[\left|\int_{\mathbb{R}^m} g(\nu)Z_X(d\nu)\right|^2\right] = \int_{\mathbb{R}^m} |g(\nu)|^2 \mu_X(d\nu). \tag{5.49}$$

The sampler is a simple WSS point process on $\mathbb{R}^m$ with intensity $\lambda \in (0, \infty)$, with point sequence $\{V_n\}_{n\geq 1}$. The sampled process and the sampler are assumed independent and wide-sense stationary. The *sample brush*

$$Y(t) = \sum_{n\geq 1} X(V_n)\delta(t - V_n) \tag{5.50}$$

is identified with the signed measure $\sum_{n\geq 1} X(V_n)\varepsilon_{V_n}$, where ε_a is the Dirac measure at a.

We define the *extended spectral measure of the sample brush* to be a locally finite measure μ_Y such that, for any $\varphi \in \mathcal{B}_Y$

$$\text{Var}\left(\int_{\mathbb{R}^m} \varphi(t) X(t) N(dt)\right) = \int_{\mathbb{R}^m} |\widehat{\varphi}(\nu)|^2 \mu_Y(d\nu), \tag{5.51}$$

where $\mathcal{B}_Y$ is a large enough vector space of functions, here also called the "*test functions*". By "large enough", we mean that there cannot be two different locally finite measures μ_Y verifying (5.51) for all $\varphi \in \mathcal{B}_Y$. Observe that,

$$\int_{\mathbb{R}^m} \varphi(t) Y(t) dt = \int_{\mathbb{R}^m} \varphi(t) \left(\sum_{n \geq 1} X(V_n) \delta(t - V_n)\right) dt$$

$$= \sum_{n \geq 1} \varphi(V_n) X(V_n) = \int_{\mathbb{R}^m} \varphi(t) X(t) N(dt),$$

so that equality (5.51) becomes, formally,

$$\text{Var}\left(\int_{\mathbb{R}^m} \varphi(t) Y(t) dt\right) = \int_{\mathbb{R}^m} |\widehat{\varphi}(\nu)|^2 \mu_Y(d\nu).$$

(This expression renders the analogy with the classical Bochner spectral measure more apparent.)

Let N be a wide-sense stationary simple point process on $\mathbb{R}^m$ with intensity $\lambda \in (0, \infty)$, Bartlett spectrum μ_N and test function space $\mathcal{B}_N$.

Theorem 5.3.1 *Suppose that the stochastic process $\{X(t)\}_{t \in \mathbb{R}^m}$ and the point process N are independent. Then, the sampled brush (5.50) admits the extended power spectral measure*

$$\mu_Y = \mu_N * \mu_X + \lambda^2 \mu_X + |m_X|^2 \mu_N. \tag{5.52}$$

If $\mathcal{B}_N$ is stable with respect to multiplications by complex exponential functions, we can take for test function space $\mathcal{B}_Y = \mathcal{B}_N$.

(This result is to be compared with that giving the spectral measure μ_Y of the product of two independent WSS stochastic processes, $Y(t) = Z(t)X(t)$: $\mu_Y = \mu_Z * \mu_X$.)

Proof We have

$$\int_{\mathbb{R}^m} \varphi(t) X(t) N(dt) = \int_{\mathbb{R}^m} \varphi(t) \left(\int_{\mathbb{R}^m} e^{2i\pi\langle \nu, t\rangle} Z_X(d\nu) + m_X\right) N(dt)$$

$$= \int_{\mathbb{R}^m} \left(\int_{\mathbb{R}^m} \varphi(t) \, e^{2i\pi\langle \nu,t\rangle} N(dt) \right) Z_X(d\nu)$$

$$+ m_X \int_{\mathbb{R}^m} \varphi(t) \, N(dt),$$

where we have formally exchanged the order of integration. Since the integrals with respect to N and Z_X are of a different nature (one is a usual infinite sum, the other is a Wiener integral), this exchange must be formally justified, which is done at the close of the proof. By the conditional variance formula, we have, denoting by $\mathcal{F}_N$ the sigma-field generated by the point process N,

$$\mathrm{Var}\left(\int_{\mathbb{R}^m} \varphi(t) \, X(t) N(dt) \right)$$

$$= E\left[\mathrm{Var}\left(\int_{\mathbb{R}^m} \left(\int_{\mathbb{R}^m} \varphi(t) \, e^{2i\pi\langle \nu,t\rangle} N(dt) \right) Z_X(d\nu) + m_X \int_{\mathbb{R}^m} \varphi(t) \, N(dt) \Big| \mathcal{F}^N \right) \right]$$

$$+ \mathrm{Var}\left(E\left[\int_{\mathbb{R}^m} \left(\int_{\mathbb{R}^m} \varphi(t) \, e^{2i\pi\langle \nu,t\rangle} N(dt) \right) Z_X(d\nu) + m_X \int_{\mathbb{R}^m} \varphi(t) \, N(dt) \Big| \mathcal{F}^N \right] \right).$$

Denote by $\alpha+\beta$ the right-hand side of this equality. Observe that, since $\varphi \in L^2(M_2)$,

$$\left| \int_{\mathbb{R}^m} \varphi(t) \, e^{2i\pi\langle \nu,t\rangle} N(dt) \right|^2 \le \left| \int_{\mathbb{R}^m} |\varphi(t)| \, N(dt) \right|^2 < \infty, \quad P-\text{a.s.} \qquad (5.53)$$

Using the fact that, when N is fixed, $m_X \int_{\mathbb{R}^m} \varphi(t) \, N(dt)$ is deterministic,

$$\alpha = E\left[\mathrm{Var}\left(\int_{\mathbb{R}^m} \left(\int_{\mathbb{R}^m} \varphi(t) \, e^{2i\pi\langle \nu,t\rangle} N(dt) \right) Z_X(d\nu) \Big| \mathcal{F}^N \right) \right]$$

$$= E\left[\int_{\mathbb{R}^m} \left| \int_{\mathbb{R}^m} \varphi(t) \, e^{2i\pi\langle \nu,t\rangle} N(dt) \right|^2 \mu_X(d\nu) \right] \quad \text{[by Eqs. (5.49) and (5.53)]}$$

$$= \int_{\mathbb{R}^m} E\left[\left| \int_{\mathbb{R}^m} \varphi(t) \, e^{2i\pi\langle \nu,t\rangle} N(dt) \right|^2 \right] \mu_X(d\nu)$$

$$= \int_{\mathbb{R}^m} \left(\operatorname{Var} \int_{\mathbb{R}^m} \varphi(t) e^{2i\pi\langle \nu,t\rangle} N(dt) + \left| E\left[\int_{\mathbb{R}^m} \varphi(t) e^{2i\pi\langle \nu,t\rangle} N(dt) \right] \right|^2 \right) \mu_X(d\nu)$$

$$= \int_{\mathbb{R}^m} \left(\int_{\mathbb{R}^m} |\widehat{\varphi}(x-\nu)|^2 \mu_N(dx) + \left| \int_{\mathbb{R}^m} \varphi(t) e^{2i\pi\langle \nu,t\rangle} \lambda dt \right|^2 \right) \mu_X(d\nu)$$

(hypothesis on $\mathcal{B}_N$)

$$= \int_{\mathbb{R}^m} \left(\int_{\mathbb{R}^m} |\widehat{\varphi}(x-\nu)|^2 \mu_N(dx) \right) \mu_X(d\nu) + \lambda^2 \int_{\mathbb{R}^m} |\widehat{\varphi}(-\nu)|^2 \mu_X(d\nu)$$

$$= \int_{\mathbb{R}^m} \left(\int_{\mathbb{R}^m} |\widehat{\varphi}(x+\nu)|^2 \mu_N(dx) \right) \mu_X(d\nu) + \lambda^2 \int_{\mathbb{R}^m} |\widehat{\varphi}(+\nu)|^2 \mu_X(d\nu)$$

$$= \int_{\mathbb{R}^m} |\widehat{\varphi}(\nu)|^2 (\mu_N * \mu_X)(d\nu) + \lambda^2 \int_{\mathbb{R}^m} |\widehat{\varphi}(\nu)|^2 \mu_X(d\nu).$$

Since $E\left[\int_{\mathbb{R}^m} \left(\int_{\mathbb{R}^m} \varphi(t) e^{2i\pi\langle \nu,t\rangle} N(dt) \right) Z_X(d\nu) \big| \mathcal{F}_\infty^N \right] = 0$,

$$\beta = \operatorname{Var}\left(m_X \int_{\mathbb{R}^m} \varphi(t) N(dt) \right) = |m_X|^2 \int_{\mathbb{R}^m} |\widehat{\varphi}(\nu)|^2 \mu_N(d\nu) \quad (\text{because } \varphi \in \mathcal{B}_N).$$

Finally,

$$\operatorname{Var}\left(\int_{\mathbb{R}^m} \varphi(t) Y(t) dt \right) = \int_{\mathbb{R}^m} |\widehat{\varphi}(\nu)|^2 \left(\mu_N * \mu_X + \lambda^2 \mu_X + |m_X|^2 \mu_N \right)(d\nu),$$

showing that $\{Y(t)\}_{t\in\mathbb{R}^m}$ admits an extended Bochner spectral measure given by Equation (5.52). $\qquad\square$

Lemma 5.3.1 *Let N be a simple locally bounded* WSS *point process defined on $\mathbb{R}^m$ and admitting a Bartlett spectral measure μ_N. Let M_2 be its second moment measure. Let $\{X(t)\}_{t\in\mathbb{R}^m}$ be a* WSS *random field with Cramér-Khinchin decomposition Z_X and power spectral measure μ_X. Then, for all $\varphi \in L^2(M_2)$*

$$\int_{\mathbb{R}^m} \varphi(t) X(t) N(dt) = \int_{\mathbb{R}^m} \left(\int_{\mathbb{R}^m} \varphi(t) e^{2i\pi<\nu,t>} N(dt) \right) Z_X(d\nu). \qquad (5.54)$$

Proof We do the proof in the univariate case. The multivariate case follows the same lines, with more cumbersome notation. The left-hand side of (5.54) is

$$A = \sum_{n \in \mathbb{Z}} \varphi(T_n) X(T_n) = \lim_{c \uparrow \infty} \sum_{n \in \mathbb{Z}} \varphi(T_n) X(T_n) 1_{[-c,+c]}(T_n) = \lim_{c \uparrow \infty} A(c)$$

where the limit is in $L^1_{\mathbb{C}}(P)$. Indeed

$$E[|A - A(c)|] \leq E\left[\int_{[-c,+c]} |\varphi(t) X(t)| N(dt)\right]$$

$$= \int_{[-c,+c]} |\varphi(t)| E[|X(t)|] \lambda dt \leq \lambda K \int_{[-c,+c]} |\varphi(t)| dt$$

where $K = \sup_t E[|X(t)|] < \infty$ $\left(E[|X(t)|] \leq E[|X(t)|^2]^{\frac{1}{2}} = E[|X(0)|^2]^{\frac{1}{2}}, \text{ by}\right.$

Schwarz's inequality$\left.\right)$. Therefore, since $\varphi \in L^1_{\mathbb{C}}(\mathbb{R})$, $\lim_{c \uparrow \infty} E[|A - A(c)|] = 0$.
The right-hand side is

$$B = \lim_{c \uparrow \infty} \int_{\mathbb{R}} \left(\int_{[-c,+c]} \varphi(t) e^{2i\pi<\nu,t>} N(dt)\right) Z_X(d\nu) = \lim_{c \uparrow \infty} B(c)$$

where the limit is in $L^2_{\mathbb{C}}(P)$. Indeed

$$E\left[|B - B(c)|^2\right] = E\left[\left|\int_{\mathbb{R}} \left(\int_{[-c,+c]} \varphi(t) e^{2i\pi\nu t} N(dt)\right) Z_X(d\nu)\right|^2\right]$$

$$= E\left[E\left[\left|\int_{\mathbb{R}} \left(\int_{[-c,+c]} \varphi(t) e^{2i\pi\nu t} N(dt)\right) Z_X(d\nu)\right|^2 \middle| \mathcal{F}^N_\infty\right]\right]$$

$$= E\left[\int_{\mathbb{R}} \left|\int_{[-c,+c]} \varphi(t) e^{2i\pi\nu t} N(dt)\right|^2 \mu_X(d\nu)\right].$$

Denote $\varphi_c(t) = \varphi(t) 1_{\overline{[-c,c]}}(t)$. Then

$$E\left[\int_{\mathbb{R}} \left|\int_{\mathbb{R}} \varphi_c(t) e^{2i\pi\nu t} N(dt)\right|^2 \mu_X(d\nu)\right] = \int_{\mathbb{R}} E\left[\left|\int_{\mathbb{R}} \varphi_c(t) e^{2i\pi\nu t} N(dt)\right|^2\right] \mu_X(d\nu).$$

But

$$
E\left[\left|\int_{\mathbb{R}} \varphi_c(t)\, e^{2i\pi\nu t}\, N(dt)\right|^2\right] \le E\left[\left(\int_{\mathbb{R}} |\varphi_c(t)|\, N(dt)\right)^2\right]
$$

$$
= \int_{\mathbb{R}\times\mathbb{R}} |\varphi_c(t)|\, |\varphi_c(s)|\, M_2\,(dt\times ds),
$$

a quantity that tends to 0 as $c \uparrow \infty$, by dominated convergence. Dominated convergence applied to the *finite* measure μ_X then yields the desired convergence in $L^2_{\mathbb{C}}(P)$. But

$$
A(c) = \sum_{n\in\mathbb{Z}} \varphi(T_n)\, X(T_n) 1_{[-c,+c]}
$$

$$
= \sum_{n\in\mathbb{Z}} \varphi(T_n)\left(\int_{\mathbb{R}} e^{2i\pi\nu T_n} Z_X(d\nu)\right) 1_{[-c,+c]}(T_n)
$$

$$
= \int_{\mathbb{R}} \left(\sum_{n\in\mathbb{Z}} \varphi(T_n)\, e^{2i\pi\nu T_n} 1_{[-c,+c]}(T_n)\right) Z_X(d\nu) = B(c),
$$

where we have used the fact that the sums involved are finite. Thus

$$
\lim_{c\uparrow\infty} A(c) = \begin{cases} A & \text{in } L^1_{\mathbb{C}}(P) \\ B & \text{in } L^2_{\mathbb{C}}(P) \end{cases}
$$

from which it follows that $A = B$, a.s. (use the fact that if a sequence of random variables converges in $L^1_{\mathbb{C}}(P)$ or $L^2_{\mathbb{C}}(P)$ to some random variable, one can extract a subsequence that converges almost surely to the same random variable). □

Example 5.3.1 (Cox sampling) Let N be a Cox process with a WSS intensity process $\{\lambda(t)\}_{t\in\mathbb{R}^m}$ with spectral measure μ_λ. Then, recalling from Example 5.2.8 the expression $\mu_N(d\nu) = \mu_\lambda(d\nu) + \lambda d\nu$, we obtain

$$
\mu_Y = \mu_\lambda * \mu_X + \lambda^2 \mu_X + |m_X|^2 \mu_\lambda + \lambda\left(\sigma_X^2 + |m_X|^2\right)\ell^m
$$

where ℓ^m is the Lebesgue measure. (In order to obtain this expression, we have used the fact that $\mu_X * \ell^m = \mu_X$. Indeed, for all $C \in \mathcal{B}(\mathbb{R}^m)$,

$$(\mu_X * \ell^m)(C) = \int_{\mathbb{R}^m} \ell^m(C - \nu)\,\mu_X(d\nu)$$

$$= \int_{\mathbb{R}^m} \ell^m(C)\,\mu_X(d\nu) = \ell^m(C)\mu_X(\mathbb{R}^m) = \ell^m(C)\sigma_X^2.)$$

Also from Example 5.2.8 and Theorem 5.3.1, $\mathcal{B}_N = L_\mathbb{C}^1(\mathbb{R}^m) \cap L_\mathbb{C}^2(\mathbb{R}^m) = \mathcal{B}_Y$. In the particular case where the sampler N is a homogeneous Poisson process and the sampled process is centered,

$$\mu_Y = \lambda^2 \mu_X + \lambda \sigma_X^2 \ell^m. \tag{5.55}$$

Example 5.3.2 (Regular sampling) Regular sampling refers to the case where the sampler is a grid (in this example: on the real line) with points separated by the distance T. We suppose for simplicity that the sampled process is centered. Applying Formula (5.52) with $\mu_N = \frac{1}{T}\sum_{n\in\mathbb{Z}\setminus\{0\}} \varepsilon_{\frac{n}{T}}$, and observing that $\mu_X * \varepsilon_{\frac{n}{T}}(C) = \mu_X(C - \frac{n}{T})$, we find in the case where there exists a power spectral density f_X for the sampled process the following expression for the spectral density of the sample comb

$$f_Y(\nu) = \left(\frac{1}{T}\right)^2 \sum_{n\in\mathbb{Z}} f_X\left(\nu - \frac{n}{T}\right).$$

Therefore, the spectral density of the sampled process can be entirely recovered from that of the sample comb provided the former is band-limited, with band width $2B < \frac{1}{T}$.

Example 5.3.3 (Poisson sampling, I) Poisson sampling refers to the case where the sampler is a homogeneous Poisson process with intensity $\lambda \in (0, \infty)$. Under the same conditions for the sampled process as in the previous example, the spectral measure of the sample comb admits the density

$$f_Y(\nu) = \lambda^2 f_X(\nu) + \lambda \sigma_X^2.$$

Therefore, whatever the sampling frequency $\nu_s = \lambda$, there is no aliasing, and the spectrum of the sampled process can be recovered from that of the sample comb.

5.3.2 Reconstruction from Samples

In the two previous examples, we were interested in the recovery of the spectral measure of the sampled process from that of the sample comb. We now consider the problem of reconstructing the process itself from its samples. More modestly, we approximate the sampled process by a filtered version of the sample comb:

$$\int_{\mathbb{R}^m} \varphi(t-s)\, Y(s)\, ds$$

where $\varphi \in L^1_{\mathbb{C}}\mathbb{R}^m) \cap L^2_{\mathbb{C}}\mathbb{R}^m)$. We also assume that $\mathcal{B}_N = L^1_{\mathbb{C}}\mathbb{R}^m) \cap L^2_{\mathbb{C}}\mathbb{R}^m)$. The difference between $X(t)$ and its approximation, that is, the *reconstruction error*, is measured by

$$\epsilon = E\left[\left|\int_{\mathbb{R}^m} \varphi(t-u)\, Y(u)\, du - X(t)\right|^2\right].$$

The following result is stated in the unidimensional case for simplicity in the writing.

Theorem 5.3.2 *The reconstruction error is, when the sampled process is centered:*

$$\epsilon = \int_{\mathbb{R}^m} |\lambda\widehat{\varphi}(\nu) - 1|^2\, \mu_X(d\nu) + \lambda \int_{\mathbb{R}^m} |\widehat{\varphi}(\nu)|^2\, (\mu_X * \mu_\lambda)(d\nu). \qquad (5.56)$$

Proof See Exercise 5.4.26.   □

We now give some examples of reconstruction error for different sampling schemes. We develop the computations for a generalized "band-limited" sampled process, filtered with a "base-band" filter of transmittance φ. More precisely, denoting by S the support (assumed of Lebesgue measure $2B < \infty$) of the spectral measure μ_X,

$$\widehat{\varphi}(\nu) = \frac{1}{\lambda}1_S(\nu)$$

where λ is the intensity of the sample comb.

Example 5.3.4 (Poisson sampling, II) The situation is that of Example 5.3.3. The spectral measure of the sample comb is then given by (5.55). The reconstruction error is

$$\epsilon = \int_{\mathbb{R}^m} |\lambda\widehat{\varphi}(\nu) - 1|^2\, \mu_X(d\nu) + \lambda\sigma_X^2 \int_{\mathbb{R}^m} |\widehat{\varphi}(\nu)|^2\, (d\nu). \qquad (5.57)$$

In the "classical" band-limited case described above, we have

$$\epsilon = \lambda\sigma_X^2 \int_{\mathbb{R}} |\widehat{\varphi}(\nu)|^2\, (d\nu) = \lambda\sigma_X^2 \int_{\mathbb{R}} \frac{1}{\lambda^2}1_S(\nu)\, d\nu,$$

that is

$$\epsilon = \sigma_X^2 \frac{2B}{\lambda}.$$

Therefore, sampling at the "Nyquist rate" $\lambda = 2B$ gives a very poor performance, not better than the estimate based on no observation at all.

This does not mean, however, that below the rate $\lambda = 2B$, there is no information (or in a sense as the result suggests "negative information") concerning the process itself contained in its samples. A better choice of a filter would indeed give a linear estimate with error less than σ_X^2. For instance, if we let $\widehat{\varphi}$ be real, we find for the reconstruction error

$$\epsilon = \int_{\mathbb{R}} \left[(\lambda \widehat{\varphi}(\nu) - 1)^2 f_X(\nu) + \lambda \sigma_X^2 \widehat{\varphi}(\nu)^2 \right] d\nu,$$

where it is assumed that the sampled process has the power spectral density f_X. A minimum occurs for

$$\widehat{\varphi}(\nu) = \frac{\lambda f_X(\nu)}{\lambda^2 f_X(\nu) + \lambda \sigma_X^2}$$

and then

$$\epsilon = \sigma_X^2 \left(1 - \int_{\mathbb{R}} \frac{\lambda \widetilde{f}_X(\nu)}{1 + \lambda \widetilde{f}_X(\nu)} \widetilde{f}_X(\nu) \, d\nu \right).$$

where

$$\widetilde{f}_X(\nu) := \frac{f_X(\nu)}{\int_{\mathbb{R}} f_X(\nu') \, d\nu'} = \frac{f_X(\nu)}{\sigma_X^2}$$

is the normalized power spectral density. This optimal filter requires the knowledge of the spectral density of the sampled process, which is in principle available from the samples, as shown in Example 5.3.3.

Example 5.3.5 (Regular sampling, II) When the sampled comb is derived by regular sampling as in Example 5.3.2, the reconstruction error is

$$\epsilon = \int_{\mathbb{R}} \left| \frac{1}{T} \widehat{\varphi}(\nu) - 1 \right|^2 \mu_X(d\nu) + \frac{1}{T} \int_{\mathbb{R}} |\widehat{\varphi}(\nu) - 1|^2 \, d\nu \qquad (5.58)$$

(where we took into account the fact that $\ell * \mu_X = \mu_X(\mathbb{R})\ell$). In the band-limited case, if we consider $T = 1/2B$, that is, $\lambda = 2B$, equation (5.58) gives an error equal to zero. Therefore, the process is perfectly reconstructed by

$$X(t) = \int_{\mathbb{R}} \varphi(t-s) X(s) N(ds) = \sum_{n \in \mathbb{Z}} X(T_n) \operatorname{sinc}(2B(t - T_n)),$$

which is the usual reconstruction formula.

Example 5.3.6 (Effects of jitter in regular Nyquist sampling) The reconstruction error from uniform samples in the presence of jitter is obtained plugging μ_Y given by (5.44) into the error formula (5.56). The previous example showed that within the "classical" sampling framework the process may be perfectly reconstructed. Now, in the presence of jitter this is not possible and the reconstruction error is given by

$$\epsilon = \frac{1}{2B} \left(\int_{-B}^{B} \sigma_X^2 \left(1 - \left(|\psi_Z|^2 * \widetilde{f}_X \right)(\nu) \right) d\nu \right), \qquad (5.59)$$

where $\widetilde{f}_X$ is the normalized power spectral density of the sampled process.

5.4 Exercises

Exercise 5.4.1 (*Point process integral*) Let N be a random measure on E and let $\varphi : (E, \mathcal{E}) \rightarrow (\overline{\mathbb{R}}, \overline{\mathcal{B}})$ be a non-negative measurable function. Prove that $N(\varphi)$ is a random variable.

Exercise 5.4.2 (*Properties of the renewal function*) The notation is that of Sect. 5.1.3. Recall that the trivial case where the interevent times are identically null is excluded.

1. Prove that $E[N(t)] < \infty$.
2. Show that $t \rightarrow E[N(t)]$ is a right-continuous function.
3. Denoting $N(\infty) = \lim_{t \to \infty} N(t)$, prove that

$$P(S_1 < \infty) < 1 \Leftrightarrow P(N(\infty) < \infty) = 1 \Leftrightarrow E[N(\infty)] < \infty.$$

Exercise 5.4.3 (*Laplace transform and independence*) Prove Theorem 5.1.4.

Exercise 5.4.4 (CF *of the forward recurrence time*) Let S_1 have the distribution F with $E[S_1] < \infty$. Prove that the characteristic function of a random variable S_0 with the CDF $F_0(x) := \frac{1}{E[S_1]} \int_0^x (1 - F(y)) \, dy$ is given by

$$E\left[e^{iuS_0}\right] = \frac{E\left[e^{iuS_1} - 1\right]}{iu \, E[S_1]}.$$

Exercise 5.4.5 (*Thinning*) Consider the operation of *thinning*, that is of randomly erasing points of a Poisson process. More precisely: Consider the situation as depicted in Theorem 5.1.6. Let I be an arbitrary index set and let $\{L_i\}_{i \in I}$ be a family of *disjoint* measurable sets of $\mathbb{R}^d$. Define for each $i \in I$ the simple point process N_i on $\mathbb{R}^m$ by

$$N_i(C) = \sum_{n \geq 1} 1_C(x) 1_{L_i}(Z_n).$$

Prove that the family N_i, $i \in I$, is an independent family of Poisson processes with respective mean measures ν_i, $i \in I$, where

$$\nu_i(dx) := Q(x, L_i)\nu(dx).$$

Exercise 5.4.6 (*On-off model*) Consider a semi-Markov process as described after Example 5.1.8, with the following specific features: $E = \{1, 2\}$, $p_{12} = p_{21} = 1$, $p_{11} = p_{22} = 0$, $G_{12} = G_1$ (mean μ_1) and $G_{21} = G_2$ (mean μ_2). Therefore, once in state 1, the process waits a random time of CDF G_1 whereupon it switches to state 2, where it waits a random time of CDF G_2 whereupon it switches back to state 1, etc. In other terms, we have an alternating renewal process: the interevent times sequence is independent, and it alternates between two distributions. Describe the stationary version of this process.

Exercise 5.4.7 (*Transport*) Consider the situation depicted in Theorem 5.1.6. Form a point process N^* on $\mathbb{R}^d$ by associating to a point $X_n \in \mathbb{R}^m$ a point $Z_n \in \mathbb{R}^d$. Formally

$$N^*(L) = \sum_{n \geq 1} 1_L(Z_n), \tag{5.60}$$

where $L \in \mathcal{B}^m$. (We then say that N^* is obtained by transporting N *via the stochastic kernel* $Q(x, \cdot)$. Prove that N^* is a Poisson process on $\mathbb{R}^d$ with mean measure ν^* given by

$$\nu^*(L) = \int_{\mathbb{R}^m} \nu(dx)Q(x, L).$$

Exercise 5.4.8 (*Translation*) Let N be a Poisson process on $\mathbb{R}^m$ with mean measure ν and let $\{V_n\}_{n \geq 1}$ be an IID family of independent random vectors of $\mathbb{R}^m$ with common distribution Q. Form the point process N^* on $\mathbb{R}^m$ by translating each point X_n of N by V_n. Formally,

$$N^*(C) = \sum_{n \geq 1} 1_C(X_n + V_n).$$

What is the nature of N^*? its mean measure?

Exercise 5.4.9 (*Jumping amplitude*) Let N be a homogeneous Poisson process on $\mathbb{R}_+$ and let $\{Z_n\}_{n \geq 0}$ be an IID sequence of integrable real random variables, centered, with finite variance σ^2, and independent of N. Define the stochastic process $\{X(t)\}_{t \geq 0}$ by $X(t) = Z_{N(t)}$, where $N(t) = N(0, t]$.

1) Show that it is a wide-sense stationary stochastic process and give its covariance function.
2) Give its power spectral density.

3) Assume that the Z_n's are absolutely continuous random variables. Compute $P(X(t_1) = X(t_2))$ and $P(X(t_1) > X(t_2))$.

Exercise 5.4.10 (*The original flip-flop*) Let N be a HPP on $\mathbb{R}_+$ with intensity λ. Define the (*telegraph* or *flip-flop*) process $\{X(t)\}_{t\geq0}$ with state space $E = \{+1, -1\}$ by

$$X(t) = Z(-1)^{N(t)},$$

where $X(0) = Z$ is an E-valued random variable independent of the counting process N. (Thus the telegraph process switches between -1 and $+1$ at each event of N.) The probability distribution of Z is arbitrary.

(1) Compute $P(X(t+s) = j|X(s) = i)$ for all t, $s \geq 0$ and all i, $j \in E$.
(2) Give for all $i \in E$ the limit of $P(X(t) = i)$ as t tends to ∞.
(3) Show that when $P(Z = 1) = \frac{1}{2}$, the process is a stationary process and give its power spectral measure.

Exercise 5.4.11 (*Another flip-flop*) Let N be a HPP on $\mathbb{R}$ with intensity $\lambda > 0$. Define for all $t \in \mathbb{R}$

$$X(t) = (-1)^{N((t,t+a])}$$

(1) Show that $\{X(t)\}_{t\in\mathbb{R}}$ is a WSS stochastic process;
(2) Compute its power spectral density;
(3) Give the best affine estimate of $X(t+\tau)$ in the terms of $X(t)$, that is, find α, β minimizing

$$E\left[|X(t+\tau) - (\alpha + \beta X(t))|^2\right], \quad \text{when } \tau > 0$$

Exercise 5.4.12 (*Jumping phase*) Let for each $t \in \mathbb{R}_+$

$$X(t) := e^{i\Phi_{N(t)}}$$

where $\{N(t)\}_{t\geq0}$ is the counting process of a homogeneous Poisson process on $\mathbb{R}_+$ with intensity $\lambda > 0$, and $\{\Phi_n\}_{n\geq0}$ is an IID sequence of random variables uniformly distributed on $[0, 2\pi]$, and independent of this Poisson process. Show that $\{X(t)\}_{t\geq0}$ is a wide-sense stationary process, give its covariance function $C_X(\tau)$ and its power spectral measure.

Exercise 5.4.13 (*Shot noises*) Let N_1, N_2 and N_3 be three independent homogeneous Poisson processes on $\mathbb{R}$ with respective positive intensities θ_1, θ_2 and θ_3. Let $\{X_1(t)\}_{t\in\mathbb{R}}$ be the shot noise constructed on $N_1 + N_3$ it with an impulse function $h : \mathbb{R} \to \mathbb{R}$ that is bounded and with compact support (null outside a finite interval). Let $\{X_2(t)\}_{t\in\mathbb{R}}$ be the shot noise constructed on $N_2 + N_3$ it with the same impulse function h. Compute the power spectral density of the wide-sense stationary process $\{X(t)\}_{t\in\mathbb{R}}$, where $X(t) = X_1(t) + X_2(t)$.

Exercise 5.4.14 (*Security distance*) Consider two independent Poisson processes N_1 and N_2 on $\mathbb{R}^m$ with respective mean measures ν_1 and ν_2. Assume that $\nu_i(\mathbb{R}^m) < \infty$, $i = 1, 2$. Compute the average number of elements in N_1 that see no point of N_2 within distance a.

Exercise 5.4.15 (*Micropulses and fractal Brownian motion*) Let $\bar{N}_\varepsilon$ be a Poisson process on $\mathbb{R} \times \mathbb{R}_+$ with the mean measure $\nu(dt \times dz) = \frac{1}{2\varepsilon^2} z^{-1-\theta} dt \times dz$, where $0 < \theta < 1$ and $\varepsilon > 0$. For all $t \geq 0$, let $S^+_{0,t} = \{(s, z) : 0 < s < t, t - s < z\}$ and $S^-_{0,t} = \{(s, z) : -\infty < s < 0, -s < z < t - s\}$, and define[7]

$$X_\varepsilon(t) = \varepsilon \left\{ \bar{N}_\varepsilon(S^+_{0,t}) - \bar{N}_\varepsilon(S^-_{0,t}) \right\}.$$

(1) Show that $X_\varepsilon(t)$ is well defined for all $t \geq 0$.
(2) Compute for all $0 \leq t_1 \leq t_2 \ldots \leq t_n$ the characteristic function of $(X_\varepsilon(t_1), \ldots, X_\varepsilon(t_n))$.
(3) Show that for all $0 \leq t_1 \leq t_2 \cdots \leq t_n$, $(X_\varepsilon(t_1), \ldots, X_\varepsilon(t_n))$ converges in distribution to $(B_H(t_1), \ldots, B_H(t_n))$ as $\varepsilon \downarrow 0$, where $\{B_H(t)\}_{t \geq 0}$ is a fractal Brownian motion (fBm) with Hurst parameter $H = \frac{1-\theta}{2}$ and variance $E[B_H(1)^2] = \theta^{-1}(1 - \theta)^{-1}$. Recall that $\{B_H(t)\}_{t \geq 0}$ is called a fBm with Hurst parameter $H, 0 < H < \frac{1}{2}$, if it is a centered Gaussian process such that $B_H(0) = 0$ with covariance function

$$E[B_H(t)B_H(s)] = \frac{1}{2} \left(|s|^{2H} + |t|^{2H} - |s - t|^{2H} \right) E\left[B_H(1)^2\right].$$

Exercise 5.4.16 (*Stationary rectangular grid*) Consider the point process on $\mathbb{R}^2$ whose points form a regular (T_1, T_2)-grid on $\mathbb{R}^2$ with random origin, that is

$$N = \left\{ (n_1 T_1 + U_1, n_2 T_1 + U_2), (n_1, n_2) \in \mathbb{Z}^2 \right\}$$

where $T_1 > 0$, $T_2 > 0$, and U_1, U_2 are independent random variables uniformly distributed on $[0, T_1]$, $[0, T_2]$ respectively. Prove that this point process is strictly stationary with average intensity $\lambda = 1/(T_1 T_2)$.

Exercise 5.4.17 (*Test function space of the rectangular grid*) In Example 5.2.7, show that

$$\mathcal{B}_N := \left\{ \varphi; \varphi \in L^1_{\mathbb{C}}(\mathbb{R}^2) \text{ and } \sum_{n_1, n_2 \in \mathbb{Z}} \left| \widehat{\varphi} \left(\frac{n_1}{T_1}, \frac{n_2}{T_2} \right) \right| < \infty \right\}.$$

---

[7] Cioczek-Georges, R., Mandelbrot, B.B.: "A class of micropulses and antipersistent fractal Brownian motion", *Stochastic Processes and their Applications*, 60, pp. 1–18, (1995).

is a test function space in the sense that if the locally finite measures μ_1 and μ_2 are such that

$$\int_{\mathbb{R}^m} |\widehat{\varphi}(\nu)|^2 \, \mu_1(d\nu) = \int_{\mathbb{R}^m} |\widehat{\varphi}(\nu)|^2 \, \mu_2(d\nu)$$

for all $\varphi \in \mathcal{B}_N$, then $\mu_1 \equiv \mu_2$.

Exercise 5.4.18 (*Second moment measure*) For a second-order point process N on $\mathbb{R}^m$, prove that the second moment measure M_2 on $\mathbb{R}^m \times \mathbb{R}^m$ is well and uniquely defined by the formula $M_2(A \times B) = E[N(A)N(B)]$ for all sets A, $B \in \mathcal{B}^m$, and prove formula (5.24).

Exercise 5.4.19 (*Second moment measure of a Poisson process*) Let N be a Poisson process on $\mathbb{R}^m$ with mean measure ν.

1. Show that in order for N to be a second-order point process, it is necessary and sufficient that the mean measure be locally finite. Show that in this case, the second moment measure M_2 is given by

$$M_2(A \times B) = \nu(A \cap B) - \nu(A)\nu(B).$$

2. Show that that for all φ, $\psi \in L^1_{\mathbb{R}^m}(\nu) \cap L^2_{\mathbb{R}^m}(\nu)$ where ν is the ,

$$E\left[N(\varphi)N(\psi)^*\right] = \int_{\mathbb{R}^m} \varphi(t)\psi^*(t)\,\nu(dt) + \int_{\mathbb{R}^m} \varphi(t)\,\nu(dt) \int_{\mathbb{R}^m} \psi^*(t)\,\nu(dt).$$

3. Prove that in this case, $L^2_N(M_2) = L^1_{\mathbb{R}^m}(\nu) \cap L^2_{\mathbb{R}^m}(\nu)$.

Exercise 5.4.20 (*An expression of wide-sense stationarity*) Let N be a wide-sense stationary point process on $\mathbb{R}$. Prove that for all non-negative measurable functions $\phi, \psi : \mathbb{R} \to \mathbb{R}$, the quantity

$$E\left[\left(\int_{\mathbb{R}} \phi(\tau + t)\,N(dt)\right)\left(\int_{\mathbb{R}} \psi(\tau + t)\,N(dt)\right)\right]$$

is independent of $\tau \in \mathbb{R}$.

Exercise 5.4.21 (*Spectrum of a thinned WSS point process*) Prove in detail the assertions in Example 5.2.2.

Exercise 5.4.22 (*A WSS cluster point process*) Consider the point process N constructed as follows. Let $T > 0$ and let U be a random variable uniformly distributed on $[0, T]$. Let L be a fixed positive integer. Let $\{V_{n,\ell}\}_{n \in \mathbb{Z}, 1 \leq \ell \leq L}$ be an IID sequence of random variables uniformly distributed on $[0, T]$, independent of U. The collection of points of N is $\{nT + V_{n,\ell}\}_{n \in \mathbb{Z}, 1 \leq \ell \leq L}$. Show that N is stationary, and give its Bartlett spectral measure (with an associated test function space).

Exercise 5.4.23 (*Cross-spectra*) Compute the cross-spectral measures between a point process and its transform by the operations of thinning, clustering, translation or jittering, as well as between the transforms themselves.

Exercise 5.4.24 (*High rate Poisson sampling*) In Example 5.3.4 show that the error ϵ tends to 0 as the sampling rate λ tends to infinity.

Exercise 5.4.25 (*Process dependent sampling*) Consider the general situation of random sampling, as in Sect. 5.3. We suppose now that the sampling rate depends on the process. More precisely, the model for the sampler is now a Cox process on $\mathbb{R}^d$ with the conditional intensity of the form

$$\lambda(t) = \lambda(t, \mathbf{X}) \,,$$

where $\mathbf{X}$ represents a stationary stochastic process $\{X(t)\}_{t \in \mathbb{R}}$. (For instance, in the univariate case, $\lambda(t) = |X(t)|^2$, $\lambda(t) = |\dot{X}(t)|^2$ where $\dot{X}$ is the derivative at t of $t \to X(t)$. More complicated functionals can be considered.)

Assume that $E\left[X(t)^2 \lambda(t, X)^2\right] < \infty$ for all $t \in \mathbb{R}^m$, and that $\{\lambda(t)\}_{t \in \mathbb{R}^m}$ is a locally integrable process. Let μ_Z be the power spectrum of the stationary process

$$Z(t) = X(t)\lambda(t)$$

assumed stationary (as is the case in the examples above). Show that

$$\mu_Y(d\nu) = \mu_Z(d\nu) + \overline{X^2\lambda}d\nu$$

where $\overline{X^2\lambda} := E\left[X(t)^2\lambda(t)\right]$ (independent of t).

Exercise 5.4.26 (*Reconstruction error*) Prove Theorem 5.3.2.

Appendix A

A.1 Review of Integration

In 1933, the Russian mathematician Kolmogorov has shown that a natural framework for probability theory was that of measure and integration theory. In fact, the return from a study of abstract integration theory is considerable for many other reasons. In particular, the integral of a function f with respect to an abstract measure μ contains a variety of mathematical objects besides the usual Lebesgue integral on $\mathbb{R}^d$

$$\int_{\mathbb{R}^d} f(x_1, \ldots, x_d) \, dx_1 \ldots dx_d.$$

For instance, an infinite sum

$$\sum_{n \in \mathbb{Z}} f(n)$$

can also be viewed as a Lebesgue integral with respect to the counting measure on $\mathbb{Z}$. The Stieltjes–Lebesgue integral

$$\int_{\mathbb{R}} f(x) \, dF(x)$$

with respect to a function F of bounded variation is again a special case of the Lebesgue integral. Most importantly in this text, the expectation

$$E[Z]$$

of a random variable Z will be recognized as an abstract integral.

© Springer International Publishing Switzerland 2014
P. Brémaud, *Fourier Analysis and Stochastic Processes*, Universitext,
DOI 10.1007/978-3-319-09590-5

The reader has a working knowledge of the Riemann integral. This type of integral is sufficient for many purposes, but it has a few weak points when compared to the Lebesgue integral. For instance:

1. The class of Riemann-integrable functions is too narrow. As a matter of fact, some functions have an obvious integral, and Riemann's integration theory denies it, whereas Lebesgue's theory recognizes it.
2. The stability properties under the limit operation of the functions that admit a Riemann integral are too weak. Indeed, it often happens that such limits do not have a Riemann integral, whereas the limit, for instance, of non-negative functions for which the Lebesgue integral is well defined also admits a well-defined Lebesgue integral.
3. The Riemann integral is defined with respect to the specific measures on euclidean spaces (length, area, volume, etc.), whereas the Lebesgue integral can be defined with respect to a general *abstract measure*, a probability for instance.

A.1.1 Sigma-Fields and Measurability

Denote by $\mathcal{P}(X)$ the collection of all subsets of an arbitrary set X.

Definition A.1 A family $\mathcal{X} \subseteq \mathcal{P}(X)$ of subsets of X is called a *sigma-field* on X if:

(α) $X \in \mathcal{X}$;
(β) $A \in \mathcal{X} \implies \bar{A} \in \mathcal{X}$;
(γ) $A_n \in \mathcal{X}$ for all $n \in \mathbb{N} \implies \cup_{n=0}^{\infty} A_n \in \mathcal{X}$.
One then says that $(X, \mathcal{X})$ is a *measurable space* .

In words, a sigma-field is a collection of subsets of X that contains X and is stable under complementation and countable union. It is also stable under countable intersection (exercise).

Two extremal examples of sigma-fields on X are the *gross* sigma-field $\mathcal{X} = \{\varnothing, X\}$, and the *trivial* sigma-field $\mathcal{X} = \mathcal{P}(X)$.

Definition A.2 The sigma-field *generated* by a non-empty collection of subsets $\mathcal{C} \subseteq \mathcal{P}(X)$ is, by definition, the smallest sigma-field on X containing all the sets in $\mathcal{C}$. It is denoted by $\sigma(\mathcal{C})$.

Recall that a set $O \in \mathbb{R}^n$ is called *open* if for any $x \in O$, one can find a non-empty open ball centered on x and contained in O.

Definition A.3 Let $X = \mathbb{R}^n$ be endowed with the Euclidean topology. The Borel sigma-field $\mathcal{B}(\mathbb{R}^n)$, also denoted by $\mathcal{B}^n$, is, by definition, the sigma-field generated by the open sets.

The next result gives a more convenient way of defining the Borel sigma-field.

Theorem A.1 $\mathcal{B}(\mathbb{R}^n)$ *is also generated by the collection* $\mathcal{C}$ *of all rectangles of the type* $\prod_{i=1}^{n}(-\infty, a_i]$, *where* $a_i \in \mathbb{Q}$ *for all* $i \in \{1, \ldots, n\}$.

For $n = 1$ one writes $\mathcal{B}(\mathbb{R}) = \mathcal{B}$. For $I = \prod_{j=-1}^{n} I_j$, where I_j is a general interval of $\mathbb{R}$ (I is then called a *rectangle* of $\mathbb{R}^n$), the Borel sigma-field $\mathcal{B}(I)$ on I consists of all the Borel sets contained in I. By definition, $\mathcal{B}(\overline{\mathbb{R}})$, or $\overline{\mathcal{B}}$, is the sigma-field on $\overline{\mathbb{R}}$ generated by the intervals of type $(-\infty, a]$, $a \in \mathbb{R}$.

Definition A.4 Let $(X, \mathcal{X})$ and $(E, \mathcal{E})$ be two measurable spaces. A function $f : X \to E$ is said to be a *measurable function* with respect to $\mathcal{X}$ and $\mathcal{E}$ if

$$f^{-1}(C) \in \mathcal{X} \quad \text{for all } C \in \mathcal{E}.$$

This is denoted in various ways, for instance:

$$f : (X, \mathcal{X}) \to (E, \mathcal{E}) \quad \text{or} \quad f \in \mathcal{E}/\mathcal{X}.$$

A function $f : (X, \mathcal{X}) \to (\overline{\mathbb{R}}, \overline{\mathcal{B}})$, where $(X, \mathcal{X})$ is an arbitrary measurable space, is called an *extended* measurable function. As for functions $f : (X, \mathcal{X}) \to (\mathbb{R}, \mathcal{B}(\mathbb{R}))$, they are called *real* measurable functions.

Definition A.5 A measurable function $f : (X, \mathcal{X}) \to (\mathbb{R}, \mathcal{B}(\mathbb{R}))$ of the type

$$f(x) = \sum_{i=1}^{k} a_i \, 1_{A_i}(x), \qquad (A.1)$$

where $k \in \mathbb{N}_+$, $a_1, \ldots, a_k \in \mathbb{R}$, $A_1, \ldots, A_k \in \mathcal{X}$, is called a *simple measurable function* (defined on X).

It seems difficult to prove measurability since most sigma-fields are not defined explicitly (see the definition of $\mathcal{B}(\mathbb{R}^n)$ for instance). However, the following result renders the task feasible.

Theorem A.2 *Let* $(X, \mathcal{X})$ *and* $(E, \mathcal{E})$ *be two measurable spaces, where* $\mathcal{E} = \sigma(\mathcal{C})$ *for some collection* $\mathcal{C}$ *of subsets of* E. *Then* $f : (X, \mathcal{X}) \to (E, \mathcal{E})$ *if and only if* $f^{-1}(C) \in \mathcal{X}$ *for all* $C \in \mathcal{C}$.

Proof We shall first make two obvious preliminary observations. Let X and E be arbitrary sets, $f : X \to E$ an arbitrary function from X to E, $\mathcal{G}$ an arbitrary sigma-field on E, and let $\mathcal{C}_1, \mathcal{C}_2$ be arbitrary non-empty collections of subsets of E. Then

(i) $\sigma(\mathcal{G}) = \mathcal{G}$
(ii) $\mathcal{C}_1 \subseteq \mathcal{C}_2 \Rightarrow \sigma(\mathcal{C}_1) \subseteq \sigma(\mathcal{C}_2)$

Now, define

$$\mathcal{G} = \{C \subseteq E; \ f^{-1}(C) \in \mathcal{X}\}.$$

One checks that $\mathcal{G}$ is a sigma-field. By hypothesis, $\mathcal{C} \subseteq \mathcal{G}$. Therefore, by (ii) and (i), $\mathcal{E} = \sigma(\mathcal{C}) \subseteq \sigma(\mathcal{G}) = \mathcal{G}$. □

An immediate application of this result and Theorem A.1 is:

Corollary A.1 *Let* $(X, \mathcal{X})$ *be a measurable space and let* $n \geq 1$ *be an integer. Then* $f = (f_1, \ldots, f_n) : (X, \mathcal{X}) \to (\mathbb{R}^n, \mathcal{B}(\mathbb{R}^n))$ *if and only if for all* i, $1 \leq i \leq n$, $\{f_i \leq a_i\} \in \mathcal{X}$ *for all* $a_i \in \mathbb{Q}$.

A function $f : \mathbb{R}^k \to \mathbb{R}^m$ is said continuous if the inverse image of an open set is open.[1] The following result is then a direct application of Theorem A.2 considering the definition of the Borel sigma-field on $\mathbb{R}^n$.

Corollary A.2 *Any continuous function* $f : \mathbb{R}^k \to \mathbb{R}^m$ *is measurable with respect to* $\mathcal{B}^k$ *and* $\mathcal{B}^m$.

Measurability is stable under composition:

Theorem A.3 *Let* $(X, \mathcal{X})$, $(Y, \mathcal{Y})$ *and* $(E, \mathcal{E})$ *be three measurable spaces, and let* $\phi : (X, \mathcal{X}) \to (Y, \mathcal{Y})$, $g : (Y, \mathcal{Y}) \to (E, \mathcal{E})$. *Then* $g \circ \phi : (X, \mathcal{X}) \to (E, \mathcal{E})$.

Proof Let $f = g \circ \phi$ (meaning: $f(x) = g(\phi(x))$ for all $x \in X$). For all $C \in \mathcal{E}$,

$$f^{-1}(C) = \phi^{-1}(g^{-1}(C)) = \phi^{-1}(D) \in \mathcal{X},$$

because $D = g^{-1}(C)$ is a set in $\mathcal{Y}$ since $g \in \mathcal{E}/\mathcal{Y}$, and therefore $\phi^{-1}(D) \in \mathcal{X}$ since $\phi \in \mathcal{Y}/\mathcal{X}$. □

Corollary A.3 *Let* $\varphi = (\varphi_1, \ldots, \varphi_n)$ *be a measurable function from* $(X, \mathcal{X})$ *to* $(\mathbb{R}^n, \mathcal{B}(\mathbb{R}^n))$, *and let* $g : \mathbb{R}^n \to \mathbb{R}$ *be a continuous function. Then* $g \circ \phi : (X, \mathcal{X}) \to (\mathbb{R}, \mathcal{B}(\mathbb{R}))$.

Proof Follows directly from Theorem A.3 and Corollary A.2. □

This corollary in turn allows us to show that addition, multiplication and quotient preserve neasurability.

Corollary A.4 *Let* $\varphi_1, \varphi_2 : (X, \mathcal{X}) \to (\mathbb{R}, \mathcal{B}(\mathbb{R}))$, *and let* $\lambda \in \mathbb{R}$. *Then* $\varphi_1 \times \varphi_2$, $\varphi_1 + \varphi_2$, $\lambda\varphi_1$, $(\varphi_1/\varphi_2)1_{\varphi_2 \neq 0}$ *are real measurable functions. Moreover, the set* $\{\varphi_1 = \varphi_2\}$ *is a measurable set.*

Proof For the first three functions, take in the previous corollary $g(x_1, x_2) = x_1 \times x_2$, $= x_1 + x_2$, $= \lambda x_1$ successively.

For $(\varphi_1/\varphi_2)1_{\varphi_2 \neq 0}$, define $\psi_2 = \frac{1_{\varphi_2 \neq 0}}{\varphi_n}$, check that the latter function is measurable, and use the just proven fact that the product $\varphi_1 \psi_2$ is then measurable.

Finally, $\{\varphi_1 = \varphi_2\} = \{\varphi_1 - \varphi_2 = 0\} = (\varphi_1 - \varphi_2)^{-1}(\{0\})$ is a measurable set since $\varphi_1 - \varphi_2$ is a measurable function and $\{0\}$ is a measurable set. □

---

[1] Of course, this definition is equivalent to the usual (ε, δ) definition of continuity, which we shall admit here.

Finally, taking the limit preserves measurability, as we shall see next. (Without otherwise explicitly mentioned, the limits of functions must be understood as *pointwise limits*).

Theorem A.4 *Let* $f_n : (X, \mathcal{X}) \rightarrow (\overline{\mathbb{R}}, \overline{\mathcal{B}})$, $n \in \mathbb{N}$. *Then* $\liminf_{n \uparrow \infty} f_n$ *and* $\limsup_{n \uparrow \infty} f_n$ *are measurable functions, and the set*

$$\{\limsup_{n \uparrow \infty} f_n = \liminf_{n \uparrow \infty} f_n\} = \{\exists \lim_{n \uparrow \infty} f_n\}$$

belongs to $\mathcal{X}$. *In particular, if* $\{\exists \lim_{n \uparrow \infty} f_n\} = X$, *the function* $\lim_{n \uparrow \infty} f_n$ *is a measurable function.*

Proof We first prove the result in the particular case when the sequence of functions is nondecreasing. Denote by f the limit of this sequence. By Theorem A.2 it suffices to show that for all $a \in \mathbb{R}$, $\{f \leq a\} \in \mathcal{X}$. But since the sequence $\{f_n\}_{n \geq 1}$ is nondecreasing, we have that $\{f \leq a\} = \cap_{n=1}^{\infty} \{f_n \leq a\}$, which is indeed in $\mathcal{X}$, being a countable intersection of sets in $\mathcal{X}$.

Now recall that, by definition,

$$\liminf_{n \uparrow \infty} f_n = \lim_{n \uparrow \infty} g_n,$$

where

$$g_n = \inf_{k \geq n} f_k.$$

The function g_n is measurable since for all $a \in \mathbb{R}$, $\{\inf_{k \geq n} f_k \leq a\}$ is a measurable set, being the complement of $\{\inf_{k \geq n} f_k > a\} = \cap_{k \geq n}\{f_k > a\}$, a measurable set (as the countable intersection of measurable sets). Since the sequence $\{g_n\}_{n \geq 1}$ is non-decreasing, the measurability of $\liminf_{n \uparrow \infty} f_n$ follows from the particular case of non-decreasing functions.

Similarly, $\limsup_{n \uparrow \infty} f_n = -\liminf_{n \uparrow \infty}(-f_n)$ is measurable.

The set $\{\limsup_{n \uparrow \infty} f_n = \liminf_{n \uparrow \infty} f_n\}$ is the set on which two measurable functions are equal, and therefore, by the last assertion of Corollary A.4, it is a measurable set.

Finally, if $\lim_{n \uparrow \infty} f_n$ exists, it is equal to $\limsup_{n \uparrow \infty} f_n$, which is, as we just proved, a measurable function. $\qquad \Box$

The basis of the construction of the Lebesgue integral is the following *fundamental approximation theorem*.

Theorem A.5 *Let* $f : (X, \mathcal{X}) \rightarrow (\overline{\mathbb{R}}, \overline{\mathcal{B}})$ *be a non-negative measurable function. There exists a non-decreasing sequence* $\{f_n\}_{n \geq 1}$ *of non-negative simple measurable functions that converges pointwise to* f.

Proof Take

$$f_n(x) = \sum_{k=0}^{n2^{-n}-1} k2^{-n} 1_{A_{k,n}}(x) + n 1_{A_n}(x),$$

where

$$A_{k,n} = \{x \in X : k2^{-n} < f(x) \le (k+1)2^{-n}\}, \qquad A_n = \{x \in X : f(x) > n\}$$

This sequence of functions has the announced properties. In fact, for any $x \in X$ such that $f(x) < \infty$, and n large enough,

$$|f(x) - f_n(x)| \le 2^{-n},$$

and for any $x \in X$ such that $f(x) = \infty$, $f_n(x) = n$ indeed converges to $f(x) = +\infty$.   □

A.1.2 Measures

Definition A.6 Let $(X, \mathcal{X})$ be a measurable space and let $\mu : \mathcal{X} \to [0, \infty]$ be a set function such $\mu(\varnothing) = 0$ and such that for any *countable* family $\{A_n\}_{n \ge 1}$ of *mutually disjoint* sets in $\mathcal{X}$

$$\mu\left(\cup_{n=1}^\infty A_n\right) = \sum_{n=1}^\infty \mu(A_n). \tag{A.2}$$

The set function μ is called a *measure* on $(X, \mathcal{X})$, and $(X, \mathcal{X}, \mu)$ is called a *measure space*.

Property (A.2) is the *sigma-additivity* property. See also Exercise A.5.

Example A.1 (The Dirac measure) Let $a \in X$ and let $\mathcal{X}$ be an arbitrary sigma-field on X. The measure ε_a defined on $(X, \mathcal{X})$ by $\varepsilon_a(C) = 1_C(a)$ is called the *Dirac measure* at $a \in X$. The set function $\mu : \mathcal{X} \to [0, \infty]$ defined by

$$\mu(C) = \sum_{i=0}^\infty \alpha_i 1_{a_i}(C),$$

where $\alpha_i \in \mathbb{R}_+$ for all $i \in \mathbb{N}$, is a measure on $(X, \mathcal{X})$ denoted by $\mu = \sum_{i=0}^\infty \alpha_i \varepsilon_{a_i}$.

Example A.2 (Weighted counting measure) Let $\{\alpha_n\}_{n \ge 1}$ be a sequence of non-negative numbers. The set function $\mu : \mathcal{P}(\mathbb{Z}) \to [0, \infty]$ defined by $\mu(C) = \sum_{n \in C} \alpha_n$ is a measure on $(\mathbb{Z}, \mathcal{P}(\mathbb{Z}))$. If $\alpha_n \equiv 1$ it is called the *counting measure* on $\mathbb{Z}$.

Example A.3 (Lebesgue measure) Measure theory tells us that there exists one and only one measure ℓ on $(\mathbb{R}, \mathcal{B}(\mathbb{R}))$ such that

$$\ell((a, b]) = b - a.$$

This measure is called the *Lebesgue measure* on $\mathbb{R}$. (See the more general result below, Theorem A.7).

Definition A.7 Let μ be a measure on $(X, \mathcal{X})$. If $\mu(X) < \infty$ the measure μ is called a *finite* measure. If $\mu(X) = 1$ the measure μ is called a *probability* measure. If there exists a sequence $\{K_n\}_{n \geq 1}$ of $\mathcal{X}$ such that $\mu(K_n) < \infty$ for all $n \geq 1$, and $\bigcup_{n=1}^{\infty} K_n = X$, the measure μ is called a *sigma-finite* measure. A measure μ on $(\mathbb{R}^n, \mathcal{B}(\mathbb{R}^n))$ such that $\mu(C) < \infty$ for all bounded Borel sets C is called a *locally finite* measure.

Example A.4 The Dirac measure ε_a is a probability measure. The counting measure ν on $\mathbb{Z}$ is a sigma-finite measure. Any locally finite measure on $(\mathbb{R}^n, \mathcal{B}(\mathbb{R}^n))$ is sigma-finite. Lebesgue measure is a locally finite measure.

The following result is the *sequential continuity theorem* for measures.

Theorem A.6 *Let $(X, \mathcal{X}, \mu)$ be a measure space. Let $\{A_n\}_{n \geq 1}$ be a sequence of $\mathcal{X}$, non-decreasing (that is, $A_n \subseteq A_{n+1}$ for all $n \geq 1$). Then*

$$\mu\left(\bigcup_{n=1}^{\infty} A_n\right) = \lim_{n \uparrow \infty} \mu(A_n). \tag{A.3}$$

Proof Write $A_n = A_0 + (A_1 - A_0) + \cdots + (A_n - A_{n-1})$ and therefore

$$\mu(A_n) = \mu(A_0) + \mu(A_1 - A_0) + \cdots + \mu(A_n - A_{n-1}).$$

Then write $\bigcup_{n \geq 0} A_n = A_0 + (A_1 - A_0) + \cdots + (A_n - A_{n-1}) + \cdots$ and therefore

$$\mu(\bigcup_{n \geq 0} A_n) = \mu(A_0) + \mu(A_1 - A_0) + \cdots + \mu(A_n - A_{n-1}) + \cdots. \qquad \square$$

Definition A.8 A function $F : \mathbb{R} \to \mathbb{R}$ is called a *cumulative distribution* function (CDF) if the following properties are satisfied:

1. F is non-decreasing;
2. F is right-continuous;
3. F admits a left-hand limit, denoted by $F(x-)$, at all $x \in \mathbb{R}$.

Example A.5 Let μ be a locally finite measure on $(\mathbb{R}, \mathcal{B}(\mathbb{R}))$ and define

$$F_\mu(t) = \begin{cases} +\mu((0, t]) & \text{if } t \geq 0, \\ -\mu((t, 0]) & \text{if } t < 0. \end{cases}$$

This is a cumulative distribution function (CDF), and moreover,

$$F_\mu(b) - F_\mu(a) = \mu((a, b]),$$
$$F_\mu(a) - F_\mu(a-) = \mu(\{a\}).$$

To show that this function is indeed a CDF is the object of Exercise A.7. The function F_μ is called the CDF of μ.

Theorem A.7 *Let $F : \mathbb{R} \to \mathbb{R}$ be a CDF. There exists an unique locally finite measure μ on $(\mathbb{R}, \mathcal{B}(\mathbb{R}))$ such that $F_\mu = F$.*

This result is easily stated, but it is not trivial, even in the case of the Lebesgue measure (Example A.3). It is typical of the existence results which answer the following type of question: Let $\mathcal{C}$ be a collection of subsets of X with $\mathcal{C} \subseteq \mathcal{X}$, where $\mathcal{X}$ is a sigma-field on X. Given a set function $u : \mathcal{C} \to [0, \infty]$, does there exist a measure μ on $(X, \mathcal{X})$ such that $\mu(C) = u(C)$ for all $C \in \mathcal{C}$, and is it unique?

Let $\mathcal{P}$ be a collection of subsets of X, and let $\mu : \mathcal{P} \to [0, \infty]$ be sigma-additive, that is for any countable family $\{A_n\}_{n\geq 1}$ of mutually disjoint sets in $\mathcal{P}$ such that $\cup_{n\geq 1} A_n \in \mathcal{P}$, we have $\mu(\cup_{n\geq 1} A_n) = \sum_{n\geq 1} \mu(A_n)$. Then μ is called a measure on $\mathcal{P}$. Let $\mathcal{C} \subseteq \mathcal{P}$ be a collection of subsets of X. A mapping $\mu : \mathcal{P} \to [0, \infty]$ is called *sigma-finite on $\mathcal{C}$* if there exists a countable family $\{C_n\}_{n\geq 1}$ of sets in $\mathcal{C}$ such that $\cup_{n\geq 1} C_n = X$ and $\mu(C_n) < \infty$ for all $n \geq 1$.

Theorem A.8 *Let μ_1 and μ_2 be two measures on $(X, \mathcal{X})$ and let $\mathcal{S}$ be a collection of subsets of X that is stable under finite intersection and that generates $\mathcal{X}$. Suppose that μ_1 and μ_2 are sigma-finite on $\mathcal{S}$. If μ_1 and μ_2 agree on $\mathcal{S}$ (that is $\mu_1(C) = \mu_2(C)$ for all $C \in \mathcal{S}$), then they are identical.*

Carathéodory's Theorem

We now give the relevant definitions for *Caratheodory's extension theorem*, one of the fundamental tools of measure theory.

Definition A.9 Let X be an arbitrary set. The collection $\mathcal{A} \subseteq \mathcal{P}(X)$ is called an *algebra* if

(α) $X \in \mathcal{A}$;
(β) $A, B \in \mathcal{A} \implies A \cup B \in \mathcal{A}$;
(γ) $A \in \mathcal{A} \implies \overline{A} \in \mathcal{A}$.

The only difference with a sigma-field is that we require stability by *finite* (instead countable) unions. Note that, similarly to the sigma-field case, $\varnothing \in \mathcal{A}$ and $\mathcal{A}$ is stable under finite intersections. This is why a sigma-field is also called a *sigma-algebra*.

On $\mathbb{R}$, the collection of finite sums of disjoint intervals is an algebra. (By interval we mean any type of intervals: open, closed, semi-open, semi-closed, infinite, etc., or, more formally, any convex subset of $\mathbb{R}$).

Definition A.10 Let X be an arbitrary set. The collection $C \subseteq P(X)$ is called a *semi-algebra* if

(α) $X \in C$;
(β) $A, B \in C \implies A \cup B \in C$;
(γ) If $A \in C$ then $\overline{A}$ can be expressed as a *finite union of disjoint sets* of C.

On $\mathbb{R}$, the collection of intervals is a semi-algebra.

Theorem A.9 *Let C be either an algebra or a semi-algebra defined on X. Let μ be a sigma-finite measure on (X, C). Then there exists a unique extension of μ to $(X, \sigma(C))$ that is a measure.*

Negligible Sets

Definition A.11 Let $(X, \mathcal{X}, \mu)$ be a measure space. A *μ-negligible set* is a set contained in a measurable set $N \in \mathcal{X}$ such that $\mu(N) = 0$. One says that some property $\mathcal{P}$ relative to the elements $x \in X$ holds *μ-almost everywhere (μ-a.e.)* if the set $\{x \in X : x \text{ does not satisfy } \mathcal{P}\}$ is a μ-negligible set.

For instance, if f and g are two measurable functions defined on X, the expression

$$f \leq g \quad \mu\text{-a.e.}$$

means that $\mu(\{x : f(x) > g(x)\}) = 0$.

Theorem A.10 *A countable union of μ-negligible sets is a μ-negligible set.*

Proof Let $A_n, n \geq 1$ be a sequence of μ-negligible sets, and let $N_n, n \geq 1$ be a sequence of measurable sets such that $\mu(N_n) = 0$ and $A_n \subseteq N_n$. Then $N = \cup_{n \geq 1} N_n$ is a measurable set containing $\cup_{n \geq 1} A_n$, and N is of μ-measure 0, by the sub-sigma-additivity property. $\square$

Example A.6 (The rationals are Lebesgue-negligible) Any singleton $\{a\}$, $a \in \mathbb{R}$, is a Borel set of Lebesgue measure 0. The set of rationals $\mathbb{Q}$ is a Borel set of Lebesgue measure 0.

Proof The Borel sigma-field $\mathcal{B}$ is generated by the intervals $I_a = (-\infty, a]$, $a \in \mathbb{R}$ (Theorem A.1), and therefore $\{a\} = \cap_{n \geq 1}(I_a - I_{a-1/n})$ is also in $\mathcal{B}$. Denoting by ℓ the Lebesgue measure, $\ell(I_a - I_{a-1/n}) = 1/n$, and therefore $\ell(\{a\}) = \lim_{n \geq 1} \ell(I_a - I_{a-1/n}) = 0$. $\mathbb{Q}$ is a countable union of sets in $\mathcal{B}$ (singletons) and is therefore in $\mathcal{B}$. It has Lebesgue measure 0 as a countable union of sets of Lebesgue measure 0. $\square$

Theorem A.11 *If two continuous functions $f, g : \mathbb{R} \to \mathbb{R}$ are ℓ-a.e. equal, they are everywhere equal.*

Proof Let $t \in \mathbb{R}$ be such that $f(t) \neq g(t)$. For any $c > 0$, there exists $s \in [t-c, t+c]$ such that $f(s) = g(s)$ (Otherwise, the set $\{t; f(t) \neq g(t)\}$ would contain the whole interval $[t-c, t+c]$, and therefore could not be of null Lebesgue measure). Therefore, one can construct a sequence $\{t_n\}_{n \geq 1}$ converging to t and such that $f(t_n) = g(t_n)$ for all $n \geq 1$. Letting n tend to ∞ yields $f(t) = g(t)$, a contradiction. $\square$

Translation Invariant Measures

Recall the following notation: For any set $A \subseteq \mathbb{R}^n$ and $h \in \mathbb{R}^n$, $A + h := \{a + h; a \in \mathbb{R}^n\}$.

Theorem A.12 *(a) The Lebesgue measure ℓ^n on $(\mathbb{R}^n, \mathcal{B}(\mathbb{R}^n))$ is invariant under translation, that is for all $h \in \mathbb{R}^n$, all $A \in \mathcal{B}(\mathbb{R}^n)$, $\ell^n(A + h) = \ell^n(A)$.*

(b) If μ is a locally finite measure on $(\mathbb{R}^n, \mathcal{B}(\mathbb{R}^n))$ that is invariant under translation, there exists a finite constant $c \geq 0$ such that $\mu = c\ell^n$.

Proof (a) Apply Theorem A.8 to $\mu_1 = \ell^n$ and μ_2 given by $\mu_2(A) = \ell^n(A + h)$ and to the π-system consisting of the rectangles $\prod_{i=1}^{n}(a_i, b_i]$.

(b) Define

$$c = \mu\left((0, 1]^n\right).$$

For any integer $k \geq 1$, the n-cube $(0, 1]^n$ is the disjoint union of k^n n-cubes, all obtained by translation of $(0, \frac{1}{k}]^n$. Therefore

$$\mu\left(\left(0, \frac{1}{k}\right]^n\right) = \frac{c}{k^n}.$$

Let now $d_1, \ldots, d_n \geq 0$. We have

$$\prod_{i=1}^{n}\left(0, \frac{\lfloor kd_i \rfloor}{k}\right] \subseteq \prod_{i=1}^{n}(0, d_i] \subseteq \prod_{i=1}^{n}\left(0, \frac{\lfloor kd_i \rfloor + 1}{k}\right]$$

and therefore

$$\left(\prod_{i=1}^{n}\lfloor kd_i \rfloor\right)\frac{c}{k^n} = \mu\left(\prod_{i=1}^{n}\left(0, \frac{\lfloor kd_i \rfloor}{k}\right]\right) \leq \mu\left(\prod_{i=1}^{n}(0, d_i]\right)$$

$$\leq \mu\left(\prod_{i=1}^{n}\left(0, \frac{\lfloor kd_i \rfloor + 1}{k}\right]\right) \leq \left(\prod_{i=1}^{n}\lfloor kd_i \rfloor + 1\right)\frac{c}{k^n}.$$

Letting $k \uparrow \infty$, we obtain

$$\mu\left(\prod_{i=1}^{n}(0, d_i]\right) = c\prod_{i=1}^{n}d_i = c\ell^n\left(\prod_{i=1}^{n}(0, d_i]\right).$$

Since both μ and ℓ^n are translation-invariant, the above equality extends to arbitrary rectangles $\prod_{i=1}^{n}(a_i, b_i]$. Theorem A.8 concludes the proof. $\square$

A.1.3 Construction of the Integral

We are now in a position to define the Lebesgue integral of a measurable function $f : (X, \mathcal{X}) \to (\overline{\mathbb{R}}, \overline{\mathcal{B}})$ with respect to μ, denoted

$$\int_X f \, d\mu, \quad \text{or} \quad \int_X f(x) \, \mu(dx), \quad \text{or} \quad \mu(f).$$

The integral is defined in 3 steps, which we recall (without proof of the details):

STEP 1. For any non-negative simple measurable function $f : (X, \mathcal{X}) \to (\mathbb{R}, \mathcal{B}(\mathbb{R}))$ of the form (A.1), one defines the integral of f with respect to μ by

$$\int_X f \, d\mu = \sum_{i=1}^{k} a_i \, \mu(A_i).$$

STEP 2. If $f : (X, \mathcal{X}) \to (\overline{\mathbb{R}}, \overline{\mathcal{B}})$ is non-negative the integral is defined by

$$\int_X f \, d\mu = \lim_{n \uparrow \infty} \uparrow \int_X f_n \, d\mu, \tag{A.4}$$

where $\{f_n\}_{n \geq 1}$ is a non-decreasing sequence of non-negative simple measurable functions $f_n : (X, \mathcal{X}) \to (\mathbb{R}, \mathcal{B}(\mathbb{R}))$ such that $\lim_{n \uparrow \infty} \uparrow f_n = f$ (Theorem A.5). This definition can be shown to be consistent, in that the integral so defined is independent of the choice of the approximating sequence. Note that the quantity (A.4) is non-negative and can be infinite.

STEP 3. One checks that if $f \leq g$, where $f, g : (X, \mathcal{X}) \to (\overline{\mathbb{R}}, \overline{\mathcal{B}})$ are non-negative, then

$$\int_X f \, d\mu \leq \int_X g \, d\mu.$$

Denoting

$$f^+ = \max(f, 0) \quad \text{and} \quad f^- = \max(-f, 0)$$

(and in particular $f = f^+ - f^-$ and $f^{\pm} \leq |f|$), we therefore have

$$\int_X f^{\pm}\, d\mu \leq \int_X |f|\, d\mu.$$

Thus, if

$$\int_X |f|\, d\mu < \infty, \tag{A.5}$$

the right-hand side of

$$\int_X f\, d\mu := \int_X f^+\, d\mu - \int_X f^-\, d\mu \tag{A.6}$$

is meaningful and defines the integral of the left-hand side. Moreover, the integral of f with respect to μ defined in this way is finite.

Definition A.12 A measurable function $f : (X, \mathcal{X}) \to (\overline{\mathbb{R}}, \overline{\mathcal{B}})$ satisfying (A.5) is called a μ-*integrable function*.

STEP 3 (ctd). The integral can be defined for some *non-integrable* functions. For example, it is defined for all non-negative functions. More generally, if $f : (X, \mathcal{X}) \to (\overline{\mathbb{R}}, \overline{\mathcal{B}})$ is such that at least one of the integrals $\int_X f^+\, d\mu$ or $\int_X f^-\, d\mu$ is finite, one defines the integral as in (A.6). This leads to one of the forms "finite minus finite", "finite minus infinite", and "infinite minus finite". The case which is rigorously excluded is that in which $\mu(f^+) = \mu(f^-) = +\infty$.

Example A.7 (Integral with respect to the weighted counting measure) Any function $f : \mathbb{Z} \to \mathbb{R}$ is measurable with respect to $\mathcal{P}(\mathbb{Z})$ and $\mathcal{B}$. With the measure μ defined in Example A.2, and with $f \geq 0$ for instance,

$$\mu(f) = \sum_{n=1}^{\infty} \alpha_n f(n).$$

We shall see how this "natural" definition in this special case fits with the general construction of the integral summarized above. It suffices to consider the approximating sequence of simple functions

$$f_n(k) = \sum_{j=-n}^{+n} f(j) 1_{\{j\}}(k)$$

whose integral is

$$\nu(f_n) = \sum_{j=-n}^{+n} f(j)\mu(\{j\}) = \sum_{j=-n}^{+n} f(j)\alpha_j$$

and to let n tend to ∞. When $\alpha_n \equiv 1$, the integral reduces to the sum of a series:

$$v(f) = \sum_{n \in \mathbb{Z}} f(n).$$

In this case, integrability means that the series is absolutely convergent.

Example A.8 (Integral with respect to the Dirac measure) Let $(X, \mathcal{X})$ be an arbitrary measurable space and let ε_a be the Dirac measure at point $a \in X$. Let $f : (X, \mathcal{X}) \to (\mathbb{R}, \mathcal{B}(\mathbb{R}))$. We shall prove formally that it is ε_a-integrable and that

$$\varepsilon_a(f) = f(a).$$

For a simple function f as in (A.1), we have

$$\varepsilon_a(f) = \sum_{i=1}^{k} a_i \, \varepsilon_a(A_i) = \sum_{i=1}^{k} a_i \, 1_{A_i}(a) = f(a).$$

For a non-negative function f, and any non-decreasing sequence of simple non-negative measurable functions $\{f_n\}_{n \geq 1}$ converging to f, we have

$$\varepsilon_a(f) = \lim_{n \uparrow \infty} \varepsilon_a(f_n) = \lim_{n \uparrow \infty} f_n(a) = f(a).$$

Finally, for any $f : (X, \mathcal{X}) \to (\mathbb{R}, \mathcal{B}(\mathbb{R}))$

$$\varepsilon_a(f) = \varepsilon_a(f^+) - \varepsilon_a(f^-) = f^+(a) - f^-(a) = f(a)$$

is a well-defined quantity.

Stieltjes–Lebesgue Integral

Let F be cumulative distribution function on $(\mathbb{R}, \mathcal{B}(\mathbb{R}))$ an let μ_F be the associated locally finite measure on $(\mathbb{R}, \mathcal{B}(\mathbb{R}))$. By definition, the *Stieltjes–Lebesgue integral* of the measurable function $g : (\mathbb{R}, \mathcal{B}(\mathbb{R})) \to (\mathbb{R}, \mathcal{B}(\mathbb{R}))$ with respect to F is the integral of g with respect to μ_F. It is denoted by $\int_{\mathbb{R}} g(x) \, dF(x)$. Therefore

$$\int_{\mathbb{R}} g(x) \, dF(x) := \int_{\mathbb{R}} g(x) \, \mu_F(dx).$$

A.1.4 Properties of the Integral

Elementary Properties

First, recall that for all $A \in \mathcal{X}$

$$\int_X 1_A \, d\mu = \mu(A),$$ (A.7)

and that the notation $\int_A f \, d\mu$ means $\int_X 1_A f \, d\mu$. We now list without proof a set basic properties of the integral:

Theorem A.13 *Let* $f, g : (X, \mathcal{X}) \to (\overline{\mathbb{R}}, \overline{\mathcal{B}})$ *be* μ-integrable functions, and let $a, b \in \mathbb{R}$. Then
- (a) $af + bg$ is μ-integrable and $\mu(af + bg) = a\mu(f) + b\mu(g)$;
- (b) If $f = 0$ μ-a.e., then $\mu(f) = 0$; If $f = g$ μ-a.e., then $\mu(f) = \mu(g)$;
- (c) If $f \leq g$ μ-a.e., then $\mu(f) \leq \mu(g)$;
- (d) $|\mu(f)| \leq \mu(|f|)$;
- (e) If $f \geq 0$ μ-a.e. and $\mu(f) = 0$, then $f = 0$ μ-a.e.;
- (f) If $\mu(1_A f) = 0$ for all $A \in \mathcal{X}$, then $f = 0$ μ-a.e.
- (g) If f is μ-integrable, then $|f| < \infty$ μ-a.e.

For a complex measurable function $f : X \to \mathbb{C}$ (i.e. $f = f_1 + if_2$, where $f_1, f_2 : (X, \mathcal{X}) \to (\mathbb{R}, \mathcal{B}(\mathbb{R}))$) such that $\mu(|f|) < \infty$, one defines

$$\int_X f \, d\mu = \int_X f_1 \, d\mu + i \int_X f_2 \, d\mu.$$

The extension to complex measurable functions of the properties (a), (b), (d) and (f) in Theorem A.13 is immediate.

The time spent in learning about the Riemann integral on $\mathbb{R}$ has not been in vain. In fact, the procedures that led to the actual computation of such and such Riemann integral (starting with simple integrands and using basic rules) can be reproduced in the framework of Lebesgue integration theory. Thus, in a sense that we need not formalize here, and for all practical purposes, "Riemann-integrable functions are Lebesgue-integrable and the integrals then coincide". The converse is not true: The function f defined by $f(x) = 1$ if $x \in \mathbb{Q}$ and $f(x) = 0$ if $x \notin \mathbb{Q}$ is a measurable function, and it is Lebesgue integrable with its integral equal to zero because $\{f \neq 0\} \equiv \mathbb{Q}$, has ℓ-measure zero. However, f is not Riemann integrable.

Example A.9 (Generalized Riemann-integral) The function $f : (\mathbb{R}, \mathcal{B}(\mathbb{R})) \to (\mathbb{R}, \mathcal{B}(\mathbb{R}))$ defined by

$$f(x) = \frac{x}{1 + x^2}$$

does not have a Lebesgue integral, because

$$f^+(x) = \frac{x}{1+x^2} 1_{(0,\infty)}(x) \quad \text{and} \quad f^-(x) = -\frac{x}{1+x^2} 1_{(-\infty,0)}(x)$$

have infinite Lebesgue integrals. However, it has a *generalized* Riemann integral

$$\lim_{A\uparrow\infty} \int_{-A}^{+A} \frac{x}{1+x^2} \, dx = 0.$$

Beppo Levi, Fatou and Lebesgue

We seek conditions allowing to interchange the order of limit and integration, that is,

$$\int_X \lim_{n\uparrow\infty} f_n \, d\mu = \lim_{n\uparrow\infty} \uparrow \int_X f_n \, d\mu. \qquad (A.8)$$

Example A.10 (The classical counterexample) For $(X, \mathcal{X}, \mu) = (\mathbb{R}, \mathcal{B}(\mathbb{R}), \ell)$, define

$$f_n(x) = (-n^2|x| + n) \, 1_{[-\frac{1}{n},+\frac{1}{n}]}$$

(the triangular "hat" of height n and base length $\frac{2}{n}$). One has

$$\lim_{n\uparrow\infty} f_n(x) = 0 \quad \text{if} \quad x \neq 0,$$

that is $\lim_{n\uparrow\infty} f_n = 0$ μ-a.e. Therefore $\mu(\lim_{n\uparrow\infty} f_n) = 0$. However, $\mu(f_n) = 1$ for all $n \geq 1$.

The *monotone convergence* theorem below is also called the *Beppo Levi theorem*.

Theorem A.14 *Let* $f_n : (X, \mathcal{X}) \to (\overline{\mathbb{R}}, \overline{\mathcal{B}})$, $n \geq 1$, *be such that*

(i) $f_n \geq 0$ μ-a.e.;
(ii) $f_{n+1} \geq f_n$ μ-a.e.

Then there exists a non-negative function $f : (X, \mathcal{X}) \to (\overline{\mathbb{R}}, \overline{\mathcal{B}})$ *such that*

$$\lim_{n\uparrow\infty} \uparrow f_n = f \quad \mu\text{-a.e.,}$$

and (A.8) holds.

The *dominated convergence* theorem below is sometimes refered to as the *Lebesgue theorem*.

Theorem A.15 *Let* $f_n : (X, \mathcal{X}) \to (\overline{\mathbb{R}}, \overline{\mathcal{B}})$, $n \geq 1$, *be such that, for some function* $f : (X, \mathcal{X}) \to (\overline{\mathbb{R}}, \overline{\mathcal{B}})$ *and some* μ-*integrable function* $g : (X, \mathcal{X}) \to (\overline{\mathbb{R}}, \overline{\mathcal{B}})$:

(i) $\lim_{n \uparrow \infty} f_n = f$, μ-*a.e.*;

(ii) $|f_n| \leq |g|$ μ-*a.e. for all* $n \geq 1$.

Then (A.8) holds.

The next result is a useful technical tool, called *Fatou's lemma*.

Theorem A.16 *Let* $f_n : (X, \mathcal{X}) \to (\overline{\mathbb{R}}, \overline{\mathcal{B}})$, $n \geq 1$, *be such that* $f_n \geq 0$ μ-*a.e. for all* $n \geq 1$. *Then*

$$\int_X (\liminf_{n \uparrow \infty} f_n) \, d\mu \leq \liminf_{n \uparrow \infty} \left(\int_X f_n \, d\mu \right).$$

Image Measure

Definition A.13 Let $(X, \mathcal{X})$ and $(E, \mathcal{E})$ be two measurable spaces, let $h : (X, \mathcal{X}) \to (E, \mathcal{E})$ be a measurable function, and let μ be a measure on $(X, \mathcal{X})$. The measure $\mu \circ h^{-1}$ on $(E, \mathcal{E})$, called the image of μ by h, is defined by

$$(\mu \circ h^{-1})(C) = \mu(h^{-1}(C)), \quad C \in \mathcal{E}.$$

(One easily checks that it is indeed a measure.)

In the proof of the following theorem, the combination of the approximation theorem for random variables and of the monotone convergence theorem is typical.

Theorem A.17 *For arbitrary non-negative measurable function* $f : (X, \mathcal{X}) \to (\overline{\mathbb{R}}, \overline{\mathcal{B}})$

$$\int_X (f \circ h)(x) \, \mu(dx) = \int_E f(y) (\mu \circ h^{-1}) \, (dy). \tag{A.9}$$

For functions $f : (X, \mathcal{X}) \to (\overline{\mathbb{R}}, \overline{\mathcal{B}})$ *of arbitrary sign either one of the conditions*

(a) $f \circ h$ *is* μ-*integrable, or*
(b) f *is* $\mu \circ h^{-1}$-*integrable,*

implies the other, and equality (A.9) then holds.

Proof The equality (A.9) is readily verified when f is a non-negative simple measurable function. In the general case one approximates f by a non-decreasing sequence of non-negative simple measurable functions $\{f_n\}_{n \geq 1}$ and (A.9) then follows from the same equality written with $f = f_n$, by letting $n \uparrow \infty$ and using the monotone convergence theorem. For the case of functions of arbitrary sign, apply (A.9) with f^+ and f^-. $\qquad \square$

Tonelli and Fubini

Let $(X_1, \mathcal{X}_1, \mu_1)$ and $(X_2, \mathcal{X}_2, \mu_2)$ be two measure spaces where μ_1 and μ_2 are sigma-finite. Define the product set $X = X_1 \times X_2$ and the *product sigma-field* $\mathcal{X} = \mathcal{X}_1 \otimes \mathcal{X}_2$, where by definition the latter is the smallest sigma-field on X containing all sets of the form $A_1 \times A_2$, where $A_1 \in \mathcal{X}_1$, $A_2 \in \mathcal{X}_2$.

Theorem A.18 *There exists an unique measure μ on $(X_1 \times X_2, \mathcal{X}_1 \otimes \mathcal{X}_2)$ such that*

$$\mu(A_1 \times A_2) = \mu_1(A_1)\mu_2(A_2) \tag{A.10}$$

for all $A_1 \in \mathcal{X}_1$, $A_2 \in \mathcal{X}_2$.

The measure μ is the *product measure* of μ_1 and μ_2, and is denoted $\mu_1 \times \mu_2$.

The above result and the following ones are stated for products of two sigma-finite measures, but extend in an obvious manner to a finite number of sigma-finite measures. The notation $\mathcal{X}^{\otimes n}$ represents the nth fold product of $\mathcal{X}$ by itself. For instance, $\mathcal{X}^{\otimes 2} = \mathcal{X} \otimes \mathcal{X}$.

The typical example of a product measure is the Lebesgue measure on the space $(\mathbb{R}^n, \mathcal{B}(\mathbb{R}^n))$: It is the unique measure ℓ^n on that space that is such that $\ell^n \left(\Pi_{i=1}^n A_i \right) = \Pi_{i=1}^n \ell(A_i)$ for all $A_1, \dots, A_n \in \mathcal{B}$.

Theorem A.19 *Let $(X_1, \mathcal{X}_1, \mu_1)$ and $(X_1, \mathcal{X}_2, \mu_2)$ be two measure spaces in which μ_1 and μ_2 are sigma-finite. Let $(X, \mathcal{X}, \mu) = (X_1 \times X_2, X_1 \otimes \mathcal{X}_2, \mu_1 \times \mu_2)$.*

(A) Tonelli . *If f is non-negative, then, for μ_1-almost all x_1, the function $x_2 \to f(x_1, x_2)$ is measurable with respect to $\mathcal{X}_2$, and*

$$x_1 \to \int_{X_2} f(x_1, x_2)\,\mu_2(dx_2)$$

is a measurable function with respect to $\mathcal{X}_1$. Furthermore,

$$\int_X f\,d\mu = \int_{X_1} \left[\int_{X_2} f(x_1, x_2)\,\mu_2(dx_2) \right] \mu_1(dx_1). \tag{A.12}$$

(B) Fubini . *If f is μ-integrable, then, for μ_1-almost all x_1, the function $x_2 \to f(x_1, x_2)$ is μ_2-integrable and $x_1 \to \int_{X_2} f(x_1, x_2)\,\mu_2(dx_2)$ is μ_2-integrable, and (A.12) is true.*

We shall refer to the global result as the *Fubini–Tonelli* Theorem. Part (A) says that one can integrate a non-negative measurable function in any order of its variables. Part (B) says that the same is true of an arbitrary measurable function if that function is μ-integrable. In general, in order to apply Part (B) one must use Part (A) with $f = |f|$ to ascertain whether or not $\int_X |f| \, d\mu < \infty$.

Example A.11 (When Fubini is not applicable) Consider the function f defined on $X_1 \times X_2 = (1, \infty) \times (0, 1)$ by the formula

$$f(x_1, x_2) = e^{-x_1 x_2} - 2e^{-2x_1 x_2}.$$

We have

$$\int_{(1,\infty)} f(x_1, x_2) \, dx_1 = \frac{e^{-x_2} - e^{-2x_2}}{x_2} = h(x_2) \geq 0,$$

$$\int_{(0,1)} f(x_1, x_2) \, dx_2 = -\frac{e^{-x_1} - e^{-2x_1}}{x_1} = -h(x_1).$$

However,

$$\int_0^1 h(x_2) \, dx_2 \neq \int_1^\infty (-h(x_1)) \, dx_1,$$

since $h \geq 0$ ℓ-a.e. on $(0, \infty)$. We therefore see that successive integrations yields different results according to the order in which they are performed. As a matter of fact, f is *not* integrable on $(0, 1) \times (1, \infty)$.

Diagonal Shift Invariant Measure

Theorem A.20 *Let μ be a locally finite measure on $(X^k, \mathcal{X}^{\otimes k})$ which is invariant for the diagonal shift, that is, such that for all $A_1, \ldots, A_k \in \mathcal{X}$, all $h \in X$,*

$$\mu((A_1 + h) \times \cdots \times (A_k + h)) = \mu(A_1 \times \cdots \times A_k).$$

Then, there exists a unique locally finite measure $\widehat{\mu}$ on $\mathcal{X}^{k-1}$ such that for all non-negative measurable functions f from X^k to $\mathbb{R}$,

$$\int_{\mathcal{X}^k} f(x_1, \ldots, x_k) \mu(dx_1 \times \cdots \times dx_k)$$

$$= \int_{\mathcal{X}} \left\{ \int_{\mathcal{X}^{k-1}} f(x_1, x_1 + x_2, \ldots, x_1 + x_k) \widehat{\mu}(dx_2 \times \cdots \times dx_k) \right\} dx_1.$$

Proof Consider the bijective linear transformation from X^k onto itself:

$$T : (x_1, \ldots, x_k) \mapsto (x_1, x_1 + x_2, \ldots, x_1 + x_k)$$

Since $\int\limits_{\mathcal{X}} f(x)\mu(dx) = \int\limits_{\mathcal{X}} f(Tx)(\mu \circ T)(dx)$, we have to prove that the measure $v := \mu \circ T$ is of the form

$$v(dx_1 \times \cdots \times dx_k) = dx_1 \times \widehat{\mu}(dx_2 \times \cdots \times dx_k)$$

for some locally finite measure $\widehat{\mu}$ on $\mathcal{X}^{k-1}$.

For all non-negative measurable functions $\varphi : X^k \to \mathbb{R}$,

$$\int\limits_{\mathcal{X}^k} \varphi(x_1, \ldots, x_k)v(dx_1 \times \cdots \times dx_k)$$

$$= \int\limits_{X^k} \varphi(x)(\mu \circ T)(dx) = \int\limits_{X^k} \varphi(T^{-1}x)\mu(dx)$$

$$= \int\limits_{X^k} \varphi(x_1, x_2 - x_1, \ldots, x_k - x_1)\mu(dx_1 \times \cdots \times dx_k)$$

$$= \int\limits_{X^k} \varphi(x_1 + h, x_2 - x_1, \ldots, x_k - x_1)\mu(dx_1 \times \cdots \times dx_k),$$

where the last equality follows from the assumed invariance property of μ. Replacing the function φ by

$$\tilde{\varphi}(x_1, x_2, \ldots, x_k) := \varphi(x_1 + h, x_2, \ldots, x_k)$$

where $h \in X$, we obtain, by definition of v, and

$$\int\limits_{X^k} \varphi(x_1 + h, \ldots, x_k)v(dx_1 \times \cdots \times dx_k)$$

$$= \int\limits_{X^k} \tilde{\varphi}(x_1, \ldots, x_k)v(dx_1 \times \cdots \times dx_k)$$

$$= \int\limits_{X^k} \tilde{\varphi}(x_1, x_2 - x_1, \ldots, x_k - x_1)\mu(dx_1 \times \cdots \times dx_k)$$

$$= \int_{X^k} \varphi(x_1 + h, x_2 - x_1, \ldots, x_k - x_1)\mu(dx_1 \times \cdots \times dx_k)$$

$$= \int_{X^k} \varphi(x_1, x_2 - x_1, \ldots, x_k - x_1)\mu(dx_1 \times \cdots \times dx_k),$$

where the last equality again follows from the assumed diagonal shift invariance property of μ. Therefore, for all $h \in X$,

$$\int_{X^k} \varphi(x_1 + h, \ldots, x_k)\nu(dx_1 \times \cdots \times dx_k) = \int_{X^k} \varphi(x_1, \ldots, x_k)\nu(dx_1 \times \cdots \times dx_k),$$

that is, ν is invariant with respect to the shift acting on the first coordinate. In particular, for all $C_1 \in \mathcal{X}$ and all $B \in \mathcal{X}^{k-1}$, all $h \in X$,

$$\nu((C_1 + h) \times B) = \nu(C_1 \times B).$$

Thus for fixed $B \in \mathcal{X}^{k-1}$, the measure $\nu(\cdot \times B)$ is shift-invariant, and therefore a multiple of the Lebesgue measure (Theorem A.12):

$$\nu(C_1 \times B) = \ell^m(C_1) \times \widehat{\mu}(B),$$

where the constant (with respect to C_1) $\widehat{\mu}(B)$ clearly defines a locally finite measure $\widehat{\mu}$ on $\mathcal{X}^k$.

The uniqueness of $\widehat{\mu}$ is left as an exercise.   □

Radon–Nikodym

Definition A.14 Let $(X, \mathcal{X}, \mu)$ be a measure space and let $h : (X, \mathcal{X}) \to (\overline{\mathbb{R}}, \overline{\mathcal{B}})$ be non-negative. Define the set function $\nu : \mathcal{X} \to [0, \infty]$ by

$$\nu(C) = \int_C h(x)\,\mu(dx). \tag{A.13}$$

Then, one easily checks that ν is a measure on $(X, \mathcal{X})$ called the product of μ with the function h.

Theorem A.21 *Let μ, ν and h be as in Definition A.14. For arbitrary non-negative $f : (x, \mathcal{X}) \to (\overline{\mathbb{R}}, \overline{\mathcal{B}})$,*

$$\int_X f(x)\,\nu(dx) = \int_X f(x)h(x)\,\mu(dx). \tag{A.14}$$

If $f : (X, \mathcal{X}) \to (\overline{\mathbb{R}}, \overline{\mathcal{B}})$ *has arbitrary sign then either one of the conditions*

(a) f is v-integrable,
(b) fh is μ-integrable,

implies the other, and the equality (A.14) then holds.

Proof Verifying (A.14) for elementary non-negative functions and approximating f by a non-decreasing sequence of such functions, the monotone convergence theorem is then used, as in the proof of (A.9). For the case of functions of arbitrary sign, apply (A.14) with $f = f^+$ and $f = f^-$.   □

Equality (A.13) is also written $v(\mathrm{d}x) = h(x)\,\mu(\mathrm{d}x)$. Observe that for all $X \in \mathcal{X}$

$$\mu(C) = 0 \implies v(C) = 0. \tag{A.15}$$

Definition A.15 (a) Let μ and v be two measures on $(X, \mathcal{X})$ such that (A.15) holds for all $C \in \mathcal{X}$. Then v is said to be *absolutely continuous with respect to* μ.
   (b) The measures μ and v on $(X, \mathcal{X})$ are said to be *mutually singular* if there exist two disjoint measurable sets $A, B \in \mathcal{X}$ such that $X = A \cup B$ and $v(A) = \mu(B) = 0$.

Absolute continuity is denoted $v \ll \mu$. Mutual singularity is denoted $\mu \perp v$.

Theorem A.22 *Let μ and v be two measures on $(X, \mathcal{X}$, with μ sigma-finite, such that $v \ll \mu$. Then there exists a non-negative function $h : (X, \mathcal{X}) \to (\overline{\mathbb{R}}, \overline{\mathcal{B}})$ such that*

$$v(\mathrm{d}x) = h(x)\,\mu(\mathrm{d}x). \tag{A.16}$$

This function h is called the Radon-Nikodym derivative of v with respect to μ and is denoted $\mathrm{d}v/\mathrm{d}\mu$.

The function h is easily seen to be μ-essentially unique in the sense that if h' is another such function then $h = h'$ μ-a.e.

From (A.16) we therefore have

$$\int_X f(x)\,v(\mathrm{d}x) = \int_X f(x)\,h(x)\,\mu(\mathrm{d}x)$$

for all non-negative $f = (X, \mathcal{X}) \to (\overline{\mathbb{R}}, \overline{\mathcal{B}})$.

In the subsequent chapters, we shall need the following:

Theorem A.23 *Let μ and v be two sigma-finite measures on $(X, \mathcal{X})$. There exists a unique decomposition (called the* Lebesgue *decomposition)*

$$\nu = \nu_a + \nu_s$$

and a μ-essentially unique non-negative measurable function $g : X \rightarrow \mathbb{R}$

$$\nu_a \ll \mu \qquad \nu_a \perp \nu_s \ \text{ and } \ d\nu_a = g \, d\mu . \tag{A.17}$$

(By "μ-essentialy unique", the following is meant: if $g' : X \rightarrow \mathbb{R}$ is another function satisfying the conditions A.17, then $g = g'$, μ-a.e.)

L^p Spaces

For a given integer $p \geq 1$, $L^p_{\mathbb{C}}(\mu)$ is, roughly speaking (see the details below), the collection of complex-valued measurable functions f defined on X such that $\int_X |f|^p \, d\mu < \infty$. We shall see that it is a complete normed vector space over $\mathbb{C}$, that is, a Banach space.

Let $(X, \mathcal{X}, \mu)$ be a measure space and let f, g be two complex valued measurable functions defined on X. The relation $\mathcal{R}$ defined by

$$f \mathcal{R} g \text{ if and only if } f = g \ \mu\text{-a.e.}$$

is an equivalence relation, and we shall denote the equivalence class of f by $\{f\}$. Note that for any $p > 0$ (using property b of Theorem A.13),

$$f \mathcal{R} g \implies \int\limits_X |f|^p \, d\mu = \int\limits_X |g|^p \, d\mu .$$

The operations $+$, $\times$, $*$, and multiplication by a scalar $\alpha \in \mathbb{C}$ are defined on the equivalence class by

$$\{f\} + \{g\} = \{f + g\}, \quad \{f\}\{g\} = \{fg\}, \quad \{f\}^* = \{f^*\}, \quad \alpha\{f\} = \{\alpha f\} .$$

The first equality means that $\{f\} + \{g\}$ is, by definition, the equivalence class consisting of the functions $f + g$, where f and g are arbitrary members of $\{f\}$ and $\{g\}$, respectively. A similar interpretation holds for the other equalities.

By definition, for a given $p \geq 1$, $L^p_{\mathbb{C}}(\mu)$ is the collection of equivalence classes $\{f\}$ such that $\int_X |f|^p \, d\mu < \infty$. Clearly it is a vector space over $\mathbb{C}$ (for the proof recall that

$$\left(\tfrac{|f|+|g|}{2}\right)^p \leq \tfrac{1}{2}|f|^p + \tfrac{1}{2}|g|^p$$

since $t \rightarrow t^p$ is a convex function when $p \geq 1$). In order to avoid cumbersome notation, in this section and in general whenever we consider L^p-spaces, we shall write f for $\{f\}$. This abuse of notation is harmless since two members of the same equivalence class have the same integral if that integral is defined. Therefore, using this loose notation,

$$L_{\mathbb{C}}^p(\mu) = \left\{ f : \int_X |f|^p \, d\mu < \infty \right\}. \tag{A.18}$$

When the measure is the counting measure on the set $\mathbb{Z}$ of relative integers, the traditional notation is $\ell_{\mathbb{C}}^p(\mathbb{Z})$. This is the space of random complex sequences $\{x_n\}_{n \in \mathbb{Z}}$ such that

$$\sum_{n \in \mathbb{Z}} |x_n|^p < \infty.$$

The following is a simple and often used observation.

Theorem A.24 *Let p and q be positive real numbers such that $p > q$. If the measure μ on $(X, \mathcal{X}, \mu)$ is finite, then $L_{\mathbb{C}}^p(\mu) \subseteq L_{\mathbb{C}}^q(\mu)$. In particular, $L_{\mathbb{C}}^2(\mu) \subseteq L_{\mathbb{C}}^1(\mu)$.*

Proof From the inequality $|a|^q \leq 1 + |a|^p$, true for all $a \in \mathbb{C}$, it follows that $\mu(|f|^q) \leq \mu(1) + \mu(|f|^p)$. Since $\mu(1) = \mu(\mathbb{R}) < \infty$, $\mu(|f|^q) < \infty$ whenever $\mu(|f|^p) < \infty$. □

Theorem A.25 *Let p and q be positive real numbers different from 1 such that*

$$\frac{1}{p} + \frac{1}{q} = 1$$

(p and q are then said to be conjugate), and let $f, g : (X, \mathcal{X}) \mapsto (\mathbb{R}, \mathcal{B}(\mathbb{R}))$ be non-negative. Then, we have Hölder's inequality

$$\int_X fg \, d\mu \leq \left[\int_X f^p \, d\mu \right]^{1/p} \left[\int_X g^q \, d\mu \right]^{1/q}. \tag{A.19}$$

In particular, if $f, g \in L_{\mathbb{C}}^2(\mathbb{R})$, then $fg \in L_{\mathbb{C}}^1(\mathbb{R})$.

Proof Let

$$A = \left(\int_X f^p \, d\mu \right)^{1/p}, \quad B = \left(\int_X g^q \, d\mu \right)^{1/q}.$$

We may assume that $0 < A < \infty$, $0 < B < \infty$, because otherwise Hölder's inequality is trivially satisfied. Define $F = f/A$, $G = g/B$, so that

$$\int_X F^p \, d\mu = \int_X G^q \, d\mu = 1.$$

The inequality

$$F(x)G(x) \le \frac{1}{p} F(x)^p + \frac{1}{q} G(x)^q. \tag{A.20}$$

is trivially satisfied if x is such that $F(x) = 0$ or $G(x) = 0$. If $F(x) > 0$ and $G(x) > 0$ define

$$s(x) = p \ln(F(x)), \qquad t(x) = q \ln(G(x)).$$

From the convexity of the exponential function and the assumption that $1/p + 1/q = 1$,

$$e^{s(x)/p + t(x)/q} \le \frac{1}{p} e^{s(x)} + \frac{1}{q} e^{t(x)},$$

and this is precisely inequality (A.20). Integrating this inequality yields

$$\int_X (FG)\, d\mu \le \frac{1}{p} + \frac{1}{q} = 1,$$

and this is just (A.19). For the last assertion of the theorem, take $p = q = 2$. □

Theorem A.26 *Let $p \ge 1$ and let $f, g : (X, \mathcal{X}) \mapsto (\mathbb{R}, \mathcal{B}(\mathbb{R}))$ be non-negative and such that*

$$\int_X f^p\, d\mu < \infty, \qquad \int_X g^p\, d\mu < \infty.$$

Then, we have Minkowski's inequality

$$\left[\int_X (f+g)^p \right]^{1/p} \le \left[\int_X f^p\, d\mu \right]^{1/p} + \left[\int_X g^p\, d\mu \right]^{1/p}. \tag{A.21}$$

Proof For $p = 1$ the inequality (in fact an equality) is obvious. Therefore, assume $p > 1$. From Hölder's inequality

$$\int_X f(f+g)^{p-1}\, d\mu \le \left[\int_X f^p\, d\mu \right]^{1/p} \left[\int_X (f+g)^{(p-1)q} \right]^{1/q}$$

and

$$\int_X g(f+g)^{p-1}\, d\mu \le \left[\int_X g^p\, d\mu \right]^{1/p} \left[\int_X (f+g)^{(p-1)q} \right]^{1/q}.$$

Adding up the above two inequalities and observing that $(p-1)q = p$, we obtain

$$\int_X (f+g)^p \, d\mu \leq \left(\left[\int_X f^p \, d\mu \right]^{1/p} + \left[\int_X g^p \, d\mu \right]^{1/p} \right) \left[\int_x (f+g)^p \right]^{1/q}.$$

One may assume that the right-hand side of (A.21) is finite and that the left-hand side is positive (otherwise the inequality is trivial). Therefore $\int_X (f+g)^p \, d\mu \in (0, \infty)$. We may therefore divide both sides of the last display by $\left[\int_X (f+g)^p \, d\mu \right]^{1/q}$. Observing that $1 - 1/q = 1/p$ yields the desired inequality (A.21).   □

Theorem A.27 *Let $p \geq 1$. The mapping $v_p : L^p_{\mathbb{C}}(\mu) \mapsto [0, \infty)$ defined by*

$$v_p(f) = \left(\int_X |f|^p \, d\mu \right)^{1/p} \tag{A.22}$$

is a norm on $L^p_{\mathbb{C}}(\mu)$.

Proof Clearly, $v_p(\alpha f) = |\alpha| v_p(f)$ for all $\alpha \in \mathbb{C}$, $f \in L^p_{\mathbb{C}}(\mu)$.

Also, $v_p(f) = 0$ if and only if $\left(\int_X |f|^p \, d\mu \right)^{1/p} = 0$ which is in turn equivalent to $f = 0$, μ – a.e. Finally, $v_p(f+g) \leq v_p(f) + v_p(g)$ for all $f, g \in L^p_{\mathbb{C}}(\mu)$, by Minkowski's inequality.   □

We shall denote $v_p(f)$ by $\|f\|_p$. Thus $L^p_{\mathbb{C}}(\mu)$ is a normed vector space over $\mathbb{C}$, with the norm $\|\cdot\|_p$ and the induced distance

$$d_p(f, g) := \|f - g\|_p. \qquad \qquad \square$$

Theorem A.28 *Let $p \geq 1$. The distance d_p makes of $L^p_{\mathbb{C}}(\mu)$ a complete normed vector space.*

In other words, $L^p_{\mathbb{C}}(\mu)$ is a *Banach space* for the norm $\|\cdot\|_p$.

Proof To show completeness one must prove that for any sequence $\{f_n\}_{n \geq 1}$ of $L^p_{\mathbb{C}}(\mu)$ that is a Cauchy sequence (that is such that $\lim_{m,n \uparrow \infty} d_p(f_n, f_m) = 0$), there exists $f \in L^p_{\mathbb{C}}(\mu)$ such that $\lim_{n \uparrow \infty} d_p(f_n, f) = 0$.

Since $\{f_n\}_{n \geq 1}$ is a Cauchy sequence, one can select a subsequence $\{f_{n_i}\}_{i \geq 1}$ such that

$$d_p(f_{n_{i+1}} - f_{n_i}) \leq 2^{-i}. \tag{A.23}$$

Let

$$g_k = \sum_{i=1}^{k} |f_{n_{i+1}} - f_{n_i}|, \quad g = \sum_{i=1}^{\infty} |f_{n_{i+1}} - f_{n_i}|.$$

By (A.23) and Minkowski's inequality we have $\|g_k\|_p \leq 1$. Fatou's lemma applied to the sequence $\{g_k^p\}_{k \geq 1}$ gives $\|g\|_p \leq 1$. In particular, any member of the equivalence class of g is finite μ-almost everywhere, and therefore

$$f_{n_1}(x) + \sum_{i=1}^{\infty} \left(f_{n_{i+1}}(x) - f_{n_i}(x) \right)$$

converges absolutely for μ-almost all x. Call the corresponding limit $f(x)$ (set $f(x) = 0$ when this limit does not exist). Since

$$f_{n_1} + \sum_{i=1}^{k-1} \left(f_{n_{i+1}} - f_{n_i} \right) = f_{n_k}$$

we see that

$$f = \lim_{k \uparrow \infty} f_{n_k} \quad \mu\text{-a.e.}.$$

One must show that f is the limit in $L^p_{\mathbb{C}}(\mu)$ of $\{f_{n_k}\}_{k \geq 1}$. Let $\epsilon > 0$. There exists an integer $N = N(\epsilon)$ such that $\|f_n - f_m\|_p \leq \epsilon$ whenever $m, n \geq N$. For all $m > N$, by Fatou's lemma we have

$$\int_X |f - f_m|^p \, d\mu \leq \liminf_{i \to \infty} \int_x |f_{n_i} - f_m|^p \, d\mu \leq \epsilon^p.$$

Therefore $f - f_m \in L^p_{\mathbb{C}}(\mu)$, and consequently $f \in L^p_{\mathbb{C}}(\mu)$. It also follows from the last inequality that

$$\lim_{m \to \infty} \|f - f_m\|_p = 0.$$

In the proofs of Theorems A.28 we have obtained the following result.

Theorem A.29 *Let* $\{f_n\}_{n \geq 1}$ *be a convergent sequence in* $L^p_{\mathbb{C}}(\mu)$, *where* $p \geq 1$, *and let* f *be the limit. Then, there exists a subsequence* $\{f_{n_i}\}_{i \geq 1}$ *such that*

$$\lim_{i \uparrow \infty} f_{n_i} = f \quad \mu\text{-a.e.}. \tag{A.24}$$

Note that the statement in (A.24) is about functions and is not about equivalence classes. The functions thereof are *any* members of the corresponding equivalence class. In particular, when a given sequence of functions converges μ-a.e. to two functions, these two functions are necessarily equal μ-a.e.,

Theorem A.30 *If $\{f_n\}_{n\geq 1}$ converges both to f in $L_{\mathbb{C}}^p(\mu)$ and to g μ-a.e., then $f = g$ μ-a.e.*

Riesz's Theorem

Denote by $C_0(\mathbb{R}^d)$ the set of continuous functions from $\varphi : \mathbb{R}^d \to \mathbb{R}$ that vanish at infinity ($\lim_{|x|\uparrow\infty} \varphi(x) = 0$), endowed with the norm of uniform convergence $\|\varphi\| := \sup_{x\in\mathbb{R}^d} \|\varphi(x)\|$. Let $C_c(\mathbb{R}^d)$ be the set of continuous functions from $\mathbb{R}^d$ to $\mathbb{R}$ with compact support. In particular $C_c(\mathbb{R}^d) \subset C_0(\mathbb{R}^d)$.

Definition A.16 A linear form $L : C_c(\mathbb{R}^d) \to \mathbb{R}$ such that $L(f) \geq 0$ whenever $f \geq 0$ is called a positive *Radon linear form*.

We quote without proof this fundamental result of Riesz[2]:

Theorem A.31 *Let $L : C_c(\mathbb{R}^d) \to \mathbb{R}$ be a positive Radon linear form. There exists a unique locally finite measure μ on $(\mathbb{R}^d, \mathcal{B}(\mathbb{R}^d))$ such that for all $f \in C_c(\mathbb{R}^d)$,*

$$L(f) = \int_{\mathbb{R}^d} f \, d\mu .$$

A.2 Review of Probability

Probability theory is from a purely formal point of view, a particular chapter of measure and integration theory. Since the terminologies of the two theories are different, we shall proceed to the "translation" of the theory of measure and integration into the theory of probability and expectation.

A.2.1 Expectation as Integral

Recall the probabilistic trinity, the triple $(\Omega, \mathcal{F}, P)$, where P, the probability, is a measure on the measurable space $(\Omega, \mathcal{F})$ with total mass $P(\Omega) = 1$.

For probability measures, Theorem A.8 reads as follows:

Theorem A.32 *Let P_1 and P_2 be two probability measures on $(\Omega, \mathcal{F})$, and let S be a collection of events that is stable under finite intersections and that generates $\mathcal{F}$. If P_1 and P_2 agree on S, then they are identical.*

---

[2] See for instance (Rudin 1986), Theorem 2.14.

Definition A.17 A measurable function X from $(\Omega, \mathcal{F})$ to a measurable space $(E, \mathcal{E})$ is called a *random element* with values in $(E, \mathcal{E})$ (or in E for short, when the context is unambiguous).

If $(E, \mathcal{E}) = (\mathbb{R}, \mathcal{B}(\mathbb{R}))$, or if $(E, \mathcal{E}) = (\overline{\mathbb{R}}, \mathcal{B}(\overline{\mathbb{R}}))$, X is called a *random variable* (R.V.) (*real* r.v. if $X \in \mathbb{R}$, *extended* r.v. if $X \in \overline{\mathbb{R}}$). If $(E, \mathcal{E}) = (\mathbb{R}^n, \mathcal{B}(\mathbb{R}^n))$, X is called a *random vector* (of dimension n), and then $X = (X_1, \ldots, X_n)$ where the X_i are random variables. A *complex random variable* is a function $X : \Omega \to \mathbb{C}$ of the form $X = X_R + iX_I$ where X_R and X_I are real random variables.

If X is a random element with values in $(E, \mathcal{E})$ and if g is a measurable function from $(E, \mathcal{E})$ to $(\overline{\mathbb{R}}, \mathcal{B}(\overline{\mathbb{R}}))$, then $g(X)$ is, by the composition theorem for measurable functions (Theorem A.3), a random variable.

Since a random variable X is a measurable function, we can define, under certain circumstances, its integral with respect to the probability measure P, called the *expectation* of X. Therefore

$$E[X] = \int_\Omega X(\omega)P(d\omega).$$

The general theory of integration summarized in the previous section yields the following results. First, if $A \in \mathcal{F}$,

$$E[1_A] = P(A).$$

More generally, if X is a simple random variable, that is,

$$X(\omega) = \sum_{i=1}^N \alpha_i 1_{A_i}(\omega),$$

where $\alpha_i \in \mathbb{R}$, $A_i \in \mathcal{F}$, then

$$E[X] = \sum_{i=1}^N \alpha_i P(A_i).$$

For a non-negative random variable X, the expectation is always defined by

$$E[X] = \lim_{n\uparrow\infty} E[X_n],$$

where $\{X_n\}_{n\geq 1}$ is any non-decreasing sequence of non-negative simple random variables that converge to X. This definition is consistent, that is, it does not depend on the approximating sequence of non-negative simple random variables, as long as it is increasing and has X for limit. If X is of arbitrary sign, the expectation is defined by $E[X] = E[X^+] - E[X^-]$ if not both $E[X^+]$ and $E[X^-]$ are infinite. If $E[X^+]$ *and $E[X^-]$ are infinite, the expectation is not defined. If $E[|X|] < \infty$, X is said to be *integrable*, and then $E[X]$ is a finite number.

The basic properties of the expectation are *linearity* and *monotonicity*: If X_1 and X_2 are random variables with expectations, then for all $\lambda_1, \lambda_2 \in \mathbb{R}$,

$$E[\lambda_1 X_1 + \lambda_2 X_2] = \lambda_1 E[X_1] + \lambda_2 E[X_2],$$

whenever the right-hand side has meaning (not an $\infty - \infty$ form). Also, if $X_1 \leq X_2$, P-a.s., then

$$E[X_1] \leq E[X_2].$$

It follows from this that if $E[X]$ is well-defined, then

$$|E[X]| \leq E[|X|].$$

We now state in terms of expectation the Lebesgue theorems, which give general conditions under which the limit and expectation symbols can be interchanged, that is,

$$E\left[\lim_{n \uparrow \infty} X_n\right] = \lim_{n \uparrow \infty} E[X_n]. \tag{A.25}$$

They are just a rephrasing of the Lebesgue theorems for integrals with respect to an arbitrary measure. We first restate the *monotone convergence* theorem (Theorem A.14) in the context of expectations.

Theorem A.33 *Let $\{X_n\}_{n \geq 1}$ be a sequence of random variables such that for all $n \geq 1$,*

$$0 \leq X_n \leq X_{n+1}, \ \text{P-a.s.}$$

Then (A.25) holds.

Next, we restate the *dominated convergence* theorem (Theorem A.15).

Theorem A.34 *Let $\{X_n\}_{n \geq 1}$ be a sequence of random variables such that for all ω outside a set $\mathcal{N}$ of null probability there exists $\lim_{n \uparrow \infty} X_n(\omega)$ and such that for all $n \geq 1$*

$$|X_n| \leq Y, \ \text{P-a.s.,}$$

where Y is some integrable random variable. Then (A.25) holds.

Example A.12 We recall that (A.25) is not always true when $\lim_{n \uparrow \infty} X_n$ exists. Indeed, take the following probabilistic model: $\Omega = [0, 1]$, and P is the Lebesgue measure on $[0, 1]$ (the probability of $[a, b] \subset [0, 1]$ is the length $b - a$ of this interval). Thus, ω is a real number in $[0, 1]$, and a random variable is a real function defined on $[0, 1]$. Take for X_n the function whose graph is a triangle with base $[0, \frac{2}{n}]$ and height n. Clearly, $\lim_{n \uparrow \infty} X_n(\omega) = 0$ and $E[X_n] = \int_0^1 X_n(x) dx = 1$, so that $E[\lim_{n \uparrow \infty} X_n] = 0 \neq \lim_{n \uparrow \infty} E[X_n] = 1$.

A.2.2 Distribution of a Random Element

Definition A.18 Let X be a random element with values in $(E, \mathcal{E})$. Its *distribution* is, by definition, the probability measure Q_X on $(E, \mathcal{E})$, image of the probability measure P by the application X from $(\Omega, \mathcal{F})$ to $(E, \mathcal{E})$, that is: for all $C \in \mathcal{E}$, $Q_X(C) = P(X \in C)$.

By Theorem A.17, we have

Theorem A.35 *If g is a measurable function from $(E, \mathcal{E})$ to $(\overline{\mathbb{R}}, \mathcal{B}(\overline{\mathbb{R}}))$*

$$E[g(X)] = \int_E g(x) Q_X(dx),$$

this formula requiring that one of the sides of the equality be well-defined, in which case the other is also well-defined.

In the particular case where $(E, \mathcal{E}) = (\mathbb{R}, \mathcal{B}(\mathbb{R}))$, taking $C = (-\infty, x]$, we have

$$Q_X((-\infty, x]) = P(X \leq x) = F_X(x),$$

where F_X is the *cumulative distribution function* (*c.d.f*) of X, and

$$E[g(X)] = \int_{\mathbb{R}} g(x) dF(x),$$

by definition of the Stieltjes-Lebesgue integral.

In the particular case where $(E, \mathcal{E}) = (\mathbb{R}^n, \mathcal{B}(\mathbb{R}^n))$, and where the random vector X admits a probability density f_X, that is if Q_X is the product of the Lebesgue measure on $(\mathbb{R}^n, \mathcal{B}(\mathbb{R}^n))$ by the function f_X, Theorem A.21 tells us that

$$E[g(X)] = \int_{\mathbb{R}^n} g(x) f_X(x) \, dx.$$

The mean m_X and the variance σ_X^2 of an integrable real-valued random variable are defined by:

$$m_X = E[X],$$
$$\sigma_X^2 = E[(X - m_X)^2] = E[X^2] - m_X^2.$$

The characteristic function of a real-valued random variable X is the function $\varphi : \mathbb{R} \to \mathbb{C}$ defined by

$$\varphi_X(u) = E[e^{iuX}],$$

where $E[e^{iuX}] = E[\cos(uX)] + i E[\sin(uX)]$. The *characteristic function* of a random vector $X = (X_1, \ldots, X_n) \in \mathbb{R}^n$ is the function $\varphi_X : \mathbb{R}^n \to \mathbb{C}$ defined by

$$\varphi_X(u) = E[e^{i \sum_{k=1}^n u_k X_k}],$$

where $u = (u_1, \ldots, u_n) \in \mathbb{R}^n$.

Theorem A.36 *The distribution of a random vector of $\mathbb{R}^n$ is uniquely determined by its characteristic function.*

The proof is given in Theorem 2.1.5.

A.2.3 Independence

Two events A and B are said to be *independent* events if

$$P(A \cap B) = P(A)P(B).$$

More generally, a family $\{A_i\}_{i \in I}$ of events, where I is an arbitrary index, is called an *independent family of events* if for every *finite* subset $J \in I$

$$P\left(\bigcap_{j \in J} A_j\right) = \prod_{j \in J} P(A_j).$$

Two random elements $X : (\Omega, \mathcal{F}) \to (E, \mathcal{E})$ and $Y : (\Omega, \mathcal{F}) \to (G, \mathcal{G})$ are called independent if for all $C \in \mathcal{E}, D \in \mathcal{G}$

$$P(\{X \in C\} \cap \{Y \in D\}) = P(X \in C)P(Y \in D).$$

More generally, a family $\{X_i\}_{i \in I}$, where I is an arbitrary index, of random elements $X_i : (\Omega, \mathcal{F}) \to (E_i, \mathcal{E}_i), i \in I$, is called an independent family of random elements if for every finite subset $J \in I$

$$P\left(\bigcap_{j\in J}\{X_j \in C_j\}\right) = \prod_{j\in J} P(X_j \in C_j)$$

for all $C_j \in \mathcal{E}_j$ $(j \in J)$.

The following is an immediate consequence of the definition of independent random elements.

Theorem A.37 *If the random elements X and Y, taking their values in $(E, \mathcal{E})$ and $(G, \mathcal{G})$ respectively, are independent , then so are the random elements $\varphi(X)$ and $\psi(Y)$, where $\varphi : (E, \mathcal{E}) \to (E', \mathcal{E}')$, $\psi : (G, \mathcal{G}) \to (G', \mathcal{G}')$.*

Proof For all $C' \in \mathcal{E}'$, $D' \in \mathcal{G}'$, the sets $C = \varphi^{-1}(C')$ and $D = \psi^{-1}(D')$ are in $\mathcal{E}$ and $\mathcal{G}$ respectively, since φ and ψ are measurable. We have

$$\begin{aligned} P\left(\varphi(X) \in C', \psi(Y) \in D'\right) &= P\left(X \in C, Y \in D\right) \\ &= P\left(X \in C\right) P\left(Y \in D\right) \\ &= P\left(\varphi(X) \in C'\right) P\left(\psi(Y) \in D'\right). \quad \square \end{aligned}$$

The above result is stated for two random elements for simplicity, and it extends in the obvious way to a finite number of independent random elements.

In order to prove that two sigma-fields are independent, it suffices to prove that certain subclasses of these sigma-fields are independent. More precisely:

Theorem A.38 *Let $(\Omega, \mathcal{F}, P)$ be a probability space, and let $\mathcal{S}_1$ and $\mathcal{S}_2$ be two collections of events that are stable under finite intersections. If $\mathcal{S}_1$ and $\mathcal{S}_2$ are independent, then so are $\sigma(\mathcal{S}_1)$ and $\sigma(\mathcal{S}_2)$.*

Proof The proof is omitted. $\square$

Corollary A.5 *Let $(\Omega, \mathcal{F}, P)$ be a probability space on which are given two real random variables X and Y. For these two random variables to be independent, it is necessary and sufficient that for all $a, b \in \mathbb{R}$, $P(X \leq a, Y \leq b) = P(X \leq a)P(Y \leq b)$.*

Proof This follows from Theorem A.38, using the fact that the collection $\{(-\infty, a]; a \in \mathbb{R}\}$ is stable under finite intersection and generates $\mathcal{B}$. $\square$

The independence of two random variables X and Y is equivalent to the factorisation of their joint distribution:

$$Q_{(X,Y)} = Q_X \times Q_Y,$$

where $Q_{(X,Y)}$, Q_X, and Q_Y are the distributions of (X, Y), X, and Y, respectively. Indeed, for all sets of the form $C \times D$, where $C \in \mathcal{E}, D \in \mathcal{G}$,

$$Q_{(X,Y)}(C \times D) = P((X, Y) \in C \times D) = P(X \in C, Y \in D)$$
$$= P(X \in C)P(Y \in D) = Q_X(C)Q_Y(D),$$

and this implies that $Q_{(X,Y)}$ is the product measure of Q_X and Q_Y.

In particular, by the Fubini–Tonelli theorem

Theorem A.39 *Let X and Y be independent random elements taking their values in $(E, \mathcal{E})$ and $(G, \mathcal{G})$ respectively, Then for all $g : (E, \mathcal{E}) \to (\mathbb{R}, \mathcal{B}(\mathbb{R}))$, $h : (G, \mathcal{G}) \to (\mathbb{R}, \mathcal{B}(\mathbb{R}))$ such that $E\left[|g(X)|\right] < \infty$ and $E\left[|h(Y)|\right] < \infty$, or $g \geq 0$ and $h \geq 0$, we have the product formula for expectations*

$$E\left[g(X)h(Y)\right] = E\left[g(X)\right] E\left[h(Y)\right].$$

A.2.4 Convergences

Definition A.19 A sequence $\{Z_n\}_{n \geq 1}$ of random variables with values in $\mathbb{C}$ (resp. in $\overline{\mathbb{R}}$), is said to *converge P-almost surely* (P-a.s.) to the random variable Z with values in $\mathbb{C}$ (resp. in $\overline{\mathbb{R}}$) if

$$P(\lim_{n \uparrow \infty} Z_n = Z) = 1. \tag{A.26}$$

Paraphrasing the definition: For all ω outside a set N of null probability, $\lim_{n \uparrow \infty} Z_n(\omega) = Z(\omega)$. In the case where the sequence takes values in $\overline{\mathbb{R}}$, the limit may be infinite. Otherwise, when $P(Z < \infty) = 1$, we shall add the precision: "converges to a *finite* limit".

A basic tool of the theory of almost-sure convergence, is the Borel–Cantelli lemma below. Consider a sequence of events $\{A_n\}_{n \geq 1}$. We are interested in the probability that A_n occurs infinitely often, that is, the probability of the event

$$\{\omega;\ \omega \in A_n \text{ for an infinity of indices } n\},$$

denoted by $\{A_n\ i.o.\}$, where *i.o.* means *infinitely often*. We have the *Borel–Cantelli lemma*:

Theorem A.40 *For any sequence of events $\{A_n\}_{n \geq 1}$,*

$$\sum_{n=1}^{\infty} P(A_n) < \infty \implies P(A_n\ i.o.) = 0.$$

Proof We first observe that

$$\{A_n \text{ i.o.}\} = \bigcap_{n=1}^{\infty} \bigcup_{k \geq n} A_k.$$

The set $\bigcup_{k \geq n} A_k$ decreases as n increases, so that by the sequential continuity property of probability,

$$P(A_n \text{ i.o.}) = \lim_{n \uparrow \infty} P\left(\bigcup_{k \geq n} A_k\right). \tag{A.27}$$

But by sub-σ-additivity,

$$P\left(\bigcup_{k \geq n} A_k\right) \leq \sum_{k \geq n} P(A_k),$$

and by the summability assumption, the right-hand side of this inequality goes to 0 as $n \uparrow \infty$. $\qquad \square$

Recall *Markov's inequality*.

Theorem A.41 *Let Z be a nonnegative real random variable, and let $a > 0$. We then have*

$$P(Z \geq a) \leq \frac{E[Z]}{a}. \tag{A.28}$$

Proof It suffices to take expectations in the inequality $a \, 1_{\{Z \leq a\}} \leq Z$. $\qquad \square$

Theorem A.42 *Let $\{Z_n\}_{n \geq 1}$ and Z be complex random variables. If*

$$\sum_{n \geq 1} P(|Z_n - Z| \geq \varepsilon_n) < \infty \tag{A.29}$$

for some sequence of positive numbers $\{\varepsilon_n\}_{n \geq 1}$ converging to 0, then the sequence $\{Z_n\}_{n \geq 1}$ converges P-a.s. to Z.

Proof If for a given ω, $|Z_n(\omega) - Z(\omega)| \geq \varepsilon_n$ finitely often (or *f.o.*; that is, for all but a finite number of indices n), then $\lim_{n \uparrow \infty} |Z_n(\omega) - Z(\omega)| \leq \lim_{n \uparrow \infty} \varepsilon_n = 0$. Therefore

$$P(\lim_{n \uparrow \infty} Z_n = Z) \geq P(|Z_n - Z| \geq \varepsilon_n \quad f.o.).$$

On the other hand, $\{|Z_n - Z| \geq \varepsilon_n \quad f.o.\} = \overline{\{|Z_n - Z| \geq \varepsilon_n \quad i.o.\}}$. Therefore

$$P(|Z_n - Z| \geq \varepsilon_n \quad f.o.) = 1 - P(|Z_n - Z| \geq \varepsilon_n \quad i.o.).$$

Hypothesis (A.29) implies (Borel–Cantelli lemma) that

$$P(|Z_n - Z| \geq \varepsilon_n \ \ i.o.) = 0.$$

By linking the above facts, we obtain $P(\lim_{n\uparrow\infty} Z_n = Z) \geq 1$, and of course, the only possibility is $= 1$. $\square$

The following is a *criterion* of almost-sure convergence.

Theorem A.43 *The sequence $\{Z_n\}_{n\geq 1}$ of complex random variables converges P-a.s. to the complex random variable Z if and only if for all $\epsilon > 0$,*

$$P(|Z_n - Z| \geq \epsilon \ i.o.) = 0. \tag{A.30}$$

Proof For the necessity, observe that

$$\{|Z_n - Z| \geq \epsilon \ i.o.\} \subseteq \overline{\{\omega; \lim_{n\uparrow\infty} Z_n(\omega) = Z(\omega)\}},$$

and therefore

$$P(|Z_n - Z| \geq \epsilon \ i.o.) \leq 1 - P(\lim_{n\uparrow\infty} Z_n = Z) = 0.$$

For the sufficiency, let N_k be the last index n such that $|Z_n - Z| \geq \frac{1}{k}$ (let $N_k = \infty$ if $|Z_n - Z| \geq \frac{1}{k}$ for an infinity of indices $n \geq 1$). By (A.30) with $\epsilon = \frac{1}{k}$, we have $P(N_k = \infty) = 0$. By sub-σ-additivity, $P(\cup_{k\geq 1}\{N_k = \infty\}) = 0$. Equivalently, $P(N_k < \infty, \ \text{for all } k \geq 1) = 1$, which implies $P(\lim_{n\uparrow\infty} Z_n = Z) = 1$. $\square$

Recall the *strong law of large numbers*:

Theorem A.44 *Let $\{X_n\}_{n\geq 1}$ be an IID sequence of integrable random variables. Then,*

$$P\left(\lim_{n\uparrow\infty} \frac{S_n}{n} = E[X_1]\right) = 1. \tag{A.31}$$

Definition A.20 A sequence $\{Z_n\}_{n\geq 1}$ of variables is said to *converge in probability* to the random variable Z, if for all $\varepsilon > 0$,

$$\lim_{n\uparrow\infty} P(|Z_n - Z| \geq \varepsilon) = 0. \tag{A.32}$$

Theorem A.45 *(A) If the sequence $\{Z_n\}_{n\geq 1}$ of complex random variables converges almost surely to some complex random variable Z, it also converges in probability to the same random variable Z.*

(B) If the sequence of complex random variables $\{X_n\}_{n\geq 1}$ converges in probability to the complex random variable X, one can find a sequence of integers $\{n_k\}_{k\geq 1}$, strictly increasing, such that $\{X_{n_k}\}_{k\geq 1}$ converges almost-surely to X.

(B says, in other words: From a sequence converging in probability, one can extract a subsequence converging almost-surely).

Proof (A) Suppose almost sure convergence. By Theorem A.43 , for all $\varepsilon > 0$,

$$P(|Z_n - Z| \geq \varepsilon \; i.o.) = 0,$$

that is

$$P(\cap_{n \geq 1} \cup_{k=n}^{\infty} (|Z_k - Z| \geq \varepsilon)) = \lim_{n \uparrow \infty} P(\cup_{k=n}^{\infty} (|Z_k - Z| \geq \varepsilon))0,$$

which implies that $\lim_{n \uparrow \infty} P(|Z_n - Z| \geq \varepsilon) = 0$.

(B) By definition of convergence in probability, for all $\varepsilon > 0$,

$$\lim_{n \uparrow \infty} P(|X_n - X| \geq \varepsilon) = 0$$

Therefore one can find n_1 such that $P\left(|X_{n_1} - X| \geq \frac{1}{1}\right) \leq \left(\frac{1}{2}\right)^1$. Then, one can find $n_2 > n_1$ such that $P\left(|X_{n_2} - X| \geq \frac{1}{2}\right) \leq \left(\frac{1}{2}\right)^2$, and so on, until we have a strictly increasing sequence of integers $\{n_k\}$, $k \geq 1$ such that

$$P\left(|X_{n_k} - X| \geq \frac{1}{k}\right) \leq \left(\frac{1}{2}\right)^k.$$

It follows from Theorem A.42 that $\lim_{k \uparrow \infty} X_{n_k} = X$ a.s.   □

Exercise A.18 gives an example of a sequence converging in probability, but not almost-surely. Thus, convergence in probability is a weaker notion than almost-sure convergence.

Definition A.21 A sequence $\{Z_n\}_{n \geq 1}$ of square-integrable complex random variables is said to *converge in quadratic mean* to the square-integrable complex random variable Z, if for all $\varepsilon > 0$,

$$\lim_{n \uparrow \infty} E[|Z_n - Z|^2] = 0. \tag{A.33}$$

Statement (A.33) is the same as the following:

$$\lim_{n \uparrow \infty} d(Z_n, Z) = 0, \tag{A.34}$$

where d is the distance associated with the inner product of $L^2_{\mathbb{C}}(P)$. Convergence in quadratic mean is therefore the same as convergence in $L^2_{\mathbb{C}}(P)$. Therefore, by the L^p-completeness theorem (Theorem A.28),

Theorem A.46 *For the sequence* $\{Z_n\}_{n\geq 1}$ *of square-integrable complex random variables converges in quadratic mean to some square-integrable complex random variable* Z, *it is necessary and sufficient that*

$$\lim_{n,m\uparrow\infty} E[|Z_n - Z_m|^2] = 0. \tag{A.35}$$

In other words: $L^2_{\mathbb{C}}(P)$ is a Hilbert space. We now recall the property of *continuity of the inner product* of Theorem 1.3.3 in terms of square integrables random variables.

Theorem A.47 *Let* $\{X_n\}_{n\geq 1}$ $\{Y_n\}_{n\geq 1}$ *be two sequences of square-integrable complex random variables that converge in quadratic mean to the square-integrable complex random variables* X *and* Y *respectively. Then,*

$$\lim_{n,m\uparrow\infty} E[X_n Y_m^*] = E[XY^*]. \tag{A.36}$$

Theorem A.48 *If the sequence* $\{Z_n\}_{n\geq 1}$ *of square-integrable complex random variables converges in quadratic mean to the complex random variable* Z, *it also converges in probability to the same random variable.*

Proof It suffices to observe that, by Markov's inequality, for all $\varepsilon > 0$,

$$P(|Z_n - Z| \geq \varepsilon) \leq \frac{1}{\varepsilon^2} E[|Z_n - Z|^2]. \qquad \square$$

A.3 Approximation Results

A.3.1 Stone–Weierstrass Theorem

An *associative algebra* A on a commutative field K is a vector space on this field endowed with a bilinear multiplicative operation $(x, y) \in A \times A \rightarrow x.y \in A$, such that for all $x, y, z \in A$, $x.(y.z) = (x.y).z$. If A is a Banach space with norm $|| \cdot ||$ such that for all $x, y \in A$, $||x.y|| \leq ||x|| \times ||y||$, A is called a *Banach algebra*. A vector subspace B of A that is stable with respect to the multiplicative operation (that is $x.y \in B$ whenever x and y are in B) is called a Banach subalgebra of A.

Example A.13 The following facts are easy to check:

(a) The space $\mathcal{C}([a, b])$ of continuous complex functions on the interval $[a, b] \in \mathbb{R}$ with the sup norm $||f|| = \sup_{t\in[a,b]} |f(t)|$ is a Banach algebra on $\mathbb{C}$, and the polynomial functions $p : [a, b] \rightarrow \mathbb{C}$ with complex coefficients form a Banach subalgebra of $\mathcal{C}([a, b])$.

(b) The space of trigonometric polynomials on the *torus*[3] $T = [0, 2\pi]$, that is, of functions $g : T \to \mathbb{C}$ of the form

$$g(x) = \sum_{k=-n}^{+n} c_k e^{ikx}$$

where $n \in \mathbb{N}$ and $c_k \in \mathbb{C}$ for all $-n \leq k \leq +n$ is a Banach subalgebra of the space of continuous functions on T, a Banach algebra with respect to the sup norm.

We have the *Stone–Weierstrass theorem* :

Theorem A.49 *Let X be a compact Haussdorf topological space with at least two elements and let B be a Banach subalgebra of* $C(X)$, *the space of continuous functions* $f : X \to \mathbb{C}$ *with the sup norm. If B contains a constant non identically null function and separates the points of X (that is for all distinct points* $x, y \in X$, *there exists* $f \in B$ *such that* $f(x) \neq f(y)$, *then B is dense in* $C(X)$.

Example A.14 From the facts given in Example A.13 and the above theorem:

(a) The set of trigonometric polynomials on the torus $T = [0, 2\pi]$ is dense in $C(T)$.
(b) The set of polynomial functions $p : [a, b] \to \mathbb{C}$ with complex coefficients is dense inf $C([a, b])$.

A.3.2 Density Theorems in L^p

Let (X, d) be a metric space and let $\mathcal{X}$ be the Borel sigma-field on it (generated by the open sets). A measure μ on $(X, \mathcal{X})$ is called *externally regular* if for all $A \in \mathcal{X}$,

$$\mu(A) = \sup\{\mu(O); \ O \text{ open} \supseteq A\}.$$

If moreover (X, d) is locally compact separable, a measure μ on $(X, \mathcal{X})$ is called *regular* if it is externally regular and if moreover for all $A \in \mathcal{X}$,

$$\mu(A) = \inf\{\mu(K); \ K \text{ compact} \subseteq A\}.$$

We shall admit without proof that:

Theorem A.50 *Any locally finite measure (in particular, any finite measure)* μ *on* $(\mathbb{R}^d, \mathcal{B}(\mathbb{R}^d))$ *is regular.*

---

[3] The difference between the *interval* $[0, 2\pi]$ and the *torus* $[0, 2\pi]$ lies in the fact that in the latter, 0 and 2π are considered to be the same point. In particular, a function f defined on the torus necessarily satisfies the equality $f(0) = f(2\pi)$.

Theorem A.51 *(a) The space of simple functions in $L_{\mathbb{R}}^1(\mu)$ is dense in $L_{\mathbb{R}}^p(\mu)$ for all $p \in [1, \infty)$.*

(b) Suppose that (X, d) is a metric space and let $\mathcal{X}$ be the Borel sigma-field on it. Let μ be externally regular measure on $(X, \mathcal{X})$. Then the space of bounded Lipschitzian functions that are in $L_{\mathbb{R}}^p(\mu)$ where $p \in [1, +\infty)$ is dense in $L_{\mathbb{R}}^p(\mu)$.

(c) Suppose that (X, d) is a locally compact separable metric space and let $\mathcal{X}$ be the Borel sigma-field on it. Let μ be a regular measure on $(X, \mathcal{X})$. Then the space of Lipschitzian functions with compact support that are in $L_{\mathbb{R}}^p(\mu)$ where $p \in [1, +\infty)$ is dense in $L_{\mathbb{R}}^p(\mu)$.

Proof (a) It suffices to prove that any *non-negative* function $f \in L_{\mathbb{R}}^p(\mu)$ can be approached in $L_{\mathbb{R}}^p(\mu)$ by a sequence of simple functions in $L_{\mathbb{R}}^1(\mu)$. By the approximation theorem (Theorem A.5), there exists a non-decreasing sequence of non-negative simple functions $\{f_n\}_{n \geq 1}$ with pointwise limit f. Since $f_n \leq f$, f_n is in $L_{\mathbb{R}}^p(\mu)$ and in particular in $L_{\mathbb{R}}^1(\mu)$ (for simple functions, this is equivalent; exercise) and moreover since $|f - f_n|^p \leq |f|^p$, we have by dominated convergence that $\lim_n \int |f - f_n|^p \, d\mu = 0$.

(b) It suffices to prove that any simple non-negative function f that is in $L_{\mathbb{R}}^1(\mu)$ is the limit in $L_{\mathbb{R}}^p(\mu)$ of bounded Lipschitz functions. It is in fact enough to prove this for any function $f = 1_A$ where $A \in X$, $\mu(A) < \infty$ as a simple argument shows. Since μ is externally regular, there exists for any $\varepsilon > 0$ an open set $U \supseteq A$ such that $\mu(U \backslash A) < \left(\frac{\varepsilon}{2}\right)^p$, and therefore $||1_U - 1_A||_p := ||1_U - 1_A||_{L_{\mathbb{R}}^p(\mu)} < \frac{\varepsilon}{2}$. For all $k \geq 1$, the function $f_k(x) := (kd(x, \overline{U})) \wedge 1$ is lipschitzian, bounded, and such that the sequence $\{f_k\}_{k \geq 1}$ is non-decreasing and with pointwise limit 1_A. Also, by dominated convergence, $\lim_k ||1_U - f_k||_p = 0$. Therefore, for large enough k, $||1_U - f_k||_p < \frac{\varepsilon}{2}$, and

$$||1_A - f_k||_p \leq ||1_U - f_k||_p + ||1_U - 1_A||_p < \varepsilon.$$

(c) The proof is omitted. $\qquad\qquad\qquad\qquad\qquad\qquad\qquad\qquad\qquad\qquad\quad\square$

Corollary A.6 *Let μ be a locally finite measure on $(\mathbb{R}, \mathcal{B}(\mathbb{R}))$. The step functions with compact support form a dense set in $L_{\mathbb{R}}^p(\mu)$ where $p \in [1, +\infty)$.*

Proof This follows from Theorem A.50, (b) of Theorem A.51, and the facts that (i) for any a function $f \in C_c(\mathbb{R})$,

$$f(x) = \lim_n \left(\sum_{n \in \mathbb{Z}} f\left(\frac{k}{n}\right) 1_{[\frac{k}{n}, \frac{k+1}{n})}(x) \right),$$

and (ii) any indicator function of the type $1_{(a,b]}$ is approximable by a Lipschitzian function in $L_{\mathbb{R}}^p(\mu)$ where $p \in [1, +\infty)$. $\qquad\qquad\qquad\qquad\qquad\qquad\square$

A.4 Exercises

Exercise A.1 (*Sigma-field generated by a collection of sets*)

1. Let $\{\mathcal{F}_i\}_{i\in I}$ be an arbitrary nonempty family of sigma-fields on some set Ω. Show that the family $\mathcal{F} := \cap_{i\in I}\mathcal{F}_i$ is a sigma-field ($A \in \mathcal{F}$ if and only if $A \in \mathcal{F}_i$ for all $i \in I$).
2. Let $\mathcal{C}$ be an arbitrary family of subsets of some set Ω. Show the existence of a smallest sigma-field $\mathcal{F}$ containing $\mathcal{C}$. (This means, by definition, that $\mathcal{F}$ is a sigma-field on Ω containing $\mathcal{C}$, such that if $\mathcal{F}'$ is a sigma-field on Ω containing $\mathcal{C}$, then $\mathcal{F} \subseteq \mathcal{F}'$).

Exercise A.2 (*Gross sigma-field*) Show that a function $f : X \to \mathbb{R}$ that is measurable with respect to the gross sigma-field on X and the Borel sigma-field on $\mathbb{R}$ is a constant (takes only one value).

Exercise A.3

(A) Let X and E be arbitrary sets, $f : X \to E$ an arbitrary function from X to E, $\mathcal{C}$ be an arbitrary collection of subsets of E. Prove that $\sigma(f^{-1}(\mathcal{C})) = f^{-1}(\sigma(\mathcal{C}))$.
(B) Let $f : X \to E$ be a function. Let $\mathcal{E}$ be a given sigma-field on E. What is the smallest sigma-field on X such that f is measurable with respect to $\mathcal{X}$ and $\mathcal{E}$?

Exercise A.4 (*Modulus of a function*) Let $f : X \to \mathbb{R}$ be a function. Let b $\mathcal{X}$ be a sigma-field on X. Is it true in general that if $|f|$ is measurable with respect to $\mathcal{X}$ and $\mathcal{B}(\mathbb{R})$, then so is f itself?

Exercise A.5 (*Elementary properties of measure*) Let μ be a measure on $(X, \mathcal{X})$. Show that

$$A \subseteq B \text{ and } A, B \in \mathcal{X} \implies \mu(A) \leq \mu(B)$$

(monotonicity), and (sub-sigma-additivity)

$$A_n \in \mathcal{X} \text{ for all } n \in \mathbb{N} \implies \mu\left(\bigcup_{n=0}^{\infty} A_n\right) \leq \sum_{n=0}^{\infty} \mu(A_n).$$

Exercise A.6 (*Sequential continuity*) Let $\{B_n\}_{n\geq 1}$ be a sequence of $\mathcal{X}$, non-increasing (that is, $B_{n+1} \subseteq B_n$ for all $n \geq 1$), and such that $\mu(B_{n_0}) < \infty$ for some $n_0 \in \mathbb{N}_+$.

(a) Show that $\mu\left(\bigcap_{n=1}^{\infty} B_n\right) = \lim_{n\uparrow\infty} \mu(B_n)$.
(b) Show by means of a counterexample the necessity of the condition $\mu(B_{n_0}) < \infty$ for some n_0.

Exercise A.7 (*Cumulative distribution function*) Prove the satements of Example A.5.

Exercise A.8 (*Scheffé's lemma*) Let μ be a measure on the measurable space $X, \mathcal{X}$. Let f and f_n, $n \geq 1$ be μ-integrable *non-negative* measurable functions from $(X, \mathcal{X})$ to $(\mathbb{R}, \mathcal{B}(\mathbb{R}))$, such that $\lim_{n \uparrow \infty} f_n = f$ μ-a.e. and $\lim_{n \uparrow \infty} \int_X f_n \, d\mu = \int_X f \, d\mu$. Show that $\lim_{n \uparrow \infty} \int_X |f_n - f| \, d\mu = 0$. (Hint: $|a - b| = a + b - \inf(a, b)$ and Fatou).

Exercise A.9 (*Integrals and series*) Prove that for all $a, b \in \mathbb{R}$,

$$\int_{\mathbb{R}_+} \frac{t \, e^{-at}}{1 - e^{-bt}} \, dt = \sum_{n=0}^{+\infty} \frac{1}{(a + nb)^2} .$$

Exercise A.10 (*Differentiating under the integral*) Let $(X, \mathcal{X}, \mu)$ be a measure space and let $(a, b) \subseteq \mathbb{R}$. Let $f : (a, b) \times X \to \mathbb{R}$ and, for all $t \in (a, b)$, define $f_t : X \to \mathbb{R}$ by $f_t(x) = f(t, x)$. Assume that for all $t \in (a, b)$, f_t is measurable with respect to $\mathcal{X}$, and define, if possible, the function $I : (a, b) \to \mathbb{R}$ by the formula

$$I(t) = \int_X f(t, x) \, \mu(dx) .$$

Assume that for μ-almost all x the function $t \mapsto f(t, x)$ is continuous at $t_0 \in (a, b)$ and that there exists a μ-integrable function $g : (X, \mathcal{X}) \to (\overline{\mathbb{R}}, \overline{\mathcal{B}})$ such that $|f(t, x)| \leq |g(x)|$ μ-a.e. for all t in a neighbourhood V of t_0. Prove that I is well-defined and is continuous at t_0.

(B) Assume furthermore that

(α) $t \to f(t, x)$ is continuously differentiable on V for μ-almost all x; and
(β) For some μ-integrable function $h : (X, \mathcal{X}) \to (\overline{\mathbb{R}}, \overline{\mathcal{B}})$ and all $t \in V$,

$$|(df/dt)(t, x)| \leq |h(x)| \quad \mu\text{-a.e.},$$

Prove that I is differentiable at t_0 and

$$I'(t_0) = \int_X (df/dt)(t_0, x) \, \mu(dx) .$$

Exercise A.11 (*Fubini not applicable*) Define $f : [0, 1]^2 \to \mathbb{R}$ by

$$f(x, y) = \frac{x^2 - y^2}{(x^2 + y^2)^2} \, 1_{\{(x,y) \neq (0,0)\}}$$

Compute $\int_{[0,1]} \left(\int_{[0,1]} f(x, y) \, dx \right) dy$ and $\int_{[0,1]} \left(\int_{[0,1]} f(x, y) \, dy \right) dx$. Is f Lebesgue integrable on $[0, 1]^2$?

Exercise A.12 (*Laplace transform*) Let X be a non-negative random variable. Prove that $\lim_{\theta \uparrow \infty} E\left[e^{-\theta X}\right] = P(X = 0)$.

Exercise A.13 (*Telescope formula*) Prove that for any *nonnegative* random variable X, $E[X] = \int_0^\infty [1 - F(x)]dx$ by means of the Tonelli theorem applied to the product measure $\ell \times P$.

Exercise A.14 Let $X \in \mathbb{R}^d$ be a random vector admitting the PDF f. Show that $P(f(X) = 0) = 0$.

Exercise A.15 (*A recurrence equation*) Consider the recurrence equation

$$X_{n+1} = (X_n - 1)^+ + Z_{n+1}, \; n \geq 0$$

($a^+ := \sup(a, 0)$) where $X_0 = 0$ and where $\{Z_n\}_{n \geq 1}$ is an iid sequence of random variables with values in $\mathbb{N}$. Denote by T_0 the first index $n \geq 1$ such that $X_n = 0$ ($T_0 = \infty$ if such index does not exist)

(a) Show that if $E[Z_1] < 1$, $P(T_0 < \infty) = 1$.
(b) Show that if $E[Z_1] > 1$, there exists a (random) index n_0 such that $X_n > 0$ for all $n \geq n_0$.

Exercise A.16 (*Renewal asymptotics*) Let $\{S_n\}_{n \geq 1}$ be an IID sequence of real random variables such that $P(S_1 \in (0, \infty)) = 1$ and $E[S_1] < \infty$, and let for each $t \geq 0$, $N(t) = \sum_{n \geq 1} 1_{(0,t]}(T_n)$, where $T_n = S_1 + \cdots + S_n$. Prove that $\lim_{t \to \infty} \frac{N(t)}{t} = \frac{1}{E[S_1]}$.

Exercise A.17 (*Exchanging the orders of summation and expectation*)

(A) Let $\{S_n\}_{n \geq 1}$ be a sequence of non-negative random variables. Show that

$$E\left[\sum_{n=1}^\infty S_n\right] = \sum_{n=1}^\infty E[S_n]. \tag{$*$}$$

(B) Let $\{S_n\}_{n \geq 1}$ be a sequence of real random variables such that $\sum_{n \geq 1} E[|S_n|] < \infty$. Show that the sum $\sum_{n=1}^\infty S_n$ is almost-surely well defined and that $(*)$ holds true.

Exercise A.18 (*Convergence in probability, but not almost-surely*) Let $\{X_n\}_{n \geq 1}$ be a sequence of independent random variables taking only 2 values, 0 and 1.

(A) Show that a necessary and sufficient condition of almost-sure convergence to 0 is that

$$\sum_{n \geq 1} P(X_n = 1) < \infty.$$

(B) Show that a necessary and sufficient condition of convergence in probability to 0 is that

$$\lim_{n \uparrow \infty} P(X_n = 1) = 0.$$

(C) Deduce from the above that convergence in probability does not imply in general almost-sure convergence.

Exercise A.19 Let $\alpha > 0$, and let $\{Z_n\}_{n\geq 1}$ be a sequence of random variables such that

$$P(Z_n = 1) = 1 - n^{-\alpha}, \; P(Z_n = n) = n^{-\alpha} \; (n \geq 2).$$

Show that $\{Z_n\}_{n\geq 1}$ converges in *probability* to $Z \equiv 1$. For what values of α does $\{Z_n\}_{n\geq 1}$ converge to Z *in quadratic mean*?

Exercise A.20 (*A version of the L^2-completeness theorem*) Show that for the sequence of square-integrable random variables $\{Z_n\}_{n\geq 1}$ to converge in quadratic mean, it suffices that there exists a finite limit for $E\left[Z_n Z_m^*\right]$ as $n, m \uparrow \infty$. (Note that by Theorem A.47, this is also a necessary condition.)

Exercise A.21 (*Series in the quadratic mean sense*) Prove the following theorem:
Let $\{A_n\}_{n\in\mathbb{Z}}$ and $\{B_n\}_{n\in\mathbb{Z}}$ be two sequences of centered square integrable complex random variables such that

$$\sum_{j\in\mathbb{Z}} E[|A_j|^2] < \infty \text{ and } \sum_{j\in\mathbb{Z}} E[|B_j|^2] < \infty.$$

Suppose moreover that for all $i \neq j$,

$$E\left[A_i A_j^*\right] = E\left[B_i B_j^*\right] = E\left[A_i B_j^*\right] = 0 \quad \text{for all } i \neq j.$$

Define

$$U_n := \sum_{j=-n}^{n} A_j \quad \text{and} \quad V_n := \sum_{j=-n}^{n} B_j.$$

The sum $\{U_n\}_{n\geq 1}$ (resp., $\{V_n\}_{n\geq 1}$) converges in quadratic mean to some square integrable random variable U (resp., V), and we have

$$E[U] = E[V] = 0 \text{ and } E[UV^*] = \sum_{j=1}^{\infty} E[A_j B_j^*].$$

Exercise A.22 (*Continuity of the mean and variance*) Prove the following: If the sequence $\{Z_n\}_{n\geq 1}$ of square-integrable complex random variables converges in quadratic mean to the complex random variable Z, then

$$\lim_{n\uparrow\infty} E[Z_n] = E[Z] \text{ and } \lim_{n\uparrow\infty} E\left[|Z_n|^2\right] = E\left[|Z|^2\right].$$

References

Azencott, R., Dacunha-Casteele, D.: Séries d'observations irrégulières. Masson, Paris (1984)

Beran, J.: Statistics of Long-Memory Processes. Chapman-Hall/CRC, New York (1994)

Billingsley, P.: Probability and Measure. Wiley, New York (1995)

Brémaud, P.: Markov Chains. Springer, New York (1999)

Brémaud, P.: Mathematical Principles of Signal Processing: Fourier and Wavelet Analysis. Springer, New York (2002)

Bochner, S.: Harmonic Analysis and the Theory of Probability. University of California Press, California (1955)

Brockwell, P.J., Davis, R.A.: Time Series: Theory and Methods, 2nd edn. Springer, New York (1991)

Caines, P.: Linear Stochastic Systems. Wiley, New York (1988)

Cramér, H., Leadbetter, M.R.: Stationary and Related Stochastic Processes. Wiley, New York (1967)

Dacunha-Casteele, D., Duflo, M.: Probability and Statistics, vol. II. Springer, New York (1986)

Daley, D.J., Vere-Jones, D.: An Introduction to the Theory of Point Processes, vol. 2, 2nd edn. Springer, New York (2003)

Doob, J.L.: Stochastic Processes. Wiley, New York (1953)

Gasquet, C., Witomski, P.: Fourier Analysis and Applications. Springer, New York (1998)

Helson, H.: Harmonic Analysis. Addison-Wesley, Readings (1983)

Kodaira, K.: Introduction to Complex Analysis. Cambridge University Press, Cambridge (1984)

Koopmans, L.H.: The Spectral Analysis of Time Series. Academic Press, New York (1974)

Mallat, S.: A Wavelet Tour of Signal Processing. Academic Press, San Diego (1998)

Neveu, J.: Processus ponctuels, in École d'été de Saint Flour. Lecture Notes in Mathematics, vol. 598, pp. 249–445. Springer, Heidelberg (1976)

Orfanidis, S.: Optimal Signal Processing. McMillan, New York (1985)

Priestley, M.B.: Spectral Analysis and Time Series. Academic Press, New York (1991)

Ridolfi, A.: Power Spectra of Random Spikes and Related Complex Signals: With Application to Communications and Random Sampling. VDM Verlag, Germany (2009)

Rosenblatt, M.: Stationary Sequences and Random Fields. Birkhäuser, Boston (1985)

Rozanov, Y.A.: Innovation Processes. Wiley, New York (1977)

Rudin, W.: Real and Complex Analysis, 3rd edn. McGraw-Hill, New York (1986)

Skorokhod, A.: Basic Principles and Applications of Probability Theory. Springer, New York (2000)

Shiryayev, A.N.: Probability, 3rd edn. Springer, New York (2008)

Stoica, P., Moses, R.: Spectral Analysis of Signals. Prentice-Hall, NJ (2005)

Tolstov, G.: Fourier Series, Dover edn. Prentice-Hall, NJ (1976)

Young, N.Y.: An Introduction to Hilbert Spaces. Cambridge University Press, Cambridge (1988)

Whittle, P.: Prediction and Regulation. English Universities Press, London (1963)

Wiener, N.: The Extrapolation, Interpolation, and Smoothing of Stationary Time Series with Engineering Applications. Wiley, New York (1949)

Index

Printed in the United States
By Bookmasters